Growth Mechanism of Developing Economies

Growth Mechanism of Developing Economies

Investment, Productivity and Employment

KAZUSHI OHKAWA

A Copublication
of
the International Center for Economic Growth
and
the International Development Center of Japan

Publication signifies that ICEG and IDCJ believe the work to be a competent treatment worthy of public consideration. The findings, interpretations, and conclusions of a work are entirely those of the authors and should not be attributed to ICEG, its affiliated organizations, its Board of Overseers, organizations that support ICEG, or IDCJ.

Inquiries, book orders, and catalogue requests should be addressed to :

In Japan:	International Development Center of Japan, Kyofuku Bldg., 9-11, Tomioka 2-chome, Koto-ku, Tokyo, 135 Japan. Telephone: (03) 3630-6911; FAX: (03) 3630-8120.
Other countries:	ICS Press, 243 Kearny Street, San Francisco, California 94108, U.S.A. Telephone: (415) 981-5353; FAX: (415) 986-4878.

The cover was designed by Seiichi Tagawa.

The index was provided by Katsuo Otsuka.

Library of Congress Cataloging-in-Publication Data

Ōkawa, Kazuchi, 1908-
Growth mechanism of developing economies : investment, Productivity, and employment / Kazushi Ohkawa, in collaboration with Katsuo Otsuka and Bernard Key.
p. cm.
"A copublication of the International Center for Economic Growth and the International Development Center of Japan."
Includes bibliographical references and index.
ISBN 1-55815-193-1
1. Economic development. 2. Technology transfer. I. Otsuka, Katsuo, II. Key, Bernard M. III. Title.
HD75.036 1993
338.9--dc20 92-33578
CIP

Contents

Tables and Figures

Tables

Chapter 1

Chapter 2

Chapter 4

Chapter 5

Chapter 6

Appendix 1

Appendix 2

Appendix 3

Figures

Chapter 1

Chapter 3

Chapter 6

Foreword

The topic of growth of contemporary and past developing economies is not new, and much has been written on the subject. In this book, however, Professor Kazushi Ohkawa, with the assistance of Dr. Bernard Key and Dr. Katsuo Otsuka, present a refreshingly new approach to analyzing the growth of these economies. While the basic conceptual framework and methodology was presented in an earlier work, *Lectures on Developing Economies* (published in 1989), the analysis in this volume has been refined and, a more important difference, the problem of contemporary developing economies is treated as the primary focus of analysis.

In analyzing the development process of presently developed nations and contemporary developing countries, Dr. Ohkawa and associates employ the well-recognized residual growth methodology. However, the authors go beyond an analysis of sources of growth, but link these sources of growth with a nation's social capability of absorbing technological knowledge. Thus, their findings suggest that conventional restraints to economic growth—for example, a shortage of capital—are not the underlying factors behind a nation's inability to move forward along the development path. Rather, a nation cannot follow its trend of acceleration primarily due to a lack of mechanisms for upgrading its social capability. Hence their conclusion that greater emphasis should be placed on improving a nation's social capability rather than focusing solely on increasing the country's productive capacity.

The analysis is quite comprehensive, with extensive discussion of the concept of social capability and the residual approach. The historical perspectives of both Western developed nations and Japan are compared with that of contemporary developing countries in terms of productivity growth, employment, investment and technology, price structures, and trade structures. Both cross-sectional and time-series data are employed to yield greater insights into the growth process.

We at ICEG and IDCJ are proud to publish this book. While the book is not easy reading, the insights gained are clearly worth the effort. Dr. Ohkawa's knowledge of growth theory and his application and extension of residual growth accounting to the contemporary developing economies together make this book indispensable to macroeconomic economists and other individuals interested in the growth process of developing countries today.

Furthermore, his findings and conclusions will be very useful to policymakers to better understand the processes of growth and to focus on programs that will upgrade the social capabilities of nations to assimilate knowledge and technology.

Hideo Monden
President
International Development
Center of Japan

Nicolás Ardito-Barletta
General Director
International Center for
Economic Growth

Preface

The Contents of this Report are the outcome of our task of research and training at the International Development Center of Japan (IDCJ). Collaboration with my younger colleagues has been wide and intensive in recent years, though only Dr. Katsuo Otsuka and Dr. Bernard Key are mentioned explicity. Specifically, at the discussion meetings of the study group named Developing Economies (DE) set at IDCJ, I have benefitted by receiving a number of valuable comments on my earlier drafts from the group members.

I wish to express my sincere gratitude to Mr. Hideo Monden, President, and Mr. Yasunobu Kawato, Director of Planning and Research Division, who kindly made the decision to publish the Study Series, and selected our work as its first version. We would be happy if this study contributes to the furtherance of research on DE internationally.

I owe much to the great help rendered by Dr. Bernard Key, my former colleague in research at IDCJ, who has carefully read all drafts and kindly made painstaking efforts in editing the English. Tamiko Sakatani has been of great help in typing all drafts, from the first to the final manuscript. Kenji Domoto supported me a great deal in gathering statistical materials and data computation. Without these invaluable contributions, the Report could not have been published. I am deeply grateful to all of them.

Kazushi Ohkawa

Introduction

This volume attempts to present an overview on the growth mechanism of contemporary developing economies, focussing on the aspects of Investment, Productivity and Employment. In the previous volume, *Lectures on Developing Economies,* published in 1989, the subject has been treated in its relevance to Japan's development experience. This is its extended version as the conceptual framework and methodology are essentially the same, though with revisions and refinements. However, this volume differs substantially from the previous one in that the problem of contemporary developing nations in general are taken up as the primary theme of analysis.

What is aimed at is to clarify empirically the basic conditions and elements required for achieving economic development in a long-term perspective. It is hypothesized that technological advance through the absorption of knowledge from developed nations is the basic factor in forming the growth mechanism of latecomers. Application of this hypothesis to the contemporary developing economies is the major subject of this volume, and we are convinced that evidences are found to confirm its validity. In deriving this result, our basic concept is the nation's social capability of absorbing borrowed technological knowledge and empirical observation is pursued by measuring the growth of residuals.

In examining this hypothesis, the differentials actually found between nations are wide and are supposed to stem from varying levels and composition of social capability. If its upgrading is successful, we can say that all developing economies can achieve a path of trend acceleration, in sofar as the distortion of the market mechanism operation does not prevent it. In stating this, we do not necessarily insist that a fuller privatization is the sole solution. Instead, we think adequate leadership and guidance from the government are indispensable in pursuing what we call the "interaction mechanism" for the normal development path. Actually, the potential is not realized in many cases. We judge that distortions here are essentially brought forth by lack of mechanisms for upgrading social capability, both in private and public activities. We do not merely share a view that this is due to restraints of the conventional type such as a shortage of capital supply or balance of foreign payments.

The volume consists of six chapters. The conceptual framework and the hypothesis are presented in Chapter 1. The concept of social capability, we believe, has become increasingly acceptable among economists, but a conditional use of the "residual (total factor productivity)" as its quantitative indicator is not yet conventional. This is the reason why a lengthy discussion is needed to establish the conceptual basis of growth accounting. What we aim at is to go beyond the "sources of growth" by linking it with social capability, which is composed of human and organizational components. However, more detailed analysis of its components is beyond our analytical capacity to obtain operational results. The major pattern of change identified by our measurements is the inverse U-shape in key terms. This provides the starting point for developing the subsequent analysis by group and subgroup observation: by group according to per worker product levels and by subgroup according to factors not associated with that level. We do not follow the conventional approach in terms of per capita income. Finally, the residual growth is primarily treated from the viewpoint of investment. This is not conventional in a macro-sectoral analysis, but rather analogous to the micro-project appraisal. It is our basic view that the social capability, SC, operates in its relation to the production capacity, PC, (both defined in stock terms) and that domestic capital formation is the driving force in enhancing PC. We hope readers will see substantive problems raised there in terms of the balance vs. imbalance between SC and PC as for subgroups of nations.

The historical perspective described in Chapter 2 deals with, first, the Western experience, and second, the case of Japan. These are intended to elucidate the comparative characteristics of contemporary developing economies. Viewed from the historical and analytical standpoint of experts, the treatment is sketchy, particularly in the Western case. And yet these are indispensable because no such knowledge is available systematically in terms of the residual growth approach to compare with contemporary developing nations. In particular, the trend acceleration, both real and potential, of the latecomer nations, is highlighted together with other characteristics. More generally, we quantify the effects of characteristics of contemporary developing economies such as the initial conditions, abrupt industrialization with higher rates of capital formation, much faster population-labor force growth, and the disadvantages and advantages of late latecomers which involve a greater technological gap but offer a richer backlog.

Product per worker is the basic yardstick in group observations in Chapters 1 and 2. Its trend of change by major sector—agriculture, industry and services—is examined in Chapter 3 in order to identify the pattern, convergence or divergence. The problem of "growth vs. equity" is thus dealt with not by itself, but in relation to the changes in the production structure. An overview is obtained through cross-sectional and historical observation. It is found that Kuznets' proposition "from divergence to convergence" about the historical trend of Western developed nations cannot always be applied to the contemporary developing economies. Divergence by industrialization vs. convergence by services and agricultural development—these combined effects appear differently in a broad trend from divergence to convergence and show different characteristics. It is influenced strongly by the initial large differentials of product per worker between sectors. In particular, attention is drawn to the fact that a wider differential of product per worker is still sustained in agriculture more than in industry between developing and developed nations.

Chapter 4 attempts to investigate the relationship between productivity growth and employment changes using a three-sector approach by eliminating the effects of output price changes. In treating technological advance, the "speed" of change rather than the effects of "type" is our central concern. First, sectoral characteristics of technological property are emphasized by identifying the performance as "concurrent" in industry, "reversal" in agriculture, and "mixed" in services. These are derived from investigating the relation between the speed of productivity growth and rate of employment change by subgroups. The concurrent pattern denies a view of "trade-offs". Second, development phases are demarcated in terms of changes in the structure of sectoral productivity. Development phases have often been talked about with regard to the changes in trade patterns, but the concept of phases is much more basic in our framework: upgrading the balance in the relationship between SC and PC can help an economy to make a shift from one phase to the next one. Looking at various developing economies, we found that the increase in the product per worker is rather a poor factor in making the phase shift. What we propose is the "difference indicator", representing the changes in the macro-position of productivity of each major sector. Differences of initial conditions, typology, and country size between nations can be treated as the modifying factors in interpreting this general performance. As compared to the phase in the historical path of currently developed nations, two points are found to be peculiar. One is what we newly propose as the "pre-primary phase" and the other a far more important role played by the services sector than is usually thought in shifting phases. Finally, a few comments are made on the use of cross-sectional data. In discussing the shift of development phases, by cautious treatment of subgrouped data, we avoid a mechanical dependence merely on the cross-sectional analysis.

The price structures, domestic and international, are too ambitious a topic. What we attempt in Chapter 5 is to amplify the preceding discussions by taking up aspects related to price performance. First, the earlier findings on the trend of convergence vs. divergence are decomposed into productivity effects and price effects. In so doing, the hypothesis of the long-term trend of negative association between the change in productivity and output prices is formulated in the context of the competitive mechanism. This is also applied to the discussion on the international dimension, examining the classic notion of the trend of increase of prices in agricultural products. It is particularly noted that the relative price of industrial goods tends to show a reluctant pace of decline in a number of developing nations. This does not conform to the normal path of industrialization in international competition. Second, the domestic price structure of developing economies is generally clarified using the purchasing power parity (ppp) approach, in terms of tradeables vs. non-tradables. This is because of our emphasis on the significance of the difference between domestic and international markets. From this aspect, particularly in terms of factor prices, a sharp contrast is found between an extremely high price level of producer durables vs. a very low price level of human work. Although an "equivalence" approach (analogous to a commodity for the latter) is used just to give illustrative quantification, this finding indicates a cycle in the upgrading of human capability, HC, discussed in Chapter 1.

In Chapter 6, with respect to international trade, the aspect relevant to our preceding discussion is taken up, that is, the growth mechanism under the same hypothesis. As for imports, investigation is focussed on "required" investment goods, the carrier of new technological knowledge, and its link with domestic capital formation. Concerning export expansion, the cause is found in the shifts of the supply curve by productivity growth of domestic industry. We cannot share the view that stresses the reverse relationship, although our analysis is rather confined to the supply side. The significance of an interactive sequential process is emphasized for the relationship between exports and imports.

Simultaneous activity of both in the domestic market draws attention. This long-term sequential path is interpreted through a shift in the phases of trade patterns, demarcated broadly as being consistent with those of productivity changes in the domestic economic development identified earlier.

Finally, a few words on two points. First, the paucity of data, particularly for domestic investment, prevents the application of measuring residual growth to major sectors, which is indispensable to link the macro-treatment in Chapters 1 and 2 with the sectoral productivity approach in the subsequent chapters. Appendices II and III are summaries of the interim results of our rough estimates focussing on sectoral residual growth. These are added to stimulate further studies on this area. Second, implications for international economic cooperation, in relation to national development policies, are not discussed in this volume. Readers interested in this area may draw their own judgments, in one way or the other. We think this important problem requires separate treatment, depending upon the nature of the actual issues raised. Yet, as a general suggestion derived from our analysis, we recommend more emphasis be placed on upgrading SC rather than merely promoting PC.

Chapter **1**

Growth and Investment:
A Cross-sectional Perspective

I. The Conceptual Framework

The Hypothesis: Borrowing Technological Knowledge

It is hypothesized that the varied growth performances of developing economies are essentially due to the economies' ability to grow which, in turn, depends on the nations' social capability to absorb knowledge which will permit the capital formation necessary to raise production capacity. It is not our intention, however, to ignore the significance of indigenous efforts of developing nations in making technological and organizational progress in the traditional sectors. Nevertheless, in our framework, the capability of absorbing advanced technological knowledge rather than the possibility of creating original inventions is the central theme.[1] The so-called imitative process of technological progress is, we believe, in contrast to the classic Schumpeterian notion of "innovational", as the imitative process requires an accumulation of small devices in order for the advanced knowledge to be absorbed for local use.

The factors that determine and affect a nation's social capability are numerous and complex, and it is difficult to disentangle these factors in a meaningful way. Nevertheless, two sets of factors as they relate to productivity-income levels can be distinguished: one set of factors is associated with productivity-income levels and the other set does not. As is widely known, growth and the growth-related performances of developing countries have conventionally been observed in terms of income

---

1 For a more detailed explanation, see Kazushi Ohkawa and Hirohisa Kohama, "Technology Borrowing and Human Resources," Lecture 6, *Lectures on Developing Economies: Japan's Experience and Its Relevance* (Tokyo: University of Tokyo Press, 1989).

levels (such as per capita GNP in U.S. dollars). This assumes that both the demand-expenditure and supply-production aspects of development performance can or should be explained in terms of the income level. Although the significance of the income level for demand-expenditure analysis cannot be undervalued, we do not follow this conventional approach. Because the level of a nation's social capability is enhanced by raising product per worker, when focusing on the supply-production side, product per worker rather than per capita income comes first. Subsequent to the analysis of the supply-production side and product per worker, the demand-expenditure side, which is not associated with the level of product per worker, is considered. Empirically, the growth performances of nations with more or less the same level of product per worker vary considerably: that is, the speed of technological and/or organizational progress differs considerably. Although in theory, such a categorical distinction cannot be made precisely, we believe that such a distinction is required and useful in an empirical analysis. A nation's growth performance is thus assumed to result from the combined effects of these two different types of factors: one set of factors that is "associated" and the other set that is "nonassociated" with the level of product per worker.

The operation of these factors are complex and most of them are "invisible," i.e., they are not measurable. Nevertheless, in principle, we can relate the growth performance of a nation to the level, and changes in the level, of the nation's capacity to absorb technological knowledge from economically advanced countries. Two nations with the same level of output per worker can differ in the level and composition of this absorption capability.

The upgrading of the absorption capability level of currently developed countries during their process of development is complex and hard to clarify, but as mentioned earlier, the factors behind the enhancement in absorption capability can be grouped into two categories, associated and nonassociated factors. With regard to the former, i.e., factors associated with product per worker, the interaction mechanism[2] is assumed to operate in a cumulative fashion between the conventional factors (i.e., land, labor, and

---

2 This notion was initially developed, though more narrowly, in Kazushi Ohkawa and Henry Rosovsky, "The Economics of Trend Acceleration," Chapter 8, *Japanese Growth: Trend Acceleration in the Twentieth Century* (Stanford: Stanford University Press, 1973). In that study, the interaction between capital formation and shifts in the production function was the problem that was examined.

capital) and the absorbing capability of the economy through the growth process. Note that the absorptive capability depends not only on individual human capability, but also depends upon the institutions and organizations under which human beings operate. Depending on these social factors, the mechanism may work less effectively; that is, a nation's sociocultural system and policies, as derived from its political scheme, are also important factors.

For a better understanding of the implications of our approach, the performances of postwar developing economies may be compared with the historical growth performances of the currently developed nations or the so-called industrial economies. It is our view that development of these industrial nations was achieved essentially through the absorption of advanced technological knowledge from the United States, which was the leading country in technology at least since the turn of the century.[3] In this context of followers and latecomers, a common characteristic can be found between the industrial nations of old and the contemporary developing countries—both groups were followers and acquired technological knowledge from abroad. It should be noted, however, that although the industrial nations relied primarily on the United States, in the case of contemporary developing countries today there are several leaders.

There are other significant differences between these two groups of followers. For example, for most, if not all, of the contemporary developing nations, the growth process was introduced abruptly following political independence. This experience is very different from that of the industrial economies which were politically independent for some time. Furthermore, the population explosion occurred very suddenly, while in the developed economies, growth was an extended, continuous process accompanied by moderate rates of population increase. Thus, different characteristics have emerged between the two groups of followers: (1) the gap in the technological level between the leaders and the followers was wider in the case of the contemporary developing economies, and (2) the dilemma over the choice between technology and employment (simply put, capital-intensive vs. labor-intensive processes) is far more severe for the postwar developing nations.

These differences, however, do not necessarily imply unilateral disadvantages for the postwar followers. If the social capability of absorbing advanced technological knowledge (that may have

---

3 Angus Maddison, *Phases of Capitalist Development* (London: Oxford University Press, 1982).

occurred from a richer backlog) can be sufficiently augmented, successful growth acceleration can be achieved. If this augmentation cannot be achieved, however, these nations will fall behind and the discrepancies between the developing and developed economies may be made more severe.

The notion of social capability as it relates to the backlog and inflow of foreign technology in the industrial context was originally derived by Ohkawa and Rosovsky:

> By backlog we mean the difference between the average state of technology within an industry or in the aggregate of all industries and international best practice. Thus, the existence of a backlog within a country represents an opportunity for a particular industry or for all industries to bring the level of technology closer to international best practice. The bigger [the] backlog, the greater and more obvious the opportunity. For most industries and countries . . . the main way to exploit this opportunity lies in the importation of more advanced techniques from the few leaders of world technology. This process is called inflow. . . . What needs to be explained is why inflow varies in magnitude from time to time, and this will eventually be considered in terms of social capability, i.e., the capability of various sectors of the society to absorb technical advances.[4]

A nation's social capability is composed of two elements—a "general" element and a "sector-specific" element. The level of education of the people is an example of a general element in that the level of human capability of a society affects all sectors. In contrast, differences between a nation's agricultural and industrial sectors' capability to absorb borrowed technology is an example of a sector-specific element. Changes in the sector-specific components are particularly important to the nation's growth performance over time. The shift in the level and magnitude of the nation's sectors, say from agriculture to light manufacturing and to heavy industry, will have significant impacts on the long-run growth of the nation. This can naturally be conceived in light of our phasing approach, which will be developed later.

With respect to advances in the social capability of a nation, it is assumed that the interaction mechanism that was described earlier at the macro-level is essentially applicable to both the general and the sector-specific elements. Both factors, the

---

4 Ohkawa and Rosovsky, *Japanese Economic Growth,* 91.

associated and the nonassociated factors, are relevant to the two elements.

However, in terms of the method of operation, Kuznets' comments on the "concept of relevant stock of technology" are important to keep in mind for the sector-specific elements.[5] In the process of economic development, the backlog-inflow thesis that was described earlier can be applied with substantive qualification. Much of the stock of technological knowledge that has accumulated in developed countries may not be relevant to the needs of the latecoming developing countries. Highly advanced knowledge, for example, is beyond the reach of developing economies, and while some of the stock of technical knowledge may not be completely beyond reach, it may not be easily borrowable because it requires substantial modifications before it can be assimilated to local conditions. Thus, Kuznets notes that the relevance of the existing stock of unexploited technology may change over time in the process of growth itself. That is, for a developing country, the relevance of the stock of technology in developed countries may increase as the developing country acquires a greater capacity for modification, or as its scale of priorities shifts toward the more advanced components of technology.

This is the relevance-focused aspect of what we call the over-time process of the interaction mechanism for upgrading the social capability of developing countries. We believe this aspect is particularly important for understanding the sector-specific elements of a nation's social capability. The factors that augment the sector-specific capabilities or that impact on the direction and application of adopted technologies are many and the relations are complex. Yet, the two dimensions presented earlier, i.e., the associated and nonassociated factors, are useful in understanding the over-time process. Of particular concern are those factors not associated with productivity-income levels, i.e., the nonassociated factors. For example, if we assume a higher capability for modifying borrowed advanced technology at the dimension associated with the upgrading of Y/L, there may be no essential difference between the capability of the sector-specific components. This, however, is far from what actually occurs: to a considerable extent, the sector-specific capabilities of nations differ

---

5 Simon Kuznets, "Notes on Japan's Economic Growth," in Lawrence Klein and Kazushi Ohkawa, eds., *Economic Growth: The Japanese Experience Since the Meiji Era* (Hornewood, Illinois: Richard D. Irwin, 1968), 391.

independently of the Y/L level. Among the relevant factors involved, three are of particular concern: the natural resource endowment, factor prices, and the institutional and/or organizational framework (including government policies).

The relevance of an existing stock of unexploited technological knowledge will differ from one developing economy to another due to differences in their development phases. However, typological features are also influential; for example, the comparative advantages of agriculture and industry vary between developing countries depending upon whether the nation has a rich or poor natural resource endowment. For instance, knowledge of agricultural technology of the scale-dependent type is not relevant to those developing nations with limited land resources for which technology that is of the scale-neutral type is desired. The development of an agriculture-specific capability of borrowing technology thus will be conditioned by the given natural resources.

Factor prices affect the relevance of a stock of technical knowledge not only at the macro-level in terms of a low vs. high land price, but also at the industry level. For example, from the viewpoint of developed nations, much of the existing stock of industrial technological knowledge that they possess is irrelevant to the development of small-scale industry, which uses labor-intensive technology, in developing economies. As a matter of fact, differences in factor prices have often been treated as the most influential factor determining technology borrowing. However, the sector-specific capability of developing nations can be upgraded to increase the relevance of existing technologies. The postwar experience of the developing nations provides evidence for this possibility.

Lastly, it goes without saying that the institutional/organizational framework is sector-specific with respect to technological change. That is, governments may act positively to intensify the relevance of existing technologies by upgrading certain subsector-specific capabilities.

Social Capability: Towards Quantification

There are problems in quantifying social capability. As cited by Malinvaud:

> However, said Professor Abramovitz having defended himself so far, he must say again that his main reaction to Professor Ohkawa's comments was desired to join him in stressing the need to work out the operational content of social capability so that we should at some time be able to measure differences

> among countries and over time. It would be an invaluable advance if we could learn to treat that important component of growth quantitatively and, therefore, give it the weight it deserves.[6]

In addition, Abramovitz made an important point:

> The trouble with absorbing social capability into the catch-up hypothesis is that no one knows just what it means or how to measure it.

And after explaining his way of identifying it, he concludes:

> The combination of technological gap and social capability defines a country's potentiality for productive advance by way of catch-up.[7]

We share the same view that social capability is indispensable to interpreting the productivity growth performances of countries as they progress along the development path. Yet measurement of social capability, though highly desirable, has not been attempted, and we are confronted with the problem of how to proceed in order to make the results of our research operational. We propose the conditional use of the residual, which is measurable, as a measure of social capability (details are given in Section II of this Chapter). Because no study, including previous works by Ohkawa, has attempted to measure social capability, we would like to elaborate on the reasons for measuring social capability in the context of relevant notions that were developed by Kuznets.[8]

First, recognizing that the contribution of conventional inputs to productivity (as based on Denison's measurement) was of an

---

6 Edmond Malinvaud, ed., *Economic Growth and Resources,* Vol. I, The Major Issues, Proceedings of the Fifth World Congress of the International Economic Association held in Tokyo, Japan (London: Macmillan, 1979).

7 Moses Abramovitz, *Thinking About Growth,* Part II, 7 (Cambridge: Cambridge University Press, 1989).

8 Simon Kuznets, *Modern Economic Growth: Rate, Structure, and Spread* (New Haven: Yale University Press, 1966). In particular, note Chapter 2, "Growth of Efficiency," and Chapter 4, "Allocation of Income of Entrepreneurs and the Shares of Capital and Labor."

unexpectedly small magnitude, Kuznets emphasized the significance of the "secondary effects" in terms of both quantity and quality. That is, conventional inputs had both "indirect effects" and "direct effects" on an economy's productivity. Thus, the residual was treated in the context of an interwoven mechanism of these two effects. In view of the simple notion of "technological progress" that was assigned to the residual around that time, we believe that this notion deserves attention. Kuznets did not explicitly use the term "social capability," but he was no doubt aware of the interaction mechanism between production capacity and social capability. In proceeding along this line, we believe the first step is to measure the quantitative relationship between the two and their contribution to productivity growth. This step is particularly important for this study as the residual measures that were available at that time are narrower in scope and cannot be applied to developing economies in general.

A second point to be made is that measurements cannot be done without making certain analytical assumptions. In this regard, Kuznets' argument can be represented by the concept of "residual efficiency." This term is used to distinguish it from the more general one which allowed for inputs of labor and capital in simple units disregarding quality changes. After estimating the long-term changes in labor's share in the United States, Kuznets concludes:

> Whether or not the magnitudes used here are valid, the general implication of the argument is obvious. The price per man-hour of labor relative to price per unit of capital service rose not only because of quality improvements credited to labor man-hours (but not to capital) but also because of the gains in residual efficiency . . . which are indirect components of the rise in total output not accounted for by measured inputs [emphasis is added].[9]

The implication of this debate seems far-reaching and profound, and a correct understanding is not necessarily easy. But we believe an explicit treatment of quality can be made using two different approaches: (1) a market approach and (2) a nonmarket approach. Our proposal follows the conventional methodology of evaluating input prices prevailing in a competitive market as the opportunity costs. This implies that nonconventional factors, the contribution of which is embedded in the residual, cannot be

---

9 Kuznets, *Modern Economic Growth*, 184.

evaluated by the market mechanism. (This is essentially valid, although it is not usually stated explicitly.) Market-determined prices of labor and capital inputs reflect the differences in quality, so that measurements "not accounted for by inputs" specifies the quality which lies beyond. The specification of the residual is done by employing the concept of residual efficiency that was mentioned earlier. (The reason for stating this again stems from its general methodological importance which is beyond the treatment of changes in the labor share.) Technology and the organizational/institutional framework have a common property in that their contributions to productivity growth cannot be directly assessed by a market evaluation. This is the technical reason why the residual measurements are of a specified nature: their evaluation and quantification depend upon the market evaluation of inputs and outputs.

The residual growth thus measured is a complex mixture of the contribution of the so-called nonconventional factors, which are supposed to operate not independently but interactively. Despite the recognition of this basic property, it is nevertheless desirable to try and clarify each of the factors and their contribution, at least qualitatively if not quantitatively. To proceed along this line, preliminary discussions are needed with regard to the relationship between productivity growth and technological change.

In the conventional approach, "technological knowledge" has often been treated as an independent variable, but we believe that such an approach is not realistic. The measurement of the effects of technological change is not meant to imply that the effects as measured in the residual can be separated from the effects of other factors such as changes in organizations and institutions. We are convinced that an overall measurement of productivity growth is the only approach that can be adopted in order to clarify the real path of development. However, we do not mean to downgrade the significance of technological advance. To the contrary, our hypothesis of borrowed technology provides the basic framework for our analysis. In other words, we believe that a test of the hypothesis cannot be carried out effectively by merely attempting to measure the contribution of upgraded technological knowledge directly.

Another approach which treats the residual from the viewpoint of investment is possible. Conceptually, this implies that innovations are carried out by enterprises through new investment. In these enterprises, cost-revenue accounting is essentially based on the operation of the market mechanism. The behavior of enterprises has been analyzed in depth and no attempt is made here to further review their behavior. However, we want to draw attention to the social conditions which affect innovational behavior. The expected magnitude of the residual may vary according to these conditions: when social conditions are favorable, the residual will be greater, and when the conditions are unfavorable, the residual will be smaller. In the ex post measurement, the residual is treated as a product of the individual enterprise, but in reality, it is also influenced by social conditions, positively or negatively. While the term, social conditions, may not be adequate, it is used to point out all of the factors which operate beyond the scope of the enterprises' command.[10]

Having gone through the preliminary discussion, let us discuss the major issue at hand, i.e., the linking of the growth in the residual with the concept of social capability.

First, all factors external to enterprises do not necessarily belong under the rubric of social conditions. Examples of such factors are economies of scale and effects of resource reallocation. While these factors are external to the individual enterprises, they are no doubt included in the measured residual but should be excluded from the category of factors that make up the social conditions. Rather these factors have a common property which pertains to the macro-sectoral economic performance itself, i.e., growth and structural change, and hence have nothing to do with social conditions. Quantitative exclusion of the effects of these factors from the residual, however, can be attempted. Measurement of the effects of resource reallocation will be attempted for labor migration in Appendix II. However, the effects of scale economies cannot be measured in a reliable way without making rigid assumptions and no attempt is made to measure it explicitly. Nevertheless, we contend that the effects of scale economies are largely included in the residual that is measured by sector.[11]

---

10 Both private and public enterprises can be treated in this way insofar as they can both operate competitively.

11 The formula, $Gy = a+bG_Y$, following the "Verdoorn Law" appears attractive in this regard, but cannot be applied consistently with our formula. Kaldor's approach, however, provides empirical evidence that economies of scale are characterized sectorwise rather than macro-economywise. (Nicholas Kaldor, *Causes of the Slow Rate of Economic Growth of the United Kingdom: An Inaugural Lecture,* Cambridge, England, 1966).

Second, what about the components which form social capability? The answer to this central question does not need to be comprehensive in descriptive detail,[12] but it does require conceptual clarification. In short, the social conditions which pertain to the activities that absorb advanced technological knowledge form social capability. (Absorption in this context implies borrowing and assimilation.) As stated earlier, social capability chiefly consists of (1) the capability of individual human beings, and (2) organizations and institutions. The problem is to establish consistency and conformity between the concepts and the measurement of these concepts. Since the component effects of social conditions have been defined by their nonmeasurability through market evaluation, the residual growth thus measured must be the sum of contributions of the nonmeasurable factors.

As mentioned earlier, the effects of human capability (HC), which are not evaluated by market factors, are measured as a component of the residual. Artificial (nonmarket) evaluation of human capital and/or of the effects of education and training, including the effects of learning by doing, can be useful if the analyses are carried out with caution. While we do not deny the analytical benefits of such an approach, we propose a framework in which measurement yardsticks are not mixed up with one another. With respect to the effects that stem from changes in organizations and institutions, there is no danger of mixing up the measurement yardsticks since no market values are available and since artificial evaluation is beyond the scope of this analysis. By definition, the effects of these changes can only be measured by the residual. The term "social innovations" often used by Kuznets[13] is relevant to the framework of this study. We are convinced that the major components of Kuznet's "social innovations" pertain to the changes in the capability of human beings and the rearrangement of organizations and institutions which operate in relation to technological innovations, although he did not explicitly link these components with the residual.

The term social conditions covers an extremely wide field. The way in which the components of social conditions are specified

---

12 See Abramovitz, *Thinking About Growth,* in particular, Chapters 6 and 7 in Part II. Emphasis differs between his view and ours in that we focus on the problem of developing economies, while his major concern is developed economies in the catch-up process.

13 Kuznets, *Modern Economic Growth*

depends upon the analytical purpose. Abramovitz used a much wider approach than that used in this study.[14] Under the heading, "Conditions Controlling the Pace of Development," Abramovitz discusses (1) improved facilities for technological innovations, diffusion, and adaptation; (2) conditions facilitating structural change; and (3) conditions encouraging and sustaining capital investment. A sharp conceptual distinction is made between potential and realized productivity growth. Social capability is treated in the context of potential growth while the three components mentioned above are discussed in the context of realized growth.

Abramovitz's conceptual distinction between potential and realized productivity growth is useful in dealing with the problem of developing economies. The "realization" process is primarily dealt with in subsequent pages where the focus is on investment activities and the factors responsible for differences in investment efficiency. This is the "investment approach." The conventional technique is called the "output approach" and includes both functional and accounting measurements. It goes without saying that both production capacity and social capability are stock concepts, whereas in the investment approach, residual measurements depend solely upon flow data. Although the gap between the two approaches appears serious, the task of linking the flow measurements to the stock measurements can conditionally be done through our methodology, the details of which are described later.

II. Patterns of Residual Growth: Measurements

Methodology: An Investment Approach

In the investment approach, the measurement of the residual is not conventional, but for reasons stated earlier, this approach is carried out at the macro-level. The capital stock data are not available for most of the developing economies and this prevents measurements of the residual being made.[15] Nevertheless, even

---

14 Abramovitz, *Thinking About Growth,* Part II, Chapter 6, "Rapid Growth Potential and Its Realization: The Experience of Capitalist Economies in the Postwar Period."

15 Application of the residual approach to developing economies is still very limited, despite Kuznets' earlier stress on the significance of research in this area. This was for the purpose of clarifying the magnificent contribution of technological advance. See Simon Kuznets, *Economic*

without capital stock data, this approach can be operational. This is one reason for adopting the investment approach, i.e., for the sake of technical convenience. The basic rationale behind this approach is that varied effects of different stock levels of social capability can be assessed by measuring the residual at the flow level.

The definition of the residual is in fact not a different one but the conventional one, i.e., the output produced over and above the contribution of conventional inputs. We do not attempt to decompose the residual into various items, following the procedure of Denison. This is too ambitious a task to attempt for developing economies. Instead the effects of technological and/or organizational changes, and also to a certain extent the effects of quality improvement of conventional inputs, economies of scale, and sectoral shifts, are all assumed to be contained in the residual at the macro-level. In other words, we are concerned about the variance of the aggregate magnitude of the residual. Our methodology may be referred to as a specifically extended version of the conventional growth accounting framework. The index number formula is conventionally used for measuring the total productivity growth in the growth accounting framework. In our version, the weights are specified by the factors' marginal contributions (assuming a near-competitive market for the factors). To apply this specification to the dynamic process, our model shares the assumption originally made by J. Kendrick[16] and which can be explained in a simplified form as follows.

The value added in the base period is compared with the value added in another time period. The increase in value added is compared with changes in the sum of opportunity costs (r for capital, K; w for labor, L) of conventional inputs, ΔC, where $\Delta C = r\Delta K + w\Delta L$. A central assumption is that in the base year, the cost evaluation of ΔY^* is conceivable and is equal to ΔC for a hypothetical case of no technological and/or organizational change between the base period and the period that is being compared. Under this assumption, an incremental increase of the residual, ΔR, is defined as the observed difference between the two kinds of output increases. That is,

(1) $\Delta R = \Delta Y - \Delta Y^*$,

Growth of Nations: Total Output and Production Structure (Cambridge, Mass.: Harvard University Press, 1971).

16 For a theoretical background, see J. W. Kendrick and Ryuzo Sato, "Factor Prices, Productivity and Growth," *American Economic Review* (December 1967).

where $\Delta Y^* = r\Delta K + w\Delta L$.

Because of its simplicity, this formula is operational and there is no need for capital stock data. At the same time, because of its simplicity, application of the formula to the actual development process must be done with cautious qualification and reservation. For example, issues such as the appropriateness of devices for linking ΔK with gross domestic investment and the treatment of "opportunity costs" of capital and labor in near-competitive factor markets must be dealt with through analyses of developing economies.

In applying the formula to actual measurement, our primary goal is to explain the residual performance in relation not to output but to investment, I.[17] Thus, equation (1) is transformed to

$$(2) \qquad \Delta Y/I = \Delta R/I + r\Delta K/I + w\Delta L/I.$$

The first term $\Delta Y/I$ can be called the incremental output-capital ratio (IOCR) following the convention of calling $I/\Delta Y$, its reciprocal, the incremental capital-output ratio (ICOR). Equation (2) is, therefore, a device for decomposing the IOCR into three terms: the contribution of capital ($r\Delta K/I$), the contribution of labor ($w\Delta L/I$), and the residual ($\Delta R/I$). Derived from original theories presented by Harrod and Domar, the ICOR has been used widely not only in theoretical analyses but also in empirical studies of growth and development. In both types of analyses, the direction of change in the ICOR, whether it be constant, increasing, or decreasing, is of important concern. In this study, however, an a priori assumption of the direction of change in the ICOR would impose an undue constraint, for example, in building models to be used for development analysis or plan formulation. We believe that at the macro-level, production function analysis is too difficult to be operational. However, we do not believe that a simplified use of ICOR is a useful substitute for such an analysis either.

Given such a situation, a decomposition of ICOR on an accounting basis will be useful in enriching our empirical knowledge of growth performance. Using this approach, actual measurements of these terms and an investigation of their relationship is made to

---

17 This formula was developed earlier in Ohkawa and Kohama, *Lectures on Developing Economies.* It was often accepted as merely a rough substitute of the total factor productivity (TFP) approach. It is valid in the sense that capital stock is substituted with capital formation. However, we believe this formula has its own positive merit of measuring the residual as it relates to the investment behavior of enterprises, both private and public.

provide insights into the problems of growth and development of contemporary developing economies. Technical issues that were mentioned earlier in the text regarding the actual procedure of applying equation (2) will be explained later. At this point, we wish to explain the conceptual implications of equation (2).

First, if we assume a constant value for r for simplicity, then changes in $\Delta R/I$ correspond to changes in $\Delta Y/I$ and $w\Delta L/I$. The residual per unit of investment would be greater (smaller) if the IOCR, i.e., $\Delta Y/I$, is greater (smaller) and/or if the labor term, $w\Delta L/I$, is smaller (greater). This relationship is best understood from the viewpoint of the investor. In reality, project appraisal at the micro-level takes this viewpoint although the benefit-cost treatment differs from our case. Viewed from the social standpoint, however, the labor term is not a "cost," but a "revenue" item as an increase in its value raises the welfare level of workers. A clear distinction between the two different viewpoints should be kept in mind. This pertains to the problem of the criteria of investment efficiency. Yet, we can see essentially the same concept of the residual in both cases.

Second, the labor term, $w\Delta L/I$, is an incremental expression of Lw/K (for simplicity, I is replaced by ΔK) which we call the factor inputs ratio, FIR.[18] The ratio indicates the type of technology that is adopted including the quality of labor, and thus is more than just a "labor term." Therefore, in our formula, no rigid assumption is made with respect to the type of technologies that are adopted. Rather, various combinations of production factors are treated in combination with $\Delta Y/I$ and $\Delta R/I$, and the varied magnitudes of the terms $\Delta Y/I$ and $\Delta R/I$ are relevant. The residual for investment, $\Delta R/I$, is measured as a flow concept, but it can be used as a rough indicator of the changes in the contribution of the stock of social capability subject to the rate of utilization.

Finally, we would like to briefly discuss the possible similarities and dissimilarities between the accounting method presented in this study and the functional approach which has been used chiefly for measuring shifts in the production function or total factor productivity (TFP) following the pioneering work of Solow

---

18 For a detailed discussion on this, see Ohkawa and Kohama, *Lectures on Developing Economies*. The term reflects the actual significance of the technological requirement for factor combinations, although substitution effects due to factor price changes are not rejected.

and Abramovitz.[19] Most of these studies were undertaken with the assumptions of neutral technological change in Hicks' sense and of disembodied technical change under near-competitive factor market conditions. Although there appears to be no essential difference between the two approaches, the TFP approach is much more inclined to focus on identifying the magnitude of technological progress. For example, the results of the functional approach are expressed by $\Delta R/Y$ (in our notation) which represents the contribution of a residual increase (G_R) to output growth. This is, of course, useful and we will calculate this residual value together with $\Delta R/\Delta Y$ in subsequent discussions. However, the focus of this study is on measuring the factors relevant to forming social capability in a broader context. As will be discussed in the following pages, estimates of the magnitude of key terms such as r and w are likely to be less than absolutely reliable for developing economies. Because there is no practical possibility of measuring the TFP for all developing economies in such a way that the measures are internationally comparable, it is our conviction that the residuals thus measured can provide useful information with regard to the different functions of an individual nation's capability. Thus, our study intends to be operational by making comparisons of growth performances of nations in a relative sense, rather than directly aiming for measurements of the absolute magnitude of the residual.

The last topic concerns the use of cross-sectional and time-series data. Although the purpose of this study is to examine the characteristics of postwar economic development with a long-term perspective, and although the concepts and methods that are used are primarily in reference to changes over time in a historical perspective, two approaches are used: (1) a cross-sectional approach with observations of nations at a point in time to provide an international perspective (Chapter 1), and (2) a time-series approach that examines changes over time to provide a historical perspective (Chapter 2). In the pursuit of a better understanding of the growth performances of developing economies, the multicountry

---

19 For a comprehensive review of the approaches, see M. Ishaq Nadiri, "Some Approaches to the Theory and Measurement of Total Factor Productivity: A Survey," *Journal of Economic Literature* 8 (December 1970). The functional approach is treated as the major methodology. A number of TFP studies for developing countries are contained in Hollis Chenery, Sherman Robinson, and Moshe Syrquin, *Industrialization and Growth* (London: Oxford University Press, 1986).

cross-sectional approach has been the methodology most often used thus far. The popularity of the cross-sectional approach stems from the data limitations involved carrying out a comparative historical approach. Today, however, the postwar records of economic development cover twenty to thirty years and provide us with much information. The postwar records are first used for the cross-sectional analysis, and observations over time are used later. Naturally, our study is supplemented and strengthened by comparative observations of the historical patterns of currently developed nations.

The cross-sectional approach is introduced in Chapter 1 not because it is of primary importance, but because of its provision of useful information that can enrich our knowledge on the historical perspectives. The income level is the conventional procedure used to bridge the two approaches, i.e., the cross-sectional variations and changes over time. This procedure basically assumes that economic development can be interpreted with changes in the income level where the income level is the common yardstick. However, as stated earlier, the significance of nonassociated factors makes it impossible to apply such a common yardstick. Variance of these factors suggests that there will be conformity and nonconformity between the cross-sectional and time-series data.

Estimates of Key Terms

The results of the estimated measurements are summarized in Table 1.1. Sixty-four countries were selected. For each of these countries, data are available continuously for the entire interval of twenty-five years, 1960–1985 (except for a few countries noted in the remarks). The countries are classified into five groups, groups I–V, by the level of product per worker, Y/L. Product is evaluated in U.S. dollars.[20] The year 1970 is selected for ranking Y/L in order to be closer to the mid-point of the interval under review. The border level of Y/L is fixed so as to obtain a fairly even distribution of countries among the five groups. Group V consists of industrial economies while groups I, II, III, and IV consist of developing economies. Each group is divided into two subgroups of an equal number of countries, subgroups A and B, which are used to test the

---

20 Although the use of official exchange rates for calculating gross domestic product may be suspect, we are convinced that this causes no serious problem for treating the terms. One may suggest that use of purchasing power parities are better, but we do not necessarily agree. For a detailed explanation, see Chapter 5.

Table 1.1
Growth Performance of Nations Using the Investment Approach [a]***,***
1960–1985 (percentage)

Group[b]		G_Y	G_L	G_Y-G_L	I/Y	ΔY/I	wΔL/I (1)	rΔK/I (2)	(1)+(2)	ΔR/I	I/ΔK	r
I	(12)	3.1	2.4	0.7	17.6	17.6	6.4	8.6	15.0	2.6	1.54	13.1
	A (6)	3.9	2.5	1.4	19.2	20.4	6.3	8.4	14.7	5.8	1.52	12.8
	B (6)	2.4	2.2	0.2	15.9	15.1	6.6	8.8	15.4	-0.3	1.57	13.8
II	(14)	4.0	2.4	1.6	19.2	21.1	5.7	8.7	14.4	6.6	1.50	13.1
	A (7)	5.0	2.6	2.4	20.0	25.5	5.9	9.9	15.8	9.7	1.53	15.1
	B (7)	3.0	2.1	0.9	18.3	16.6	5.4	7.9	13.3	3.3	1.48	11.7
III	(12)	5.5	3.1	2.4	23.5	23.8	5.8	9.9	15.7	8.1	1.43	14.1
	A (6)	6.5	3.0	3.5	22.5	29.9	5.9	10.5	16.4	13.5	1.42	14.9
	B (6)	4.4	3.1	1.3	24.6	17.7	5.6	9.5	15.1	2.6	1.43	13.6
IV	(12)	4.0	1.9	2.1	22.9	16.9	3.8	8.2	12.0	4.9	1.47	12.0
	A (6)	4.7	2.1	2.6	21.8	21.1	4.6	10.5	15.1	6.0	1.42	14.9
	B (6)	3.2	1.6	1.6	24.0	13.6	3.2	6.6	9.8	3.8	1.52	10.1
V	(14)	3.6	1.2	2.4	24.1	15.1	3.0	5.2	8.2	6.9	1.87	9.7
	A (7)	4.1	1.2	2.9	22.6	18.1	3.2	5.5	8.7	9.4	1.86	10.3
	B (7)	3.1	1.1	2.0	25.9	12.2	2.7	4.9	7.6	4.6	1.87	9.2

Y = GDP
L = Labor force
I = Gross domestic capital formation (investment)
K = Capital stock
w = Wages
r = Rate of return on capital
R = Residual
G = Rate of growth in terms of annual average
Δ = Incremental increase

Notes: a. The average values of the terms under review are calculated as follows: For the first five columns of data, the simple averages of country values were used. Averages of three subintervals—1960–1970, 1970, and 1980–1985—are calculated for each country; then the grand averages are obtained for the entire interval with weights reflecting the time duration. The values for the groups and subgroups are calculated by their simple averages without using any weights concerning country size. Possible differences between the calculating procedures of average annual rate of change between G_Y and G_L (these were provided by the World Bank and ILO data) for each subinterval is not expected to matter substantially.

The investment ratio, I/Y, is simply calculated by taking averages of the values for the two years, the beginning and the end of each subinterval.

The original ILO data for G_L refer to "economically active population" and is defined as all employed and unemployed persons including those seeking work for the first time. It covers "employers, persons working on their own account, salaried employees, wage earners, unpaid family workers, members of producer's cooperatives and members of the armed forces." The data were used because of the international consistency, though for analytical purposes, the coverage is no doubt too broad. In addition, the data for 1980–1985 are projections and are used as a substitute for real data since no other internationally comparable data are available at present. Because of this, care will be taken in the interpretation of the results of measurements of $w\Delta L/I$ and related terms.

For columns 6, 7, and 8, indirect adjustment is tried by using the average values obtained for the entire periods in the first five columns. Income shares for labor (β) are assumed to be 0.53 (I), 0.51 (II), 0.49 (III), 0.51 (IV) and 0.69 (V) where $1 = \alpha+\beta$ and α is the share of capital. By using the formula, $\beta G_L = w\Delta L/Y$, $w\Delta L/I$ is estimated as $w\Delta L/Y * Y/I$. Since income shares are in terms of income at factor cost, Y*, adjustment to the GDP term at market prices, Y, is made by a constant conversion ratio, Y*/Y, of 90 for groups I–IV and 0.85 for group V. $I/\Delta K$ is estimated using data on provision for capital consumption. With regard to the estimate of r, see the Technical Notes at the end of the chapter.

b. Country selection depends upon data availability, but the countries for which the influence of rich mineral resources is great are excluded. The countries selected are as follows (in reverse order of the level of Y/L in 1970):

Group I (less than US$300)
- A: Burkina Faso (Upper Volta) (116), Mali (142), Malawi (148), India (240), Burma (177), Tanzania (179)
- B: Ethiopia (124), Niger (158), Nepal (168), Benin (176), Somalia (253), Madagascar (272)

Group II (US$300–750)
- A: Kenya (324), Thailand (366), Cameroon (473), Pakistan (519), Philippines (524), Ivory Coast (542), Bolivia (737)

B: Sierra Leone (365), Mozambique (388), Sudan (431), Senegal (452), Sri Lanka (459), Congo (520), Liberia (729)

Group III (US$750–1,200)
A: Korea, Rep. of (754), Turkey (787), Paraguay (802), Morocco (977), Malaysia (1,087), Syrian Arab, Rep. of (1,146)
B: Egypt (767), Tunisia (786), El Salvador (869), Honduras (913), Colombia (1,195), Nicaragua (1,255)

Group IV (US$1,200–3,000)
A: Dominica (1,284), Brazil (1,450), Peru (1,780), Portugal (1,829), Mexico (2,455), Greece (2,920)
B: Yugoslavia (1,475), Costa Rica (1,855), Jamaica (1,975), Uruguay (2,209), Argentina (2,496), Singapore (2,606)

Group V (Over US$3,000)
A: Japan (3,822), France (6,532), Denmark (6,649), Netherlands (7,047), Belgium (7,098), Canada (9,476), United States (11,335)
B: United Kingdom (4,792), Finland (4,853), New Zealand (5,743), Switzerland (6,771), Germany, Fed. Rep. of (6,893), Norway (6,956), Sweden (9,390)

For each group and subgroup, the average level of Y/L is as follows:

(US$, 1970)

Group	Subgroup A	Subgroup B	Average
I	155	204	180
II	499	477	488
III	794	954	874
IV	1,953	2,103	2,028
V	7,423	6,506	6,965

Sources: World Bank, *World Development Report,* 1982 and 1987 issues, and *World Tables,* 1978 issue; International Labour Organization, *Economically Active Population, 1950–2025,* Volumes 1–5.

hypothesis. The rank order of the yardstick for subgrouping is incremental output per investment, ΔY/I.

The Inverse U-shape Pattern

In order to identify the patterns of growth performance in a cross-sectional perspective,[21] let us begin by looking at the most popular term, the rate of output growth (G_Y). Over the entire range of countries—from group I with the lowest level of Y/L to group V with the highest level of Y/L—we see no clear linear trend of either acceleration or deceleration. Instead, there appears to be a pattern of an "inverse U-shape" with group III being the pivot point: that is, the trend accelerates from group I to group III and then decelerates towards group V. This pattern is of utmost concern for it suggests that the factors determining the rate of output growth operate differently between the two intervals: the factors affect output growth in one manner for low- to middle-Y/L level countries and in another manner for middle- to high-Y/L level countries. Regularity is observed in the two subgroups, A and B, without exception suggesting that the way in which factors operate is uniform through subgroups A or B.

The rate of increase in the economically active population, G_L, generally shows a pattern of similar shape, peaking at group III at both the group and subgroup levels. However, the trend of G_L differs from the trend of G_Y in that the level tends to decrease much more distinctly from the pivot point to the group of industrial economies. Therefore, the rate of increase in the product per worker, shown approximately by (G_Y–G_L), presents a different pattern: instead of an inverse U-shape, an increasing trend is observed throughout the entire range of countries from group I to group V, although the variance in the latter's interval is very slight. Regarding subgroups, the regularity is a bit blurred. One could argue that this is to be expected because of the grouping of the countries in terms of Y/L. No affirmative supposition is possible, however, as the level and the rate of change are quite different matters. What is important to note, however, is that the tendency of acceleration in the rate of output growth at the lower levels of Y/L, i.e., groups I–III, also operates well with respect to the rate of increase in output per worker, while at the higher levels of Y/L, i.e., groups III–V, a remarkable slowdown of the rate of labor increase tends to overcome

---

21 Usually, for cross-sectional analysis, a larger number of country groups is statistically desirable and adopted. In this case, a smaller number is supplemented by subgrouping. We believe that this procedure is better in identifying the patterns rather than merely adopting a greater number of groups using a simple yardstick.

the deceleration effects of output growth. The demographic and other factors determining labor force performance is treated here as given. Thus, we can only point out that the rate of increase in labor supply seems to be associated with changes in the level of Y/L, with a distinct difference between two groups of countries: groups I–III and groups III–V. For the first group, i.e., groups I–III, the accelerating pace of rising product per worker is witnessed despite the sustained pressure of increasing growth of labor force supply.

Attention is also drawn to (G_Y–G_L) which is greater for subgroup A and smaller for subgroup B. For subgroup A, (G_Y–G_L) is of substantial magnitude before the pivot point, and in fact is greater than the average value of (G_Y–G_L) for all groups. The results in the table suggest that nonassociated factors are influential, although the difference at the subgroup level tends to narrow at higher Y/L levels.

Let us turn our attention to the performance of investment and output-related terms; that is, the share of investment to GDP, I/Y, which is the basic term, and the ratio of incremental change in output to investment, $\Delta Y/I$. The latter term is often referred to as the "productivity or efficiency" of investment and is a crucial term in our formula. The reciprocal of this term, $I/\Delta Y$, is the more familiar incremental capital-output ratio (ICOR).

The investment ratio tends to be greater for groups with higher Y/L levels, but its range of variance is rather narrow. This pattern is also generally found for subgroups A and B. That is, for both subgroups, higher levels of Y/L are associated with higher levels of I/L (except for group IV where the investment ratio was lower than that of group III for both subgroups). What draws attention, however, is the fact that the investment ratio is not necessarily smaller for countries in subgroup B as compared to those in subgroup A. In three groups—groups III, IV, and V—the investment ratio is greater for subgroup B than subgroup A. This may be surprising as the rate of output growth is definitely smaller for subgroup B. This pattern has a deep significance in understanding the factors determining growth performance. Because of the various big-push theories which have long been influential, people tend to believe that increases in the investment ratio is the primary factor augmenting the output growth rate. While no one would deny the importance of increases in investment as a factor for growth, the evidence suggests that a mere increase in the investment ratio does not necessarily lead to a greater rate of output growth.

More importantly, the behavior of the term $\Delta Y/I$ (i.e., IOCR) corresponds well with the pattern of output growth. The IOCR shows an inverse U-shape pattern, with the pivot point occurring at group III. This is also identified in subgroups A and B. Thus, investment efficiency, as measured by the ratio of the incremental increase in output to investment, or IOCR, exhibits a regular pattern without any distortions. This quantitative finding is of great significance and is in contrast to the usual approach which has more or less been influenced by the conventional use of ICOR which is assumed to be constant. Between the peak value of group III and the lowest value of group V, the difference in the magnitude of $\Delta Y/I$ is as large as 8.7 percentage points, amounting to about 40 percent of the average. The differences between subgroups A and B are sizeable as well. For the country groups at the lower Y/L levels, the differences tend to widen up to group III, and for higher groups, the difference narrows.

It is tempting to explain the inverse U-shape pattern of output growth with a corresponding pattern of investment efficiency. However, we do not think this can be an "explanation" since the latter pattern is not yet explained. To give a real explanation, the factors responsible for the variation in investment efficiency must be understood. Discussion on this will be developed in the following section, but for the time being we suggest that this pattern is closely relevant for the testing of our hypothesis that the basic factor is the variance in a nation's social capability of borrowing and assimilating technological and/or organizational knowledge borrowed from predecessors.

The term $w\Delta L/I$ presents an incremental increase in the labor costs per investment as viewed from the enterprises' standpoint. In growth accounting, this element is taken when the residual is estimated. (We will touch on this treatment later.) However, this term has another significant implication for the discussion on the performances of the various country groups.

As stated earlier, $w\Delta L/\Delta K$ which is the incremental form of the term Lw/K or the factor-input ratio (FIR) can imply the type of technology adopted in terms of the intensity of the factor combination. That is, the greater the value of $w\Delta L/\Delta K$, the more labor-intensive is the technology, and the smaller the value of $w\Delta L/\Delta K$, the more capital-intensive is the technology. Note that w is included by definition. The trend of $w\Delta L/I$, which is shown in the table, suggests the expected pattern; that is, as the Y/L level rises, the technology adopted appears to be more capital-intensive. However, this suggestion cannot be simply accepted because there are differences between gross investment, I, and the incremental increase in the capital stock (i.e., new investment), ΔK, that are not taken into account. If the ratio $I/\Delta K$, which indicates the rate of capital replacement, can be accurately estimated, we can legitimately talk about the term $w\Delta L/\Delta K$ which represents the incremental change in the FIR. (This term will be discussed in detail later.) However, deriving estimates of $I/\Delta K$ is a difficult task; nevertheless, an approximate value of the term is given in the table (see Technical Notes).

The measures of $r\Delta K/I$, which measures the contribution of the capital input, exhibit a pattern associated with the inverse U-shape: the term increases from group I to group III and decreases sharply from group III to group V. Moreover, with the exception of group I, $r\Delta K/I$ is clearly larger for subgroup A than subgroup B within each group, suggesting that the contribution of capital inputs is superior in subgroup A than in subgroup B. (This term is also significant in its relation to the labor term performance, but this is discussed later.)

The rate of return on capital term, r, requires special explanation. As is explained in detail in the Technical Notes, the value of r is estimated as a byproduct of our procedure. Nevertheless, its performance deserves attention. Through groups I through IV, the rate of return on capital follows the familiar inverse U-shape pattern, increasing up to group III and then declining for the last two groups. With respect to the subgroups, the value of r is greater for subgroup A than subgroup B with the exception of group I (however, data reliability for this subgroup is doubtful). We are convinced of the relative significance of the term's absolute value and argue that the marginal productivity of capital tends to be greater in subgroup A than in subgroup B.

Thus, a further explanation of special conventional inputs does not appear to be necessary: by analyzing the various country groups, a regular pattern of the inverse U-shape is discerned. However, if attention is merely paid to the country groups, the analysis would be biased; the difference between the subgroups exhibits another regular pattern as well and casts a challenge for further analysis.

The Significance of the Residual Growth

Let us look at the behavior of $\Delta R/I$ which is derived from the formula $\Delta Y/I-(w\Delta L+r\Delta K)/I$. The range of variation of this term is rather wide between the country groups, but a regular pattern is discernible. The inverse U-shape pattern is exhibited in the developing economies, i.e., groups I–IV. In the lower-Y/L level country groups, the magnitude of the residual growth per investment tends to increase and from group III, the residual declines. However, residual growth per investment increases again from group IV to group V. This is an important conclusion in light of which all the previous output and input patterns can be interpreted.

The turning point at a specific Y/L level (i.e., at group III), which represents countries with Y/L ranging from US$750 – 1,250, is of great interest. Two different patterns can be distinguished for developing economies. Up to this Y/L level, the activities of developing economies tend to augment the residual per investment by accelerating innovational activities, while at higher Y/L levels (i.e., group IV) the development activities are less vigorous. However, this trend is not sustained to group V, i.e., the developed economies, suggesting that there is a revival of activities.

The different magnitudes of $\Delta R/I$ between subgroups is sizeable and deserve attention. In each country group, the magnitude of $\Delta R/I$ is greater for subgroup A and smaller for subgroup B without exception. For group I, residual growth per investment is even negative for subgroup B. The performance of subgroup A as observed through the various groups distinctly exhibits the pattern identified earlier, while that of subgroup B does not. The sizeable variance in the growth of the residual between subgroups cannot be explained in the same way as was the pattern between the various groups. Nonassociated factors are certainly a significant issue. This point, as well as the inverse U-shape pattern, is discussed in depth in the section that follows. However, several points deserve attention.

First, at the pivot point (group III), the average $\Delta R/I$ is extremely high in subgroup A. This suggests that among the individual nations in the subgroup an even higher magnitude of the residual must have been realized. For subgroup B, on the other hand, the average $\Delta R/I$ is extremely low and for individual countries in this subgroup, even negative values can be expected. The differential between the two subgroups is widest in group III. Thus, there appears to be a trend where the differential widens in the lower Y/L groups and narrows at the higher Y/L groups.

Second, at higher Y/L levels, why does the declining trend in $\Delta R/I$ not continue? While the effects of some distortions should be allowed, we are convinced that the pattern is generally valid and are inclined to argue that the dip in the residual growth that occurs in group IV is a distortion from a continuous pattern of residual performance from groups III to V. This issue is discussed further in Section III.

In order to elucidate upon the significance of the residual, further discussion on two points is necessary here. The first discussion focuses on a comparison of the pattern of residual growth between the two measurements: ΔR per investment and ΔR per output which is the conventional approach. The second discussion looks at the treatment of labor employment in our measurement, including some conceptual problems that exist.

The residual has usually been measured as a ratio with respect to output (i.e., $\Delta R/Y$). The growth rate of the residual (where Y is the output in the base year) is often expressed as G_R to show its contribution to G_Y. There are several conceptual differences between this output approach and the investment approach. In the latter approach, the residual pertains specifically to the criteria of capital formation, whereas the residual as measured by the output approach is a measure viewed from a national standpoint, including labor. The labor term per investment, i.e., $w\Delta L/I$, represents the "costs" to enterprises, whereas $w\Delta L/Y$, the corresponding term in the output approach, refers to the earnings of workers. In other words, the output approach is neutral in the sense that it makes no distinction between capital and labor. To put it simply, the output approach aims at measuring the contribution of total factor productivity (TFP) as the source of aggregate output growth, where the contribution of conventional inputs are given by $r\Delta K/Y$ and $w\Delta L/Y$.

Estimates of the residual measurements using the output approach are shown in Table 1.2 together with the incremental ratios. The output approach yields virtually the same inverse U-shape pattern for the growth rate of conventional inputs, $r\Delta K/Y$ and $w\Delta L/Y$, and the residual, $\Delta R/Y$ (columns 1–3 in the table). Therefore, a comprehensive discussion of these terms is not given. It should be noted that the similarity in the patterns of the terms using the two approaches occurs because the effect of an I/Y variation is limited in the present case under review. If the variation in I/Y was stronger, different patterns would likely result from the two approaches.[22]

Let us turn our attention to the changes in the incremental ratios. The term $\Delta R/\Delta Y$ (i.e., G_R/G_Y) exhibits an increasing trend for all groups with the exception of group IV. In group I, $\Delta R/\Delta Y$ is less than 15 percent, but rises to more than 45 percent in group V. The

---

22 These patterns appear to be broadly similar to those described in H. Chenery and M. Syrquin, "Typical Pattern of Transformation." What appears dissimilar is the level of the pivot point particularly with respect to the TFP (residual) growth. In their study, the pivot point is much higher ($2,100 per capita GNP) (see Figure 3–8: 71), although differences in the procedures make it difficult to compare the two studies.

Table 1.2
Conventional Inputs and the Residual, 1960–1985: Using the Output Approach [a] ***(percentage)***

Group [b]		ΔR/Y	rΔK/Y	wΔL/Y	ΔR/ΔY	rΔK/ΔY	wΔL/ΔY	ΔR/wΔL
I		0.46	1.56	1.13	14.8	48.3	36.9	0.41
	A	1.11	1.61	1.21	28.4	41.1	30.9	0.92
	B	-0.05	1.40	1.05	-3.3	58.2	43.7	-0.05
II		1.28	1.94	1.09	30.0	41.2	27.0	1.13
	A	1.94	2.28	1.18	38.0	38.8	23.1	1.64
	B	0.60	1.67	0.99	19.9	47.6	29.8	0.61
III		1.95	2.92	1.36	34.0	41.6	23.9	1.43
	A	3.04	2.97	1.32	45.2	35.1	17.9	2.30
	B	0.86	2.93	1.38	14.7	53.7	46.3	0.62
IV		1.12	3.11	0.88	28.3	48.5	22.4	1.26
	A	1.31	3.12	1.00	29.4	49.8	21.8	1.30
	B	0.91	3.08	0.77	27.9	48.5	23.4	1.19
V		1.66	1.52	0.71	45.7	34.4	19.4	2.33
	A	2.12	1.52	0.72	51.9	30.4	17.7	2.94
	B	1.19	1.53	0.70	37.7	40.2	22.1	1.71

Notes: See Table 1.1.

Sources: See sources of Table 1.1.

major reason for the sustained increasing trend, as opposed to the declining trend in the last two groups that was identified earlier in the investment approach, can be found in the increases in ICOR. From Table 1.1, ICOR is estimated to be 6.6 for group V, which is far greater than the 4.2 found for group III.

With respect to differences between subgroups A and B, the residual per output is definitely higher in subgroup A than in subgroup B, a pattern that is similar to the one that was observed in the investment approach. The contribution of conventional inputs, as indicated by rΔK/ΔY and wΔL/ΔY, is the other side of the coin. The contribution of labor, wΔL/ΔY, tends to decrease throughout the entire interval. The contribution of capital, rΔK/ΔY, on the other hand, exhibits an increasing trend from group II to group IV, and its value falls in group V. (Figure 1.1 illustrates these patterns.)

With respect to the second topic, labor employment, our formula requires data on labor input. But the estimates of "economically active population" do not exactly meet this requirement as the data pertain to labor supply. Nevertheless, use of the economically

Figure 1.1
The Contribution of Conventional Inputs and Residual to Output Growth

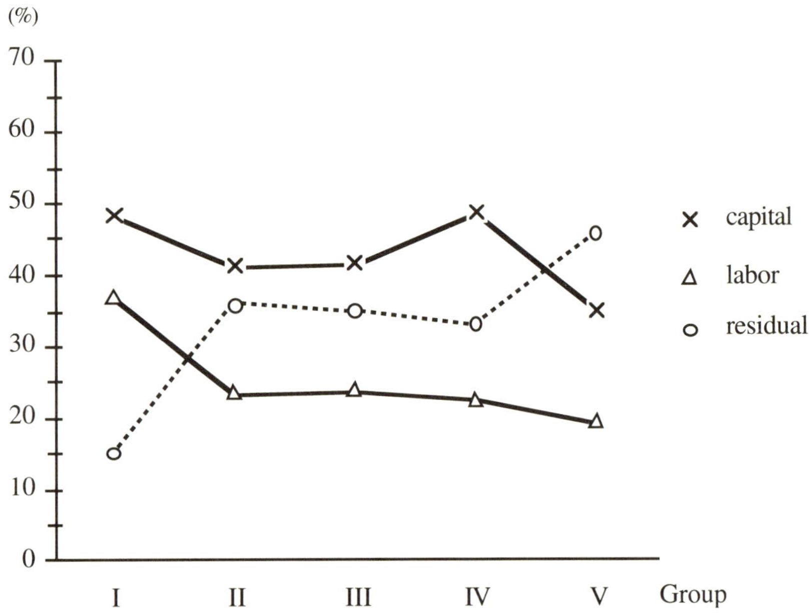

active population as a substitute can be justified for the purpose of obtaining an overview, but the insights gained must be viewed with reservation and qualification. First, it must be recognized that the characteristics of the employment problem for developing economies differ from that of the industrial economies. A discussion of these differences which directly pertain to the application of our formula follows.

The existence of self-employed and family workers mostly in the agricultural and services sectors presents a problem of underemployment in developing economies, as these types of workers are characterized by either low productivity or low working hours. Unfortunately, the data provide no information on this problem. In conventional project appraisal, the opportunity costs of labor for modern enterprises are counted not by prevailing wages but by their marginal product of labor (MP). If we follow this procedure of shadow pricing, then the value of wΔL would be lower and the incremental residual, ΔR, would accordingly be higher in our macro-accounting estimate to the extent that w>MP, even though individual enterprises pay labor the prevailing wage rates. This treatment is analogous to the macro-analytical concept and procedure of estimating the productivity effects of labor shifts from the sector of lower productivity to the sector of higher productivity. The productivity increase measured at the macro-level is higher than the weighted average of the increase in productivity of sectors. This higher productivity is referred to as the "structural effects" or "reallocation effects." We prefer this treatment insofar as output is concerned because it is operational at the macro-level, and the structural effects can be interpreted as one of the important components of the residual. This interpretation is based on the understanding that the incremental increase in output here is considered without the technological and/or organizational changes that were elucidated upon in the preceding section, and the change in output pertains to the individual sector. It is our view that at lower output per worker levels, i.e., the first part of the inverse U-shape path, structural effects play a considerable role in accelerating the growth of the residual, while its influence tends to decrease as the level of output per worker approaches the level of developed economies. (For details, see Appendix II.)[23]

III. Towards an Interpretation of the Growth Mechanism

Because of the simplicity and roughness of the measurements, the patterns of growth of the postwar developing economies must be viewed with reservation and qualification. Nevertheless, we believe that the overview obtained from the cross-sectional analysis is useful for understanding the postwar growth performance

---

23 On the labor-cost side, one may wonder whether overestimates can be avoided by using the term $w\Delta L$. As suggested earlier by the example of shadow pricing, the existing inequality, $w>MP$, is not taken into account and thus the residual is underestimated for labor of self-employed farmers and their family workers. In practice, prevailing wages are imputed for labor of self-employed and family workers in measuring the relative income shares, β (this is explained in detail in the Technical Notes). To this extent, the difference is not counted. We admit that certain overestimates of real labor input are involved, but no adjustment is possible at this stage of data availability. For the lower groups, the degree of such overestimates of the labor term may be greater. However, because of the assumption, $\alpha+\beta=1$, this would lead to underestimating α.

of developing nations. Along the lines of the hypothesis set out in Section I, let us try to interpret its characteristics, focusing on the growth mechanism and the problems involved.

The Interaction Mechanism

Of primary interest here is that the differences in the growth performances of nations are sizeable and that the factors responsible for these differences are clarified by specifying the underlying mechanism. In the preceding section, two simplified approaches of comparative observation were employed: the measurements were analyzed first by the various levels of product per worker between the five groups, and second by the two subgroups. The first approach can be interpreted as representing the "vertical axis" so to speak through which the pattern of the inverse U-shape and others relevant to it were identified. The second approach can be interpreted as representing the "horizontal axis" where the variance within each group was dealt with. We would like to explain the different growth performances of the nations combining these two axes. It is convenient to begin the analysis through observation of the variance within each group. From the measured magnitudes of the terms listed in Tables 1.1 and 1.2, let us first examine two key terms broken down into two subgroups.

First, let us begin with $\Delta Y/I$, the term for which no artificial assumption is made. The range of variance of this term is far greater between the subgroups than between the groups. The difference between subgroups A and B as a ratio of the average of each group is 34.1, 40.8, 43.5, 31.2, and 38.9 for groups I through V, respectively, while the percentage change in the average value of $\Delta Y/I$ between the groups is an increase of 14.9 from group I to group II, and 5.4 from group II to group III, and a decrease of 17.1 from group III to group IV and 18.5 from group IV to group V. The range of variance is no doubt much greater between subgroups than between groups, though changes are witnessed between the two intervals before and after the pivot point. An important point is that the difference between subgroups tends to increase towards the pivot point and is sustained at a considerable magnitude even after it. Thus, if the performance of investment efficiency is observed only in relation to changes in the Y/L level (i.e., along the vertical axis), the analysis would not reflect reality.

Another observation can be made by careful inspection of the performance of the crucial term $\Delta R/I$. As mentioned earlier, the effects of technological and/or organizational advances tend to increase with successive levels of Y/L even with respect to subgroup B. However, the differences in $\Delta R/I$ between subgroups A and B are of considerable magnitude, particularly in the first three intervals: 6.1 (I), 6.4 (II), 10.9 (III), 2.2 (IV), and 4.8 (V). The factors augmenting the nation's capability of absorbing advanced technological knowledge are suggested to operate more intensively in subgroup A than in subgroup B. Why is there such a difference? A simple answer may be that the level of social capability is higher for subgroup A than for subgroup B, where the variance is presumed to be "given" historically for each nation. Rejecting such a deterministic view, one may say that the government is responsible for the nation's capability to absorb technological knowledge through its strategy and policymaking, particularly those policies that affect human resources enhancement. Although this may partly explain the differences in the social capability of nations, it does not provide a complete and systematic explanation.

This discussion began with analyses of the differentials between subgroups because of its quantitative significance. But differences in the Y/L level also occur between groups as well, and this should also be treated in a common framework. What we would like to propose is that the level of social capability is upgraded through what we call the "interaction mechanism" which operates in a cumulative way along the nation's growth path. That is, upgraded capability raises the level of product per worker and the higher level that is realized in turn brings forth a higher level of capability. In the process of economic development, this interaction mechanism operates in an extremely complex manner, involving supply-production, demand-expenditure, and income distribution within a given external environment, even if our concern is narrowly confined to economic phenomena. Nevertheless, it is possible to trace the major paths by which the mechanism operates.

First, an increased Y/L level will accompany a higher level of per capita income through the income distribution system of each country. A higher income level will lead to higher consumption and savings through certain behavioral and institutional functions of each nation. What is conventionally called "consumption" includes expenditures for education, and thus with higher consumption, there results an upgrading of individual human capability.

Second, the output and/or income distributed to enterprises increases investment with accompanying improvements in the technology level through the process of assimilating borrowed knowledge. "Learning by doing" by technological experts and workers will upgrade their human capability, which links the function to the next step. It is assumed that the supply-demand relationship is involved in these endogenous routes though of course they are also affected by external factors.

Third, it is difficult to treat a nation's system of organizations and institutions. For certain dimensions, the associated factors operate through the interaction mechanism in upgrading the nation's capability level. However, nonassociated factors are also responsible for affecting a nation's growth performance. This can be said for both the private and public sectors. The government, which depends on an expenditure-revenue scheme, participates in the interaction mechanism with the private sector. The government's strategies and policies are important points of issue. However, these strategies and policies are not necessarily adopted with respect to the varied Y/L levels and cannot be treated functionally in the interaction mechanism. For example, increased revenues resulting from increased taxes due to the Y/L upgrading will not automatically lead to government action that augments the level of a nation's capability. Another example that is relevant to our concern may be illustrated by industrial policy for which the technology problem is inseparable. Human resource enhancement policies such as education and training may not necessarily be consistent with the requirement of technology advance.

Nonassociated factors actually include broader areas. In the actual process of economic development of nations, the initial socio-cultural conditions that are historically given continue to work through subsequent years. Furthermore, the political situation, whether it be stable or unstable, also greatly affects growth performance. With respect to economic factors, the functions of various public and private institutions and organizations are strongly affected by these situations.

While the description is sketchy, it may aid in the understanding of differences in the growth performance between groups and subgroups that was revealed earlier. In essence, (1) the sustained operation of economic factors in the interaction mechanism is the core to our interpretation of the rising trend of the residual along the path of increasing Y/L levels which was found for both subgroups A and B; and (2) the differences in the performances between the two subgroups stem from the operation of nonassociated factors such as government strategies and policies or other conditions within the nation. These nonassociated factors often operate in mutual relation with one another and it is beyond our ability to separate the effects of them.

Our interpretation thus assumes that the interaction mechanism operates more cohesively and more efficiently when nonassociated factors work in favor of growth performance. Conversely, less favorable effects of the nonassociated factors will make the mechanism operation weaker and less efficient.

Unfortunately, it is beyond the scope of this analysis to trace the links between the two categories of factors thus specified in a macro-framework in the individual countries. However, the difference in the magnitude of the residual between subgroup A and subgroup B conforms to the inverse U-shape pattern. Differences between subgroups in the operation of the interaction mechanism in terms of cohesion and efficiency tend to widen in the lower Y/L country groups and narrow in the higher Y/L country groups. This implies that the variance in the residual between the individual countries is not only sizeable but also changes in relative magnitude. In the lower Y/L country groups, the initial difference which is wide does not tend to narrow but rather tends to widen as we move up the Y/L groups. The cross-sectional treatment cannot tell us the development path exactly and caution is warranted until the concept of phasing is discussed in Chapter 4. Nevertheless, the operation of nonassociated factors is clearly a decisive element determining the growth performance of individual nations. In particular, the operation has cumulative effects: one goes forward and the other backward.

To the important question of why the difference between subgroups narrows, an answer cannot be given at this stage. However, what follows is closely relevant. The magnitude of the residual of subgroup B, though small, shows an increasing trend except for group III, while that of subgroup A decreases at the pivot point, i.e., group IV (see columns 6–9 in Table 1.1). In other words, in forming the pivot seen in the group average, the effects of subgroup A dominates. The interaction mechanism becomes less cohesive and less efficient with respect to subgroup A from group III to IV, while subgroup B catches up to a certain extent. What is needed is a more detailed interpretation of the performance of subgroup A. We will come back to this point later.

Constraints

With regard to the different growth performances of nations as well as the ups-and-downs in the growth rates, the notion of constraints is popular and seems acceptable insofar as the conventional view is concerned. The problems emanating from deficits of foreign payments externally and budget deficits of the government internally are well-known. These deficits surely bring forth constraints on growth. Therefore, one may be led to consider that the differences in growth behavior between subgroups A and B, for example, can be explained by the degree to which constraints operate.

While constraints of this kind appear to be greater for subgroup B than for subgroup A, we do not consider this phenomenon to be the basic constraint on the upgrading of the growth performance of each country. The interaction mechanism works not only for internal domestic relations but also for external relations. For example, given the same international environment, countries in subgroup A which have a higher level of exports would face less of a constraint to the extent that these countries are better able to quickly shift the supply curve to the right as they are more competitive in the international market than countries in subgroup B. (A more detailed discussion of this is given in Chapter 6.) But a nation's capability of realizing this cannot be gained abruptly and it takes a long time to build such an effective operation of the interaction mechanism. For subgroup B, foreign payments deficits are assumed to be greater due to their lower level capability in export promotion, and the constraints these nations face largely stem from the effects of the operation of the nonassociated factors.

Another constraint often discussed is the lack of capital provision. Indeed, a major part of international "economic cooperation" efforts has been focused on this issue, the validity of which is largely accepted. However, we wonder if we have not become too preoccupied with the notion of capital constraint. Investment induced by capital inflow can only be efficient when it is combined with appropriate human and organizational capability. In our framework, the less favorable growth performances of countries in subgroup B cannot be explained by the constraint of capital. This was suggested earlier by a higher level of investment proportion of GDP, I/Y, in subgroup B than in subgroup A. The problem at issue is not the constraint in terms of supply of capital resources but the lower efficiency of the capital that is being used.

Table 1.3 is arranged to verify these assertions using the framework presented in this study. For this purpose, two terms are selected: (1) the resource balance and (2) total long-term debt disbursed and outstanding. Both terms are expressed as ratios of GNP. (Although other terms could have been used, we believe these two are suitable for our simple purposes.) Resource balance refers to the difference between exports (X) and imports (M). (A more detailed discussion on merchandise trade, which is the major part of the resource balance, is developed in Chapter 6.) The trends in the resource balance are clear and only a few comments will suffice. A resource balance of negative magnitude is generally greater for subgroup B than for subgroup A. With the exception of 1960 and 1970 for group I, the difference is considerable. As always, such a simple comparison using averages tends to conceal the dispersion among the

Table 1.3
Resource Balance and Long-term Debts (as a percentage of GNP)

(1) Resource Balance (in terms of M–X)

Group	& Subgroup	1960	1970	1980	Average (1)	Difference (2)	(2)/(1)
I		4.9	7.1	11.7			
	A	5.7	7.1	11.4	8.1		
						0.1	22.5
	B	4.0	7.0	11.9	8.0		
II		5.8	2.0	6.7	4.8		
	A	2.0	2.0	4.9	3.0		
						3.7	79.2
	B	9.6	2.0	8.4	6.7		
III		2.4	2.0	6.7	4.0		
	A	1.2	2.5	5.2	3.0		
						2.0	50.0
	B	3.5	3.5	8.1	5.0		
IV		3.7	4.7	5.8	4.7		
	A	1.7	3.5	5.5	3.6		
						2.4	51.1
	B	5.7	6.3	6.1	6.0		

(2) Long-term Debts

Group		1970	1980	1985	Average
I		21.5	35.7	59.6	39.0
	A	27.5	40.7	62.2	43.5
	B	15.5	30.7	56.9	34.4
II		25.5	43.5	63.8	42.6
	A	23.9	40.4	65.2	41.9
	B	21.1	46.6	62.3	43.2
III		18.6	32.8	64.6	38.8
	A	18.3	22.7	56.2	32.6
	B	18.9	42.8	73.0	44.9
IV		24.4	34.8	71.8	47.3
	A	18.9	24.1	55.7	32.9
	B	29.8	45.4	87.9	54.4

Notes: See Table 1.1.

Sources: World Bank, *World Development; Reports* various issues, and *World Tables,* various issues.

individual countries: some countries' position in subgroup B may be better than that of some countries belonging to subgroup A. However, the difference can be identified broadly as being significant for groups II, III, and IV. Given the difference in the output growth rate, G_Y, which was shown in Table 1.1, the association between the two terms is confirmed and the validity of the conventional notion of a constraint appears to be substantiated. As stated earlier, however, we do not necessarily believe this to be the case. Incidentally, such a long and sustained phenomenon should be distinguished from the behavior of short-term business cycles. In the latter, greater deficits in the current account of foreign payments clearly operate as a lower ceiling. When adjustment policies become effective, a higher rate of growth would be attained. In the former case, such a "ceiling" is sustained at lower or higher levels in the long-term perspective. Therefore, this pattern can be explained by the factors operating endogenously in the countries.

Second, the term "long-term debts/GNP" (both public and private) is taken as a suitable indicator of the behavior of capital inflow over long-term intervals. There are two alternative arguments regarding the interpretation of this term. One argument is that greater capital inflow will enhance economic activity in the country. This implicitly assumes that there are no constraints on enhancing the contribution of capital formation undertaken in the economy. The other argument takes note of the possible operation of constraints of one kind or another. In an ex post observation, the actual results would reveal the answer to the question of which argument best fits reality. Although the term is not easy to interpret, the use of this term, long-term debts, appears in our view to suggest the validity of the latter argument, although the effects of capital flight cannot be checked.

On average, the ratio of long-term debts to GNP is greater for subgroup B than for subgroup A in groups II, III, and IV. An exception is found only for group I, the lowest group, where the reverse is true; we speculate that this is due to the richer contribution of natural resource endowment in the lowest group. The differences in the ratio between subgroups A and B in groups II, III, and IV do not seem to follow a regular pattern. For individual years, minor exceptions are witnessed in 1970 and 1980. Yet, we contend that for most developing economies, a larger ratio will be found for subgroup B than subgroup A.

What we are keenly concerned with, of course, is the major pattern. Despite a higher capital inflow and domestic accumulation of capital for a quarter of a century, the growth performance of subgroup B in terms of G_Y, $\Delta Y/I$ in general and $\Delta R/I$ in particular, has been distinctly less favorable when compared to subgroup A. No doubt the lack of capital is not the basic constraint. It is our view that the basic factors responsible for the difference in performance are of an endogenous nature. The real constraint is, we believe, the imbalance between the nation's production capacity and the nation's social capability.

Production Capacity and Social Capability: The Problems Involved

It may be in order to clarify the stock concept that is used in the investment approach. To link this concept with the process of capital accumulation, it is convenient to transform equation (2) in terms of new capital formation, ΔK, by using the following equation,

$$\varepsilon = I/\Delta K.$$

Equation (2) then becomes

$$(3) \quad 1/\varepsilon(\Delta Y/\Delta K = \Delta R/\Delta K + w\Delta L/\Delta K + r)$$

and apart from variation in ε we obtain

$$(3)' \quad (\Delta Y/\Delta K - r) = \Delta R/\Delta K + w\Delta L/\Delta K.$$

Given the rate of return on capital, r, the term $\Delta Y/\Delta K$ or the incremental COR, varies depending upon changes in $\Delta R/\Delta K$ and $w\Delta L/\Delta K$. If the direction of COR change is assumed *a priori* to indicate the type of technological change *a lá* Harrod, the conditions of forming the inequality $(\Delta R + w\Delta L) >< \Delta K$ must be explored. However, we do not attempt to do this because we are interested in the relation between variations in ΔR and $w\Delta L$.[24] The numerical values of the relevant terms are shown in Table 1.4.

To say "decomposition" is not precise but is conditional since the estimated rate of return on capital is shown to vary (Table 1.1). Yet, a broad pattern can be discerned. Capital formation accompanies residual growth ($\Delta R/\Delta K$) on the one hand and creates employment ($w\Delta L/\Delta K$) on the other. The term $w\Delta L$, rather than ΔL, indicates this as it implies employment with varied quality levels of labor. The relationship between the two dimensions is one of the crucial problems of developing economies. The incremental labor-output ratio, LOR which is defined as $\Delta Y/w\Delta L$ in the table, corresponds to the incremental COR and is a simple indicator of this relationship.

---

24 The idea was originally developed in Kazushi Ohkawa, "Capital Output Ratio and the 'Residuals'. Issues of Development Planning," *IDCJ Working Paper Series*, No. 28 (March 1984).

Table 1.4
Decomposition of Incremental COR by New Investment [a]

	Average			
	ΔY/ΔK	ΔR/ΔK	wΔL/ΔK	ΔY/wΔL
I	27.1	4.0	9.8	2.8
II	31.7	9.9	8.6	3.7
III	34.0	11.5	8.3	4.1
IV	24.8	7.2	5.7	4.5
V	28.3	13.0	5.6	5.0
	Subgroup A			
	ΔY/ΔK	ΔR/ΔK	wΔL/ΔK	ΔY/wΔL
I	31.0	8.5	9.6	3.2
II	39.0	14.8	9.0	4.3
III	42.4	19.2	8.4	5.1
IV	30.0	8.5	6.5	4.6
V	33.7	17.5	6.0	5.7
	Subgroup B			
	ΔY/ΔK	ΔR/ΔK	wΔL/ΔK	ΔY/wΔL
I	23.7	-0.5	10.2	2.3
II	24.6	4.9	8.1	3.1
III	25.3	3.7	8.0	3.2
IV	20.7	5.8	4.9	4.3
V	22.8	8.6	5.1	4.5

Note: a. All terms are calculated from terms in Table 1.1.

Sources: See sources of Table 1.1.

The investment approach can be used for the analysis of productivity and employment problems as they are linked with capital formation. Leaving the interpretation of this aspect, for later discussion, however, let us take up the major topic of the relationship in the stock dimension.

What we call production capacity, PC, is defined as the combined existence of capital and labor (K:Lw) with technology incorporated. As touched upon earlier, wΔL/ΔK is the incremental (reciprocal) term of K/wL or the factor-input ratio (FIR) which pertains to the nation's PC. We defined social capability (SC) as a

Table 1.5
Rate of Capital Stock Growth and Related Terms

	Group			Subgroup					
	Average			A			B		
	G_K	G_Y–G_K	G_K–G_L	G_K	G_Y–G_K	G_K–G_L	G_K	G_Y–G_K	G_K–G_L
I	3.2	-0.1	0.9	3.4	0.5	0.9	3.0	-0.6	0.8
II	4.0	0.0	1.7	4.7	0.3	2.1	3.4	-0.4	1.3
III	5.8	-0.3	2.7	5.8	0.7	2.8	5.8	-1.3	2.7
IV	5.3	-1.3	3.5	6.3	-1.6	4.2	4.4	-1.1	2.8
V	4.9	-1.3	3.8	4.9	-0.7	3.7	5.0	-1.8	3.9

stock concept and we discuss the relationship between PC and SC, though the latter is not directly measurable. The earlier discussion on the interaction mechanism, of course, pertains to the development process of upgrading these stocks. Therefore, one may say that the above explanation using the investment approach cannot be completed without extending it to include the stock approach. The answer is usually given by using capital stock both as a concept and as a measurement with regard to the nation's PC. This is along the line of the neoclassical analysis which was touched upon as the output approach earlier. To follow along the line of the neoclassicalists, we need to know the rate of capital stock accumulation (G_K) and not the level of capital stock itself. In our view, the investment approach can thus be linked with the output approach, not automatically, but by giving the magnitude of G_K conditionally.

In earlier discussions, the rate of return on capital was assumed to be given. If r is given, G_K can be measured by using the equality $r\Delta K/Y = \alpha G_K$. Using the values listed in Table 1.1, we can estimate G_K. Note that the variation of r implies differences in stock life. This is shown in Table 1.5 together with related terms.

Although the figures in Table 1.5 tell us nothing new beyond what is given in Table 1.4, they illustrate what the investment approach implies for the stock accumulation process. First, for the group, $G_Y \cong G_K$ in the first three groups (i.e., groups I, II, and III) but in groups IV and V, $G_Y < G_K$. In sharp contrast, $G_Y > G_K$ for subgroup A while $G_Y < G_K$ for subgroup B in the lower Y/L groups. Second, with regard to the term ($G_K - G_L$), there is a rising trend throughout all groups. By subgroups, the difference is rather narrow, though for subgroup A, (G_K-G_L) is larger for most of the developing economies.

Using these facts, let us now discuss the relationship between the nation's SC and its PC. K/Lw, an indicator of the nation's PC, is transformed into growth terms as $G_K-(G_L+G_W)$. In addition to the conventional term (G_K-G_L), the rate of increase in real wages (G_W) is included.[25] Therefore, what is implied in the investment approach is the relationship between G_R and $G_K-(G_L+G_W)$ in the stock dimension, though the qualification "conditional" is needed.

The performance of the term, $G_K-(G_L+G_W)$, will be discussed later. Here we are concerned with conceptual clarification. Comparison by subgroup is most important in this regard. The relationship between SC and PC is distinctly different between subgroup A and subgroup B. According to the discussion developed earlier, the operation of the interaction mechanism is superior in subgroup A countries (and inferior in subgroup B) because of the more balanced relationship between PC and SC in subgroup A. Backed up by these empirical facts, the problems involved in the conceptual framework can be further clarified. There are several points at issue.

First, let us turn our attention to the condition that the original hypothesis of borrowed technology involves a strong tendency for developing economies to develop an imbalance between PC and SC in their development path, where the growth of a nation's SC falls behind that of its PC. Usually, the difficulty of technology transfer is reflected in the differences in factor prices between the developed and developing economies. This pertains to PC formation and no one can deny its importance. It is our view, however, that the balance between the PC and the SC of a nation is significantly different between countries that are technology leaders and countries that are technology borrowers. In countries that are technology leaders, where there is a balanced relationship between the production capacity and social capability, advanced technology is created and

---

25 Using the output approach, the conventional formula is $G_y = G_R+\alpha(G_K-G_L)$, when referring to the relationship between G_R and (G_K-G_L), where G_R and $\alpha(G_K-G_L)$ are assumed to be relevant to SC and PC, respectively. Another formula, $G_R = (G_Y-G_K)+\alpha(G_K-G_L)$, may be better in showing the correspondence with the investment approach in which $\Delta Y/\Delta K$ and $w\Delta L/\Delta K$ are dealt with, assuming a given rate of capital return.

used; but in countries that are technology borrowers, this balance is not given *a priori* and must be brought forth.

However, the two dimensions, substitution and capability, do not necessarily contradict each other but can operate together in a mutually enforcing way. If we are asked to incorporate the idea of substitution into the framework, SC would be interpreted simply as the capability of substituting human work of augmented human capability, HC, for capital in absorbing advanced foreign technological knowledge.[26] This would be the important component of absorbing advanced technology.

Second, reference to social capability must be taken up because earlier discussion on the interaction mechanism pertained to the formation of a nation's SC. Social capability is used in the production activity of enterprises in our framework and is combined with the nation's PC. As explained earlier, the magnitude of the residual is determined by the combined activities of PC and SC; this is the reason why the magnitude of the residual can be used as an indicator of the effects of SC. The assumption behind this line of reasoning was that inputs and outputs could be evaluated through the operation of the market price mechanism. However, we must recognize that SC which is beyond measurement also pertains to the operation of the market mechanism: in particular, factor prices are involved in $w\Delta L/\Delta K$ and product prices are involved in $\Delta Y/\Delta K$. However, we cannot always assume a priori that there is perfect free competition but we must accept "imperfectness" in the markets.

Here we face a difficult problem; measurements are after all approximations. The more substantive issue is: to what extent can improvement of the market mechanism in developing economies contribute to upgrading the level and components of the economies' SC? In earlier discussion of the interaction mechanism, the contribution of the market mechanism was shown (although there are certain limits). In addition, the importance of government intervention was also recognized. A simple assumption that is clear and attractive is that fuller operation of the market mechanism can lead to economic development which brings forth a faster growth of the residual. In our framework which emphasizes the significance of the relationship between a nation's PC and its SC, however, such a straightforward causal linkage between the market mechanism and residual growth cannot be presumed; we believe that fuller

---

26 For details, in light of Japan's experience, see Ohkawa and Kohama, "Technology Borrowing and Human Resources," Lecture 6, *Lectures on Developing Economies*.

operation of the market mechanism by itself does not necessarily result in the economy attaining a greater capability for absorbing advanced technology. The complexity comes from the factual challenge. In Chapters 5 and 6, this problem is discussed in response to the challenge.

Third, the concept of the nation's capability has hitherto been used without making the distinction between private and public activities. This is because the government contributes substantially to forming a nation's social capability for absorbing advanced technological knowledge. However, the distinction between the two must be made by treating separately the effects of strategies and policies adopted by the government. The use of the term SC is thus confined to the activities of private and public enterprises. A better relationship between inputs and outputs is basically determined by a higher level of SC, and government policies no doubt affect it from outside.[27]

With respect to the function of the government, the problem of balancing a nation's SC and its PC should also be applied to government activity in the formation of SC and government policies which affect the use of SC. The problem of government strategies and policies can, in our view, be adequately evaluated, in light of its performance in balancing the formation of SC and its activity in influencing the pattern of SC use to production in enterprises. A critical view of an import substitution policy is that it results in the adoption of capital-intensive technology and/or industry. If this view is taken, our framework should evaluate the import substitution policy with respect to the efforts of the government in forming the level and components of social capability to balance with the formation of the nation's PC.[28] This problem is discussed in Chapter 6.

---

27 With regard to measurements, it must be recognized that the residual growth is an indicator of the combined activities. We recognize the SC as such and no attempt is made to isolate the private and public activities.

28 This would again lead us to complexities, including the problems associated with interrelated action between the public and private sectors. Full treatment of these problems is beyond the scope of our analysis. A simple theoretical model can be built assuming that the nation's SC is solely upgraded by the government, leaving its use entirely to the enterprises. However, as was noted earlier, this is far from what occurs in real life.

Fourth, we wish to clarify changes in the relationship between PC and SC that occur in a nation's long-term path of development. (This is dealt with using the concept of phases and fuller discussion is developed in Chapters 4 and 6.) The earlier finding of an inverse U-shape pattern provides the starting point for the discussion and the historical performances of presently developed nations which are analyzed in Chapter 2 will provide background knowledge. Preliminary illustration is given by the data shown in the previous two tables. Because the data are cross-sectional, they do not directly indicate the various phases of economic development, but they do provide important knowledge for demarcating phases.

Phases can be defined in various ways depending upon the analytical purpose. In our framework, the phases are defined in terms of changes in the relationship between a nation's PC and its SC in terms of the level and the components. As the economy develops along the normal development path, the level of the relationship between the two is raised. In the case where the nation faces constraints on enhancing its SC, a shift from a lower to a higher phase of development can be possible only through upgrading the level of SC; so long as the imbalance continues, the phase shift is not possible. A normal path can be identified empirically by the increasing trend in the residual growth in terms of $\Delta R/\Delta K$. As illustrated by the dip in the group average of the residual term for group IV, deviation from this path is an important issue.

The phase analysis needs a sectoral approach. Due to limited availability of data, the above-mentioned model cannot be directly applied. However, a productivity approach is developed in Chapters 4 and 6, and estimated measurements of sectoral residual growth are given in Appendix II.

Fifth, in examining the nation's PC performance, the relationship between $(G_K - G_L)$ and G_W is important. The near constancy of the term $w\Delta L/\Delta K$ for groups I, II, and III suggests only minor differences between $(G_K - G_L)$ and G_W. That is, wages tend to increase at a rate that is close to the rate of increase in capital intensity in the macro-setting. The cross-sectional results cannot be directly related to time-series outcomes, but two important points are suggested. First, the distinct performance of $(G_K - G_L) > G_W$ for the latter groups characterizes the basic pattern of factor combination during the earlier phase of development. The widely known hypothesis of "unlimited supplies of labor" implies unchanged wages for unskilled labor. This may not necessarily contradict our findings. However, rather than follow that hypothesis, we identify the significance of the performances of G_W and $(G_K - G_L)$, which tends to increase at a moderate pace in the lower-Y/L level country groups. Second, the sectoral contents of this macro-pattern require special attention. The relationship between K/L and w and its change over time differ widely between sectors. The pattern of K/L and w which was mentioned earlier is a weighted sum of these different sectoral patterns. In assimilating borrowed foreign technology, local adaptation may differ between nations in terms of the selection of which industry and/or technology to develop. The adaptation and adoption of borrowed technology is carried out under prevailing factor prices, which are basically determined by the market mechanism but influenced by government intervention. What we wish to recognize is that the regular pattern of the FIR broadly indicates the macro-framework of forming various PC structures conditioned by factor prices. The magnitude of $w\Delta L/\Delta K$ is greater somehow for subgroup A than for subgroup B, but the difference between the two groups is much narrower for $w\Delta L/\Delta K$ as compared to that of $\Delta R/\Delta K$, which is the indicator of SC.

Finally, further discussion is needed on the problem of labor employment, which is suggested by the description of the term $\Delta Y/w\Delta L$. The issue pertains to the relationship between $\Delta Y/\Delta K$ and $w\Delta L/\Delta K$. This relationship is essentially of the same nature as the relationship between SC and PC that was discussed in detail. The issue of "trade-offs" is often raised between productivity increase and enhancement of the employment situation. But this problem cannot be discussed without knowing the effects of residual growth.

As shown in Table 1.4, the difference in the group average between the lower- and higher-level country groups is broad. The term $w\Delta L/\Delta K$ remains stable for the three lower-level country groups and then decreases in the last two country groups. The term $\Delta R/\Delta K$, in contrast, increases for each successive country group, except for a dip in group IV. In the lower-level country groups, the change in $\Delta R/\Delta K$ between the groups is greater for subgroup A, while only moderate changes occur in the term $w\Delta L/\Delta K$ for both subgroups. Thus, the term $\Delta Y/w\Delta L$ tends to increase in the lower-level country groups, while in the higher-level groups, the decrease in $w\Delta L/\Delta K$ contributes to a decline. Second, comparing by subgroups, the term $\Delta R/\Delta K$ is significantly larger for subgroup A than for subgroup B across all country groups; the term $w\Delta L/\Delta K$ is also consistently larger in subgroup A, though the differences between subgroups are smaller in magnitude. The term $\Delta Y/w\Delta L$ is greater for subgroup A for all country groups. Taken together, these terms provide an explanation of the pattern of the term $\Delta Y/\Delta K$ which was shown in Table 1.4. It goes without saying that the relationship between productivity and employment tends to be more favorable for subgroup A and less favorable for subgroup B: both $w\Delta L/\Delta K$ and $\Delta Y/w\Delta L$ are lower for subgroup B than for subgroup A. This illustrates differences in the level rather than a "trade-off." Sectoral analysis is developed in Chapters 3 and 4.

IV. Technical Notes: Estimates of Conventional Inputs

In applying our formula, output is given as

$$\Delta Y/I = G_Y(\Delta Y/Y) / (I/Y)$$

by use of the given data. As a result, estimation of the residual pertains solely to the conventional inputs measurement. As mentioned in the text, the major task is concerned with the estimate of income shares, α and β, and the estimate of the average coefficient of capital replacement, $I/\Delta K$.

The Income Share of Labor

Income shares are conventionally measured by first estimating labor's share, β, with the assumption that $\alpha+\beta=1$ for distributed income at factor costs. The capital and labor inputs are estimated by using these shares, where $\alpha G_K = r\Delta K/Y^*$ and $\beta G_L = w\Delta L/Y^*$ (Y^* stands for output at factor costs net of depreciation allowance). In defining the residual using formula (1), no specification has been given, but in the measurements, the conventional approach of using Y^* instead of Y (i.e., GDP at market prices) is followed. The conversion is approximated by a flat use of the ratio $Y^*/Y = 0.90$ for all groups of developing economies. For group V, a ratio of 0.85 is used (for some background data, see Table 1.8). Thus, the residual in terms of Y^* is first measured and then converted to "the residual per gross investment" or $\Delta R/I$.

The data sources are: (1) for β in manufacturing, World Bank, "World Development Indicators," Table 8, *World Development Report 1987*, (Washington, D.C.: World Bank, 1987). Note that β is for added value which was converted to Y^*; (2) the ratio Y/L in agriculture to that of nonagriculture in 1960, 1970, and 1980 was calculated using data from World Bank, *World Development Report*, 1982 and 1987 issues, and *World Tables*, 1983; (3) the historical series of β estimates for the United Kingdom, France, Germany, and the United States came from Simon Kuznets, "Distribution of Product and Income," Chapter 4, *Modern Economic Growth;* (4) for agriculture, data are from Yujiro Hayami and

Vernon W. Ruttan, *Agricultural Development: An International Perspective,* Revised edition (Baltimore: Johns Hopkins University Press, 1985), Tables 7–8; (5) for the case of Japan, (a) for nonagriculture, Ryoshin Minami and Akira Ono, "Yoso Shotoku to Bumpairitsu no Suikei" [Estimates of factor income and shares], *Keizai Kenkyu* [Economic review] 29 (April 1978), (b) for agriculture, Saburo Yamada and Yujiro Hayami, "Agriculture," Chapter 4, Part II, in Kazushi Ohkawa and Miyohei Shinohara, eds., *Patterns of Japanese Economic Development,* and (c) for differences by scale of establishment, data were from Shokichi Motai and Kazushi Ohkawa, "Small-scale Industries: A Study on Japan's 1966 Manufacturing Census," *IDCJ Working Paper Series,* No. 11 (December 1987); and (6) for postwar developed economies, data came from Edward F. Denison, *Why Growth Rates Differ: Postwar Experience in Nine Western Countries* (Washington, D.C.: Brookings Institution, 1967) and Hollis Chenery, Sherman Robinson and Moshe Syrquin, *Industrialization and Growth* (London: Oxford University Press, 1986).

Sources in (1) provide total earnings of employees as a percentage of value added which can be used as β_M, labor's income share in manufacturing, covering almost 100 nations. For our purposes, simple averages of β_M for the countries in our specified groups are made. In order to estimate the income share of labor in the macro-economy, supporting information for β_N, labor's income share in the nonmanufacturing sector are needed. By using sectoral income shares as weights, β_M and β_N can be aggregated to β. Sources in (2) are used as follows. Developing economies are characterized by increasing differentials of productivity and wages between the modern and traditional sectors. At any given point in time, the differential in productivity between the two sectors is usually greater than the differential in wages. Thus, in the present case $\beta_N>\beta_M$ is expected with a tendency for the degree of inequality to increase as we go from lower- to higher-income country groups. The ratio of Y/L in agriculture to Y/L in the nonagricultural sector measured for our groups using source (2) provide broad information for estimating the degree of $\beta_N>\beta_M$. Sources (3) and (4) are used to check the results thus estimated in light of the historical records of currently developed economies.

Labor's income share in manufacturing (β_M), as shown in columns 1–3 in Table 1.6, presents a variation of the U-shape pattern across groups. Labor's income share in manufacturing declines from group I to group III, but begins to increase from group III to group V. In particular, note the sharp increase from group IV to group V.

Table 1.6
Income Share of Labor in Manufacturing and Related Terms [a]
(percentage)

	Share in Manufacturing, βm					Productivity Ratio, Y/L, of Agriculture/Nonagriculture			
	1970		1983–1985		Average	1960	1970	1980	Average
I	35.5	(6)	36.1	(6)	35.8	21.7	28.4	27.1	23.1
II	34.3	(8)	33.3	(10)	33.9	21.6	26.2	25.5	24.4[b]
III	31.8	(8)	30.1	(8)	31.0	27.9	34.5	39.5	33.9
V	35.1	(8)	29.4	(7)	32.3	33.6	36.1	38.1	35.9
V	48.5	(11)	46.7	(11)	47.6	55.4	54.9	62.9	57.8

Notes: a. For share in manufacturing, the number in parenthesis reflects the number of selected countries.
b. Three countries are excluded as they were considered to be exceptional.

Sources: See text.

Variation in the original data by country appears considerably wide within the same group. Yet the average pattern shown is convincing because it can be interpreted consistently with the patterns identified earlier with respect to key terms such as G_Y, $\Delta Y/I$, and $\Delta R/I$. As industrialization goes on, increases in product per worker in manufacturing is sustained at a higher pace than that of wage earnings in the lower-Y/L level country groups. At the Y/L level represented by group III, the pattern reverses. The slower rate of increase in wage earnings in the lower-Y/L level country groups is due to flexible supplies of labor; the pattern reverses in the higher-Y/L level groups as supply of labor becomes more flexible. A remarkable and drastic change is witnessed from group IV to group V, representing the final path toward developed economies. The sectoral ratio of output per worker, Y/L (the last four columns in the table), shows an increasing trend with each higher-Y/L level country group for the aggregate, as is expected. This is simple evidence of the narrowing of sectoral differentials in the macro-economy. The two patterns shown in Table 1.6 suggest that manufacturing does not necessarily accompany or is accompanied by a narrowing of differentials in the lower-Y/L level country groups, but does accompany it in the higher-Y/L level country groups. This is particularly conspicuous in the case of group IV to group V.

It may be possible to use the estimated value, and, as a matter of fact, it was used as the preliminary estimate of labor's income share in our research. However, in the final analysis, further refinements were made to fit the analysis into our three-sector approach that is used in the subsequent analysis.

First, with regard to β_M, two adjustments were made: one adjustment reflects the effects of establishment of small-scale enterprises and the other adjustment reflects the effects of differences in β between manufacturing and nonmanufacturing within the industrial sector (i.e., mining, construction, and public utilities). Because international data are not available for these values, rough adjustments were made using Japanese data (see source 5c for β in manufacturing and Ohkawa and Rosovsky, *Japanese Economic Growth*, Appendix for β in nonmanufacturing). The sum of these two effects amount to 25 percent of the original value of β_M, of which the former is about 20 percent. In source (4), it is stated that "where possible, [the data were] standardized by the coverage of establishments to a cut-off point of those with 5 or more employees" (Ohkawa and Rosovsky, 277). However, we question whether more than 5 smaller-scale establishments were excluded. Japan's census data from 1966 suggest that a drop of "under 9" affects 20 percent of total β_M .

Second, with regard to nonmanufacturing, the agricultural and services sectors are approached separately. For agriculture, the basic source used is (4). The original data pertain only to the historical series of the United States and Japan. Use of these averages for cross-sectional analysis is conditional and modification of the data is attempted using other sources. The results of the examination that was based on the preliminary estimates, however, in particular the relation of β_M (now β_I, for the industrial sector) with the performance of β_A thus obtained, appears acceptable (for details, see Appendix II). With regard to the services sector, in light of the preliminary examination of Y/L of this sector and the data listed in sources (3), it was determined that independent estimates were not possible. The term β_S is assumed to be 48 percent for all groups of developing economies, and is the average of the weighted averages of β_A and β_I by group. The use of a constant value for β_S is to avoid possible distortions to the aggregate pattern of β by using unreliable estimates of β_S. With regard to group V, β is approximated with the use of data from sources in (6) and β_S is derived as a residue.

Table 1.7
Estimates of Labor's Income Share in the Aggregate (percentage)

	Labor's Income Share			Product Share			Weighted			
Group	A	I	S	A	I	S	A	I	S	Sum
I	58.4	50.6	48.0	43.2	11.4	45.4	25.2	6.0	21.8	53.0
II	57.7	49.9	48.0	32.2	23.2	45.6	18.7	11.0	21.9	51.4
III	53.2	45.6	48.0	27.4	25.8	46.8	14.6	11.8	22.5	48.9
IV	53.6	47.5	48.0	17.3	33.2	43.5	14.5	15.8	20.9	51.2
V	49.9	61.6	73.9	5.1	38.5	56.4	2.5	23.7	41.7	69.0

Finally, β_A, β_I, and β_S are aggregated to a macro-β using as weights the average share of sectoral product for the period 1960–1980. The results are shown in Table 1.7. For the estimates of key terms and for the sum β, rounding was used.

Three comments need to be added. First, in Kuznets' estimate of β, the lowest value of β is 54 in 1860–1869 and 1905–1915 for Great Britain, 56 in 1853 for France, and 53 in 1895 for Germany. These values can serve as good reference points. Second, with regard to the postwar data for the nations in group V, we have a number of research results, although reference to these results is not made here. The value of the macro–β term ranges from 65 to 75, suggesting 70 as the average value. Third, in comparing all the data, it is important to note that our estimates of β assume that the procedure of imputation is valid for growth accounting analysis. Other procedures—for example the application of the same β of modern sectors to that of traditional sectors, agriculture, and small-scale manufacturing—may show different (much smaller) values of β.

Capital Inputs

The relationship between investment flow and capital stock raises debatable issues even when the scope of discussion is confined to the problem of measurement (i.e., excludes the basic conceptual problems). The description below simply aims to explain the procedure adopted and provide some background on the procedure. Corresponding to the case of labor, our formula is originally $\alpha G_K = K_r/Y \cdot \Delta K/K$ and $r\Delta K/Y \cdot Y/I = r\Delta K/I$. But in case of no capital stock estimates, we do not have direct information to estimate the incremental increase, ΔK, or the new addition to the capital stock. Our method is to estimate ΔK in relation to gross domestic investment, I, by using the coefficient of capital replacement, $\varepsilon = I/\Delta K$. How to assess realistic values for this term, however, is a

Table 1.8
The Capital Replacement Ratio as Determined from the Use of Depreciation Allowances [a, b] ***(percentage)***

Group			D/GDP		(I–D)/I		I/(I–D)
& Subgroup			1970	1980	1970	1980	(average)
I		(11)	6.7	5.8	42.4	64.9	1.54
	A	(5)	6.7	6.5	38.7	66.3	1.52
	B	(6)	6.3	5.1	46.1	63.5	1.57
II		(17)	8.1	7.7	54.2	66.5	1.50
	A	(12)	6.6	7.1	61.0	65.2	1.53
	B	(5)	9.5	8.2	47.4	67.8	1.48
III		(11)	5.4	7.9	68.3	71.1	1.43
	A	(8)	5.5	7.1	68.5	72.5	1.42
	B	(3)	5.2	8.7	68.0	69.7	1.43
IV		(11)	7.0	6.9	66.9	69.1	1.47
	A	(8)	5.6	5.6	66.3	75.0	1.42
	B	(3)	8.4	8.1	67.5	63.1	1.52
V		(14)	10.5	10.6	56.9	51.0	1.87
	A	(7)	10.1	9.7	57.9	50.4	1.86
	B	(7)	10.8	11.4	55.8	51.5	1.87

Notes:
a. The number in parenthesis refers to the number of countries in the group. The last column refers to the sum of the two years. For group I the figures for 1970 are not counted in the average.
b. I is gross fixed investment. Inventory changes are included in I for Mali in 1981 and Madagascar in 1979.

Sources: United Nations, *National Accounts Statistics 1985.*

problem. In this case, we used available data on the provision for capital consumption as a substitute for estimates of capital replacement.

One may wonder why no attempt was made to obtain simplified estimates of capital stock. Indeed, such simplified estimates have been attempted with the use of data on capital formation and depreciation allowance with certain assumptions of the life and length of stock durability. This approach can be used to obtain estimates of the total average capital share or even the several major components of the capital stock. Between this approach and the approach employed in this study, no difference is recognized. What is important, however, is that in most developing economies, the initial magnitude of the capital stock is difficult to assess: as a result, there is a serious need to apply the perpetual inventory method. Without knowing the initial stock value, the capital stock estimates may be subject to serious distortions. Our incremental approach avoids such distortions which can be caused by arbitrary assumptions.

In Table 1.8, the magnitude of the ratio I/ΔK is estimated and corresponds to a reciprocal of the average of the term (I–D)/I where D represents depreciation. The value of I/ΔK in Table 1.1 is obtained from this table.

The performance of I/ΔK through all five groups exhibits a U-shape pattern: that is, I/ΔK declines from group I to group III and then increases from group III to group V. A large value for group V is particularly noticeable. The factors responsible for forming this pattern are differences in life duration, variation in capital stock composition, etc. The level of Y itself is not necessarily associated with the pattern indicated by D/Y in the table. The variation of D/I actually appears to be negatively associated with the rate of growth of output in light of the historical experience of some developed nations. Depreciation allowances, however, are estimated by an accounting procedure, while upswings and downswings of G_Y take place in actual economic growth. Thus, caution must be taken when using D/Y to avoid swing effects.

With regard to estimating the rate of return on capital, r, in order to obtain the value of rΔK/I, data on the life (duration) of the capital stock are needed. If the life of the capital stock can be reasonably estimated using the formula, $\alpha \cdot Y/K = r$, an estimate of r can be obtained. In Table 1.8, data are available for D/Y; so K/Y can be obtained as $K/Y = K/D \cdot K/Y$, in which K/D is a measure of the average duration of the capital stock. However, the data are not available for most developing economies, and the historical data of Japan's case are used in combination with a smoothed ratio of D/Y.[29]

29 The values of K/D were arranged by Nobukiyo Takamatsu. The sources for the data are Kazushi Ohkawa et al., Table 7.1:108, *Long-term Economic Statistics (LTES),* Vol. 1, *National Income,* and Vol. 3, *Capital Stock* , Table 1.1:108. The values of this ratio are as follows:

Year	K/D	Year	K/D	Year	K/D	Year	K/D
1855	55.8	1900	53.2	1915	42.9	1930	35.2
1890	50.4	1905	47.8	1920	37.6	1935	34.4
1895	57.4	1910	45.8	1925	35.9	1940	31.7
(1890–95)	54.5	(1900–01)	48.9	(1915–25)	38.7	(1930–40)	33.8

The value of r in Table 1.1 (i.e., the last three columns of the table) is thus estimated. This procedure implies that r is measured gross of capital depreciation in order to estimate the residual in "net" terms.

Postscripts

These two postscripts are added despite some repetition with the main text. The inclusion of these postscripts is the result of questions and comments received from colleagues on an earlier draft. I hope that the postscripts help to clarify our ideas and style of problem recognition.

1) Relationship Between the Investment Approach and the Output Approach

Why is the residual measured with respect to investment in our analysis? The residual is basically conceived as

$$\Delta R = \Delta Y - (r\Delta K + w\Delta L).$$

If the residual is treated relative to output, Y, the formula becomes

$$(1) \quad \Delta R/Y = \Delta Y/Y - (r\Delta K + w\Delta L)(1/Y)$$

and if the residual is treated relative to investment, I, the formula changes to

$$(2) \quad \Delta R/I = \Delta Y/I - (r\Delta L + w\Delta L)(1/I).$$

The residual pertains to the rate of output growth G_Y ($\Delta Y/Y$) in equation (1), whereas in equation (2), the residual pertains to $\Delta Y/I$ which is the reciprocal of ICOR. The difference between the two terms, $\Delta R/Y$ and $\Delta R/I$, emanates from the magnitude of I/Y, the proportion of investment to output. The relationship can be presented as $G_Y \cdot \sigma = I/Y$.

The difference between the two terms pertains to the way in which decisions affecting growth performance are treated. Our approach presumes that investment is the most relevant variable affecting the decision and that equation (2) is thus most applicable. Investors are supposed to think about $\Delta Y/I$, together with its components, that is, the contribution of conventional inputs and the

---

Based on the average (in parenthesis), rounded figures are used for the groups in the following manner: Group I–55, Group II–50, Group III–43, Group IV–35, and Group V–25. While the duration may appear much longer than expected, we believe this is acceptable because its coverage is the macro-economy, including all structures and buildings including residential.

residual in expectation. The ex post measurements intend to trace these components. An alternative explanation is given by $\Delta R \Delta Y = \Delta R/I \cdot I/\Delta Y$ or $\Delta R/I = \Delta R/\Delta Y \cdot \Delta Y/I$. With the measured total factor productivity (TFP), the term G_R/G_Y is conventionally used to indicate the performance of technological and/or organizational advance. But this term is nothing more than $\Delta R/\Delta Y$ and can be decomposed into two terms at the incremental level, $I/\Delta Y$ (i.e., the ICOR) and $\Delta R/I$. The latter term, $\Delta R/I$, can be identified as a term that measures the combined effects of technological and/or organizational advance and the "output efficiency" of investment (IOCR). Thus, the investment approach does not ignore the purpose of the output-productivity approach at the incremental level. In this sense, the difference between the two terms is $\Delta Y/I$ (or IOCR). Conversely, the productivity approach measures the combined effects of $\Delta R/I$ and $I/\Delta Y$ (ICOR). Viewed from the $\Delta Y/I$ paradigm, $\Delta R/I$ takes into account the effects of $\Delta R/\Delta Y$. Viewed from the $\Delta R/\Delta Y$ paradigm, $\Delta R/I$ takes into account the effects of $\Delta Y/I$.

The numerical values of these terms are shown in Table 1.2 in the main text. An increasing trend for $\Delta R/\Delta Y$ through all groups with a slight dip for group IV is clear. As the Y/L level rises, the proportion of the residual also rises. However, the term $I/\Delta Y$ shows a U-shaped pattern where the pivot point occurs at group III. Thus, in the former groups, the two terms go in the same direction. In the latter groups, the pattern differs: the residual proportion goes up but it is accompanied with higher values of ICOR. Comparing by subgroups, $\Delta R/\Delta Y$ is greater for subgroup A than for subgroup B (an exception for group I is questionable) and $I/\Delta Y$ is smaller for subgroup A without exception. The two terms generally move parallel with one another. In groups III–V, however, the two terms do not move in a parallel fashion. Thus, use of $\Delta R/\Delta Y$ alone is not enough to elucidate the real growth performance of developing economies.

Furthermore, $\Delta R/I$ exhibits an inverse U-shape pattern through groups I–IV, i.e., the developing economies, as a result of a combined pattern of a tendency of increasing $\Delta R/\Delta Y$ and the U-shape pattern for $I/\Delta Y$ (the latter's effect is stronger). Regarding group V, however, the effect of an increase in $\Delta R/\Delta Y$ is stronger than that of an increase in the ICOR, leading to an increase in $\Delta R/I$ for the developed economies. These differences in $\Delta R/I$ across groups pose further challenges for analysis.

A critical view of the $\Delta R/I$ approach may stem from one's doubt that duplication is involved in the sense that investment is counted as the contribution of capital input ($r\Delta K$) to measure the residual. In response to this concern, we recognize that there are differences between the standpoint of investment decision-making and that of measuring capital's contribution. Investment will be made with the expectation of some contribution from conventional inputs (including $r\Delta K$) and nonconventional factors, and we believe no duplication is implied. Nevertheless, it may be worthwhile to clarify the relationship between I and $r\Delta K$ by using the term $r\Delta K/I$. This term is composed of r, the rate of return on capital, and $\Delta K/I$, the coefficient of new investment. The latter term represents a direct link between investment and capital stock. In appraising individual projects, investment is treated as an increase in the capital stock. In macro-accounting, the ΔK are aggregated and capital consumption is counted separately. A return is expected from using the capital stock through production activity. The term r is a marginal measure of this return, forming $r\Delta K$, the contribution of capital. Numerical values of $r\Delta K$ are shown in Table 1.1. The value of $r\Delta K/I$ exhibits an inverse U-shape pattern through all groups: until group III, $r\Delta K$ tends to increase relative to investment, but decreases rather sharply for group V. Since the variance of $\Delta K/I$ is moderate (except for groups IV–V), the inverse U-shape pattern of r is mainly responsible for forming the pattern for $r\Delta K/I$. Without exception, the same is true for the subgroup comparisons.

We would like to emphasize that among the various terms, only $\Delta R/I$, the incremental residual per investment, exhibits "deviated" values for group IV when viewed from the continuous path identified throughout all groups—that is, the U-shape pattern of $I/\Delta Y$ and the inverse U-shape pattern of r and $r\Delta K/I$. Actually, $\Delta R/I$ drops from 8.1 in group III to 4.9 in group IV and then rises to 6.9 in group V. This illustrates that our investment approach is useful in identifying important problems. Answers would require deeper scrutiny, but here we are concerned with the conceptual framework from which such solutions can be derived. This suggests that in shifting from group IV to group V, a difficulty arises and must be overcome. In particular, a sharp increase in $\Delta R/\Delta Y$ is observed from group IV to group V. This suggests that the pace of technological and/or organizational advance is accelerated in this last interval though it is accompanied by an increase in ICOR.

In discussing a discontinuous pattern, another interval, the initial interval between group I and group II, draws attention. The increase in the value of $\Delta R/\Delta Y$ is particularly large from group I to group II. This suggests that economies face a high hurdle to overcome; that is, the economies must realize faster technological and/or organizational progress in order to make the initial shift and this requires at the very least a sizeable increase in the ICOR.

Finally, a few words need to be said about the property of technology and its change. A suitable term for dealing with technology in the $\Delta R/I$ approach is $w\Delta L/\Delta K$, the numerical value of which is listed in Table 1.6 in the main text. This term is an incremental term of Lw/K, which we refer to as the factor intensity ratio (FIR). In the lower-Y/L level country groups, $w\Delta L/\Delta K$ shows a tendency to change slightly, but a sharp decrease is witnessed for the higher-Y/L level country groups, i.e., groups III–V (the pivot point occurs at group III). A drastic change in the performance of FIR is clearly identified. The Harrod type of technological change as represented by ICOR and the Hicks type of technological change as represented by β (or α) are taken into account in our approach in a simple sense in that the former can be represented by the term $\Delta Y \Delta K$ and the latter by the term $w\Delta L/\Delta Y$. The product of these two terms gives $w\Delta L/\Delta K$. This implies that $w\Delta L/\Delta K$ can represent the combined effects of these two elements of technological change at the conventional factor inputs level. The data reveal that for the first few intervals the slight changes in $w\Delta L/\Delta K$ are a combined result of slight changes in both $w\Delta L/\Delta Y$ and $\Delta Y/\Delta K$ and for the latter intervals, the drastic decrease in the value of the term is a result of a large decrease in $\Delta Y/\Delta K$ which overcomes a small increase in $w\Delta L/\Delta Y$. Substantively, $\Delta Y/\Delta K$ plays a much more important role in influencing FIR. This is shown in Table 1.6 in the main text. (In the case of subgroups, β is assumed to be equal between subgroups A and B; thus, FIR is determined solely by $\Delta Y/\Delta K$.) Note that in our framework the type of technology is represented by FIR, which is defined as having nothing to do with output behavior. Such an interpretation is used in this analysis in order to clarify the relationship with the orthodox approach.

2) Use of Residual Measures for Development Planning

The problems of development planning, which were simplified in our analytical treatment, contain such issues as: (1) objectives and means with regard to the longer-term development plan or perspective for a certain country under review; (2) criteria for establishing options in search of an appropriate strategy or policy; and (3) consistent links between the capital investment approach and the labor productivity approach. These three issues are mutually interrelated. The idea behind the framework employed in this study is to use the "residual" as a central term in dealing with these issues. The rationale for this approach is based on our recognition that development is essentially an innovative process. How to use the results of the productivity approach and the investment approach, based on the empirical findings in the main text, is one of the most important issues.

The following two formulas, equations (1) and (2), are essentially equivalent in growth accounting

(1) $G_Y - G_L = G_R + \alpha(G_K - G_L)$

(2) $\Delta Y/\Delta K = \Delta R/\Delta K + w\Delta L/\Delta K + r$

where ΔK is substituted for I. The term ΔK, which represents a new addition to the existing capital stock, is better in linking equation (2) with equation (1). The reason for developing these two formulas rather than merely relying on equation (1), is to get a closer linkage, if possible, between the macro-sectoral and micro (project) approaches. Although no discussion has been made of the latter approach, the economic nature of the micro approach is to identify a greater value for $\Delta R/\Delta K$, although the sum of $(\Delta R/\Delta K + r)$ is often used where r is assumed to be the prevailing rate of return on capital.[30] In this sense, the residual growth is the criterion in dealing with the innovational process and equation (2) can be converted to

(2') $\Delta R/\Delta K + r = \Delta Y/\Delta K - w\Delta L/\Delta K.$

The two terms on the right-hand side of equation (2') are familiar: $\Delta Y/\Delta K$ is the reciprocal of ICOR expressed in terms of ΔK, and $w\Delta L/\Delta K$ is an incremental term of Lw/K (or the FIR), an economic indicator of the type of technology, although it is often simply called the "labor term." We confirm that this term is treated as a "cost" in the accounting of the residual. National development plans set targets in trying to achieve the plans' objectives. By using equation (1), let us illustrate that these targets can generally be set in terms of productivity and labor employment. The equation treats these targets in terms of rates of growth, following the conventional format of development plans of contemporary developing countries. Let us set G_Y, G_L, and G_y which is the rate of productivity increase. The term G_y is determined by G_R and $\alpha(G_K - G_L)$. In terms of social capability (SC) vs. production capacity (PC), the plan must establish policy options that balance

---

30 This statement is very simple and we hope that it is not misleading. The project appraisal in "economic analysis" treats capital and output in a different manner from growth accounting, with time discounting. However, it is assumed that a bridge can be technically made between the two though this needs more careful examination.

the nation's SC and PC. No doubt, the rate of residual growth plays a crucial role and we can identify the significance of the residual term in both approaches, macro and micro. In terms of magnitude, correspondence between the two approaches appears to be consistent. A macro-plan that aims for a higher G_Y essentially needs a higher G_R and this seems to require a higher $\Delta R/\Delta K$ for individual project assessment. However, we should be cautious about this appearance of consistency. The incremental increase in the residual per output level and the incremental increase in the residual per new investment is not the same. The criteria for options in search of the optimum solution differ.

As a matter of fact, in equation (1), G_K implies ΔK but $\Delta R/\Delta K$ is not explicitly given. In order to link equations (1) and (2) in this regard, we need a term which expresses the relationship between output and investment. The ICOR (σ) is the conventional term used for this purpose.[31] As is widely known, the so-called Harrod-Domar formula, $G_Y \cdot \sigma = I/Y$, has often been used to estimate the required investment for realizing the planned level of G_Y. One may suspect that this equation can be used in place of equation (1) to link with equation (2). However, this is not correct for several reasons.

The output-investment ratio is needed but this ratio pertains to the means of fulfilling the objectives rather than the objectives themselves, which were illustrated earlier as G_Y and G_L. The importance of distinguishing between objectives and means must be emphasized. In this regard, equation (2) is now characterized as pertaining to the aspect of means. As is clearly shown in both equations (2) and (2'), we have $\Delta Y/\Delta K$ which is the converted term of σ. In other words, equation (2) is a decomposition of σ into three terms. It implies that an a priori assumption of σ's magnitude, that is separate from production performance as stated by equation (1), should not be adopted. Here, we face a difficult issue. Equation (2) when applied to the macro-growth performance itself, instead of interpreting it in linkage with project appraisal, can be used to estimate a plausible magnitude of $\sigma'(\Delta K/\Delta Y)$ as it may vary by the residual, the FIR, and the rate of return on capital. What can be said is that given the prevailing rates of return on capital or the prevailing interest rates, options will depend upon the relationship

---

31 For most contemporary developing countries, the data needed to estimate G_K are not available. Thus, equation (1) cannot be applied except by rough approximation. Data for the term $\Delta I/Y$, however, are available and σ is easily estimated.

between $\Delta R/\Delta K$ and $w\Delta L/\Delta K$. This difficult issue cannot be treated effectively without employing a sectoral approach. For the time being, however, let us touch upon the problem of employment, looking at $w\Delta L/\Delta K$ as the "labor term" insofar as the macro-framework is concerned.

As suggested earlier, improvement of the employment situation is another objective of a development plan and/or perspective. All the developing economies face this problem; that is, how to improve the employment situation given G_L in terms of supply, which is basically determined by demographic factors. Unlike developed economies, which are faced with the sustained situation of differential marginal productivity of labor, it is difficult for developing economies to operationalize the objective of "enhancing employment situation" in the macro-framework. Nevertheless, it must be made clear that in addition to improvement in the employment situation, productivity increase is another objective that is included in our discussion of the residual growth. Equation (2') shows that a larger (smaller) magnitude of $w\Delta L/\Delta K$ results in the residual being lower (higher). A sort of trade-off appears between the two. As pointed out earlier, in the case of project appraisal, this term is counted as a cost. When the formula is applied to the macro-approach, equation (2') appears to be essentially the same as equation (2) insofar as the terms $\Delta Y/\Delta K$ and r are given. However, we believe that such a presumption is too mechanical. In any real path of innovational growth, the two objectives—productivity growth and employment enhancement—can be pursued if the residual growth can be expected to be favorable. A better illustration of this is given by equation (1). Given a certain magnitude for G_L, a higher rate of productivity growth, G_y, can be expected with a higher rate of residual growth. Here again the crucial factor is G_R. The possible relationships between the two terms need to be examined further using the sectoral approach, and we will come back to this topic later in Chapters 3 and 4 of the main text.

Chapter **2**

Growth and Investment:

A Historical Perspective

The purpose of this chapter is to present another overview of growth performance in a historical perspective, by drawing on the relevant cross-sectional perspective that was developed in Chapter 1. The discussion will begin with selected developed countries (Section I) and will make use of a simplified form of the analytical frame used in Chapter 1. Although time-series data are available for individual countries and a more elaborate "residual" analysis of their performance is possible, this task is beyond our reach. Since our major aim is to deepen our knowledge of contemporary developing economies, a simple, common procedure will be applied for these selected countries with the aim of making comparative observations. A more detailed examination is made for Japan as the data are more easily accessible (Section II). Finally, in Section III, observations of the developing countries over time is made for the period since 1960 in order to make comparative interpretations in light of the knowledge gained from the preceding two sections.

The growth performances of contemporary developing economies has often been viewed in comparison with the historical pattern of presently developed countries. The former is sometimes evaluated by setting the latter as the "normal" or "standard" experience. However, another view rejects such an analogy and emphasizes the significance of the differences between the two groups. In this regard, remarks by Kuznets that were presented in the summary chapter serve to highlight the central problem at issue.

> Except for Japan, the presently developed countries are all in Europe, or are European offshoots overseas. At the time of their entry into modern economic growth they were already economically in advance of the rest of the world, with per capita product at least two to three times as high as per capita product

> today in the populous underdeveloped countries of Asia and Africa. This historical association among modern economic growth, the European affiliation of the developed countries, and their initially high per capita product (all with the conspicuous exception of Japan), makes it extremely important to distinguish, in the analysis of modern economic growth, between elements that can be specifically assigned to this historical connection and those that cannot, and that can therefore be conceived as transferable within non-European institutions and economic contexts.[32]

We are convinced that modern economic growth is spreading to countries all over the world, including economies in which non-Western social institutions and systems prevail. To the extent that this phenomenon goes beyond the historical connection, common and transferable factors need to be identified with regard to the basic growth mechanism. The special characteristics of the developing nations are also important. The relationship between PC and SC is analyzed in terms of the characteristics explained in Chapter 1. In applying this methodology, we believe two factors are crucial. The first factor is the different time dimension whereby technological knowledge spreads from advanced to latecomer countries in a dissimilar milieu of international circumstances. The other factor involves the tremendous difference between demographic patterns. In the postwar developed economies, the rate of increase in total population, and with a certain time lag that of the economically active population, exploded. In the history of the presently industrialized countries, this pattern was milder. Through such comparative observations, we should be able to shed more light on the characteristics of developing economies.

I. The Historical Pattern of Western Developed Nations

An Overview: 1901–1960

For our purposes it will be sufficient to begin the analysis from the turn of the century using seven countries. Although a larger number of countries is desirable, readily available data are limited. Moreover these seven countries are representative and provide a fair overview. The data in Table 2.1 are average figures for the sixty-year period, 1901–1960, and correspond to the data in Table 1.1.

---

32 Kuznets, *Economic Growth of Nations,* 304.

Table 2.1
The Historical Growth Performance of Western Developed Nations Using the Investment Approach, 1901–1960 [a, b, c] ***(percentage)***

Country	G_Y	G_L	G_Y-G_L	I/Y	ΔY/I	wΔY/I	(rΔK/I)	(ΔR/I)
United Kingdom	1.4	0.6	0.8	10.9	12.9	2.3	4.4	6.2
France [d]	1.9	0.4	1.5	19.3	9.6	0.2	3.2	6.4
Italy	2.3	0.5	1.8	17.6	13.3	0.7	4.1	8.5
Denmark	2.7	1.1	1.6	15.1	16.2	3.0	5.4	7.8
Norway	3.3	0.3	3.0	26.6	12.2	0.4	4.1	7.7
Sweden	3.5	0.8	2.7	17.9	19.2	2.1	6.6	10.5
United States	3.2	1.7	1.5	14.9	21.5	3.8	7.2	10.5

Notes:
a. The values reported are simple averages of the estimates for the three periods (1), (2), and (3) that are used in Table 2.2. Due to rounding, some discrepancies occur. As for the estimating procedures, see the notes to Table 1.1.
b. The data from Minami are in fixed investment and are roughly adjusted to total investment, including inventory changes, using an adjustment ratio of 1.18.
c. R. C. O. Mathews, C. H. Feinstein, and J. C. Odling-Smee, *British Economic Growth 1856–1973* (Stanford: Stanford University Press, 1983) is one of the most comprehensive studies in this area. Because of the complexities of the comparative data setting, regrettably we cannot incorporate their data directly into this table. But in discussing Table 2.3 later, we will rely on the TFP measured by their research.
d. 1896–1963.

Sources: Ryoshin Minami, *The Economic Developmentof Japan: A Quantitative Study* (London: Macmillan, 1986), Table 3.4:54 and Table 6.8:184–85; B. R. Mitchell, *European Historical Statistics, 1750–1970* (London and Basingstoke: Macmillan, 1978); John W. Kendrick, *Productivity Trends in the United States* (Princeton, N. J.: Princeton University Press, 1961); Maddison, *Phases of Capitalist Development*, Table C8:210; J. J. Carre, P. Dubois, and Edmond Malinvaud, *French Economic Growth* (Stanford: Stanford University Press, 1976), Table 1.1:21, Table 2.8, and Table 4.6:117.

What is impressive in the table is the fairly wide range of variance in the growth performances of the seven countries.[33] The

---

33 Use of other sources did not make any sizeable difference in the value of G_Y. For example, according to Kuznets' estimates for

most progressive in terms of G_Y are the United States and two Scandinavian countries, Sweden and Norway, while the United Kingdom is the least progressive and the other nations are somewhere in-between. In terms of $G_Y - G_L$ (= G_y), Norway and Sweden are the most progressive, while the United Kingdom lags the farthest behind. In terms of investment efficiency, $\Delta Y/I$, France and Norway have the lowest values and the highest investment proportion, I/Y. The residual per investment, $\Delta R/I$, is presented in parenthesis, because we cannot estimate $r\Delta K/I$ directly. The term $r\Delta K/I$ is an average for the "hypothetical" case where $G_Y = G_K$ for the four periods in Table 2.2, and $\Delta R/I$ is deduced from it. The simple average for the seven countries is 8.0. In the case of G_K/G_Y = 1.15, $\Delta R/I$ is 7.3, and in the case of G_K/G_Y = 0.85, $\Delta R/I$ is 9.1. This range, i.e., 7.3–9.1, cannot precisely specify all possible variations in the magnitude of $\Delta R/I$. However, the range is a plausible one insofar as there are no severe downswings. With respect to the magnitude of $\Delta R/I$ for individual countries, derivation of direct estimates is not possible.

From this range of variations, several points can be found. First, Sweden has the highest value for $\Delta R/I$ and the United Kingdom has the lowest value. (In the case of the United Kingdom, a qualification will need to be made later.) The value of $\Delta R/I$ is higher for the United States than that of the other European countries except Sweden. The estimates broadly reveal the features of growth performance which economic historians have elucidated on. Among these economic historians, the view of Maddison is particularly relevant.[34] According to Maddison, the United States replaced the United Kingdom in the role of lead country in advancing technological progress before the turn of the century and the United States sustained its position at least until the 1970s. In applying our hypothesis, we share Maddison's view. We should not expect the lead country to have the highest magnitude for the residual because of the possibility of a speedier pace of technology absorption of progressive latecomers. In fact, Sweden is the most

---

approximately the same interval, G_Y was 2.7 for Italy (1896/99 to 1963/67); 2.1 for France (1896 to 1963/66); and 3.1 for the United States (1910/14 to 1963/69). Kuznets' values of G_Y for countries not included in our table are 1.9 for Belgium (1900/04 to 1963/64); 2.8 for the Netherlands (1900/1996 to 1963/64), and 2.4 for Switzerland (1910 to 1963/67). (Kuznets, *Economic Growth of Nations,* Table 1:11).

34 Maddison, *Phases of Capitalist Development.*

progressive of the latecomers in Europe and its residual growth surpasses that of the United States. The United Kingdom, the former lead country, saw its growth pace of output and residual per investment weaken during the present century. For the other four latecomers, growth in output and residual performance is moderate, though that of France is close to that of the United Kingdom. The variance of the residual term ($\Delta R/I$) of the individual nations is generally associated with the differences in investment efficiency, as indicated by $\Delta Y/I$, although the influence of the wide variance in $w\Delta L/I$ cannot be ignored. For example, the level of $\Delta Y/I$ is highest for the United States, but so is the level of $w\Delta L/I$ and G_L. Thus, the estimate of $\Delta R/I$ is lower than it would be otherwise. For Norway and France, the reverse is the case. Taken together, these findings can be linked with the post-1960 performances of group V in Table 1.1.

Changes in the Four Periods Over Time

The period under review was eventful: there were two World Wars, an unstable interwar period with the Great Depression, and an early postwar period of recovery. A meaningful demarcation of phases, including the entire interval covered here, may be possible. Maddison's proposal is relevant in this regard.[35] He distinguishes four phases of longer intervals: 1870–1913, 1913–1950, 1950–1973, and 1973 onward. Due to the limits imposed from data availability, we demarcated three shorter periods: (1) 1901–1920, (2) 1921–1940, and (3) 1941–1960. Although taking two decades as a unit of observation may be too mechanical in view of analytical phasing, for the purpose of providing an overview of the Western historical growth pattern compared to the postwar growth pattern of developing countries, the choice of two decades is not likely to lead to serious distortions. A range of variation, 1.15 – 0.85, of G_K/G_Y for estimates of $\Delta R/I$ is assumed plausible for each of the two decades. The estimated results are presented in Table 2.2, in which period (4), 1960–1980, is added for comparison purposes.

The following points can be noted from the table. First, there is a considerable difference in the growth rate of output, G_Y, between the pre-1960 period (i.e., periods 1, 2, and 3) and the post-1960 period (i.e., period 4). Four patterns can be distinguished with respect to the pre-1960 period: (1) the United States, the lead country, sustains a fairly high rate of growth; (2) the United

35 Maddison, *Phases of Capitalist Development.*

Table 2.2
The Historical Growth Performance of Selected Western Nations for Four Periods in 1901–1980 [a] ***(percentage)***

		G_Y	G_L	G_y	I/Y	ΔY/I	wΔL/I	(rΔK/I)	(ΔR/I)
United Kingdom	(1)	0.8	0.5	0.3	7.9	9.6	2.7	3.7	4.2
	(2)	1.9	0.6	1.3	11.4	16.5	2.0	5.7	8.8
	(3)	1.7	0.6	1.1	13.1	12.7	2.3	3.8	6.6
	(4)	2.4	0.5	1.9	17.5	13.7	1.4	3.5	8.8
France[b]	(1)	1.9	0.2	1.7	17.8	10.7	0.7	3.9	6.1
	(2)	1.3	-0.1	1.4	20.2	6.5	-0.3	2.2	4.6
	(3)	2.5	0.1	2.4	20.7	12.1	0.3	3.6	8.2
	(4)	4.5	0.8	3.7	25.0	18.0	1.5	4.6	11.9
Italy	(1)	2.1	0.2	1.9	13.8	15.6	0.7	5.5	9.4
	(2)	1.5	0.5	1.0	19.0	8.1	1.0	2.0	5.1
	(3)	3.2	0.1	3.1	19.9	16.2	0.3	4.8	11.1
	(4)	4.2	0.3	3.9	25.0	16.6	0.7	4.3	11.7
Denmark	(1)	2.8	1.2	2.6	14.6	19.2	3.5	6.9	8.8
	(2)	1.6	1.2	0.4	12.7	12.3	3.8	4.3	4.2
	(3)	3.6	0.7	2.9	21.1	17.0	1.7	5.1	10.2
	(4)	4.0	0.9	3.1	20.5	19.1	2.2	5.0	11.9
Norway	(1)	2.3	0.3	2.0	21.4	10.1	0.5	3.9	5.7
	(2)	3.5	0.3	3.2	24.0	14.6	0.5	4.9	9.2
	(3)	3.9	0.2	3.7	32.8	12.0	0.3	3.5	8.2
	(4)	4.6	1.7	2.9	29.0	15.8	3.0	4.1	8.7
Sweden	(1)	2.5	1.3	1.2	13.6	18.4	4.0	6.7	7.7
	(2)	3.2	0.5	2.7	17.2	18.7	1.1	6.3	11.3
	(3)	4.7	0.5	4.2	20.8	20.5	1.2	6.7	12.7
	(4)	3.2	1.2	2.0	23.0	13.9	2.6	4.9	7.4
United States	(1)	3.3	2.0	1.3	18.2	18.1	4.5	6.6	7.0
	(2)	2.9	0.5	2.4	12.9	22.4	1.5	7.7	11.2
	(3)	3.3	1.5	1.8	13.6	24.4	5.4	7.2	11.6
	(4)	3.7	2.1	1.6	18.0	20.3	6.0	5.3	9.0

Notes: a. The difference between G_L data from the ILO and the World Bank Report cannot be ignored for these countries. The use of World Bank data, however, does result in some differences. However, these discrepancies do not substantially affect the interpretation in the main text. β is assumed constant for all countries within each group as follows: 0.60 for (1) and (2), 0.65 for (3) and 0.70 for (4) (Kuznets broadly assumed 0.55–0.70 for Western nations). The adjustment ratio, NNP/GDP, is assumed to be 0.90 for (1) and (2), 0.88 for (3), and 0.85 for (4).

The estimating procedure is essentially the same as for all terms in Table 1.1 with the important exception of $\Delta R/I$. Since the data to estimate K/D, $I/\Delta K$, and/or relevant terms are not available, we are obliged to simply assume that $G_Y = G_K$ on average for each period. The capital-output ratio may be increasing ($G_K > G_Y$) or decreasing ($G_K < G_Y$). To that extent, the estimates of the relevant terms in the table may change.

As is mentioned in the main text, the possible range of distortions of our crude measurements should be qualified, if possible. The TFP measured by Mathews and others for the British case is summarized as follows in Mathews et al., *British Economic Growth 1856–1973,* Table 17.2:532.

	1856–1873	1873–1913	1924–1939	1937–1951	1951–1973
(a) Rate of TFP growth	0.9	0.5	0.7	1.4	2.3
(b) Rate of TFP growth, excluding contribution from commerce and labor quality-shift	(0.5)	(0.1)	0.7	0.4	1.5
(The figures in parenthesis are less reliable.)					

Because our measures cannot take into account the contribution of labor quality change explicitly, (b) may be more comparable. The term $\Delta R/Y$ as calculated from $\Delta R/I$ in the table is

(1)	(2)	(3)	(4)
0.3	0.9	0.8	1.5

Our figures are generally close to their original measurements. The difference may be partly due to differences in the periods. An examination of our assumption that $G_K = G_Y$ is tried. Using Tables 16.1 and 16.2, the ratio of G_K/G_Y excluding 1913–1924 for which G_Y is near zero is estimated to be:

1873–1913	1924–1937	1937–1951	1951–1973
1.06	0.95	0.61	1.14

We contend that in all periods except 1937–1951, the effects of $G_K \gtrless G_Y$ are in a rather moderate range.

b. For France the periods are as follows: (1) 1896–1913; (2) 1913–1938; (3) 1938–1963.

Sources: See sources of Tables 2.1 and 1.1.

Kingdom, the former lead country, had the slowest rate of growth; (3) the two Scandinavian countries, Norway and Sweden, exhibit what we call "trend acceleration (TA),"[36] namely, G_Y is accelerated without being interrupted by any slowdown; and (4) the performance of three other countries—France, Italy, and Denmark—is in the middle with a sharp slowdown of G_Y during the interwar period. Nevertheless, in the post-1960 period, growth rates accelerated in all of these countries irrespective of their different growth patterns in the pre-1960 period. A notable exception is Sweden. With regard to G_y, because of variance in G_L, the pattern differs somewhat among nations, but note that for both Norway and Sweden, the TA is exhibited for G_y during the pre-1960 period.

Second, with respect to the investment proportion, I/Y, a general increasing trend is witnessed for the pre-1960 interval and a number of nations continued to raise their investment proportions into the post-1960 period. A notable exception is the United States. In period (1), I/Y is at a relatively high level and the increase in this ratio is moderate and slow in comparison with the postwar performances of the developing countries. The term $\Delta Y/I$ tends to increase distinctly in the lead country and along the path of trend acceleration in Sweden and Norway (though it is somewhat blurred). In the United Kingdom, France, Italy, and Denmark, the performance does not exhibit a regular increasing trend and is interrupted in either period (2) or (3). Towards period (4), however,

36 The concept was originally developed in Ohkawa and Rosovsky, *Japanese Economic Growth.*

the magnitude of $\Delta Y/I$ in these countries did increase, while for the United States and Sweden it distinctly declined.

Third, the rate of employment increase over time appears to vary widely for the United States, Sweden, and Norway, while in the United Kingdom, France, and Italy, G_L follows the so-called Western pattern of smallness throughout the entire period. The term, $w\Delta L/I$, does not necessarily appear to be small through the pre-1960 period as it is influenced by a relatively lower level of I/Y, but its variance affects considerably the incremental residual per investment. To that extent, the term $\Delta R/I$ will deviate from the performance of $\Delta Y/I$.

Fourth, in view of our hypothesis presented at the outset of Chapter 1, the trend acceleration (TA) must be typical of successful latecomers because it is an accumulated outcome of the sustained operation of technological and/or organizational advance. The TA was observed earlier in terms of G_Y for Sweden and Norway, but its genuine test must be in terms of $\Delta R/I$. Sweden provides the most affirmative evidence. The term $\Delta R/I$ increases by 7.7 in period (1), 11.3 in period (2), and 12.7 in period (3). If we look at the performances for the intervals from period (2) to period (4), three other countries—France, Italy, and Denmark—also show a trend acceleration. Under an accelerated high magnitude of $\Delta R/I$ for the United States, the potential of TA continued to exist for other latecomers, but other factors prevented its realization, particularly in period (3). Towards period (4), the residual performance differs. The term $\Delta R/I$ decelerated in Sweden, as well as in the United States, the lead country. As is noted in the table, the estimates of capital input are conditional so that these patterns should be interpreted with caution. Nevertheless, the proposal on the trend acceleration may be broadly accepted.[37]

---

37 In the case of Sweden, the value of $\Delta R/I$ is as follows according to the variance of K/Y:

		(1)	(2)	(3)
(a)	$G_K = G_Y$	6.1	9.9	11.0
(b)	$G_K/G_Y = 1.15$	6.0	8.9	9.9
(c)	$G_K/G_Y = 0.85$	7.2	11.0	12.2

The trend acceleration is interrupted distinctly only from period (2) to period (3) by the increase in the capital output ratio as illustrated by the difference between (b) and (c).

Our attention is drawn further to the fact that the country differential of residual growth narrows over time for the entire period under review. For example, on average residual growth is 7.8 percent in period (2); the magnitude of residual growth is small, 4–5 percent, for France, Italy, and Denmark, and large, 9–11 percent, for the United States, Sweden, and Norway. In period (4), however, the magnitude of residual growth is closer to the average 9.9 percent, ranging between 7–12 percent. Because of a possible distortion due to our "conditional" estimates of the residual, these changes in the magnitude of residual growth among nations should be confirmed with further analysis. The existence of an unfavorable vs. favorable international environment is, of course, relevant, but we are concerned with the long-term outcome of the international spread and diffusion of advanced knowledge about technology. The narrowing-down of the residual growth differentials may be an expected phenomenon so long as the technological advance proceeded under United States leadership through the entire interval under consideration.

Comparisons with Developing Nations

In deriving the characteristics of contemporary developing economies, to what extent will earlier observations be useful? The small number of developed nations observed and the limitation of existing data may limit our comparisons between the developed and developing countries. Yet, we suggest some summary observations as shown in Table 2.3. We know that within each group, the growth performance differs considerably with respect to the period and the Y/L level; thus, the table is designed so as not to miss these aspects.

Initially, let us compare periods (4) and (4') in Panel (B) to test the representativeness of the selected six countries. The magnitude of almost all of the terms are fairly representative although they are somewhat biased towards the countries of better growth performance and this should be qualified when required. In Panel (A), two types of averages are shown: the average for all groups of developing nations, and the average for groups II and III. The latter grouping is intended to show the performance of developing economies that have exhibited a vigorous growth path.

As is shown in the table, the output growth rate, G_Y, is definitely greater for developing than developed economies in almost all comparisons. Even subgroup B of the group of developing nations has a higher level of G_Y than the developed economies. The rate of increase in the labor force, G_L, is, as is well known, much greater for the former than the latter in all cases. However, the

Table 2.3
Comparison of Growth Performances of Developing and Developed Nations (percentage)

	G_Y	G_L	G_y	I/Y	ΔY/I	rΔK/I	wΔL/I	ΔR/I	ΔY/wΔL
(A) Developing nations:									
a) Developing nations									
Total	4.2	2.1	2.1	20.8	19.8	8.8	5.4	5.6	3.7
Subgroup A	5.0	2.6	2.4	20.6	24.3	9.8	5.8	8.7	4.2
Subgroup B	3.3	2.2	1.2	21.0	15.7	8.3	5.1	2.3	3.1
b) Developing nations									
Group II, III	4.8	2.5	2.3	21.4	22.5	9.3	5.7	7.5	3.9
Subgroup A	5.8	2.8	3.0	21.3	27.3	10.2	5.9	11.2	4.6
Subgroup B	3.7	2.6	1.1	21.4	15.2	8.7	5.5	1.0	2.8
(B) Developed nations:[a]									
Period (1)	2.3	0.9	1.4	14.9	15.2	5.6	2.7	7.1	5.6
(2)	1.9	0.6	1.3	16.2	15.6	5.2	1.6	8.3	9.8
(3)	3.4	0.6	2.8	20.2	17.1	4.6	2.3	10.1	7.4
(4)	3.7	1.1	2.6	22.2	16.6	4.0	2.8	9.9	5.9
(4') Group V	3.6	1.2	2.4	24.1	15.2	5.2	3.0	7.0	5.1
Subgroup A	4.1	1.2	2.7	22.6	18.1	5.5	3.2	9.4	5.7
Subgroup B	3.1	1.1	2.0	25.9	12.2	4.9	2.7	5.1	4.5

Note: a. For periods (1), (2), and (3), France is not included due to differences in periodization.

Sources: See sources of Tables 1.1 and 2.1.

rate of output growth per labor, G_y, does not appear to be uniform as a result of the canceling out of the two. In contrast, G_y for developed nations did distinctly increase from 1.3–1.4 in 1901–1940 to 2.6–2.8 in 1940–1960. In the developing nations, higher rates of increase occur in subgroup A vs. subgroup B. In fact, in comparison with historical records of developing nations, G_y is relatively low for subgroup B, while for subgroup A, particularly subgroup A in subpanel (b), G_y is close to or even surpasses the high rates realized in later periods.

Capital investment performance is more active and vigorous in contemporary developing nations as compared to the historical path of developed nations. As mentioned earlier, on average, the investment proportion, I/Y, of developing nations is higher (20–21 percent) than that of developed nations (15–20 percent) which showed a gradual pace of increase over the long run. As a result of these trends, the levels of I/Y for the two country groups for the period since 1960 are similar. As was shown in Table 1.1, even the lowest level of I/Y for group I of the developing nations (18 percent) is greater than that of period (3) of currently developed nations. In addition, as mentioned earlier, between the country groups in terms of Y/L levels, the tendency for I/Y to increase is noticeable for the country groups of lower Y/L. This pattern difference draws attention.

Explanations of this pattern, however, may differ. In view of our hypothesis of the spreading-over effect of borrowed technology, we contend that differences in I/Y suggest differences in the willingness of economies to absorb advanced technologies. We do not hold to the idea of "embodied capital," but recognize the fact that technology transfer to latecomers necessitates enhancement of domestic capital formation. The types of technologies vary and sectoral resource allocation of capital formation cannot be uniform. Therefore, we cannot assume any quantitative relationship between the two. This is more so when we consider the varied levels of the nations' capabilities in transferring advanced technologies. Even allowing for these conditional qualifications, it is possible to argue that the greater I/Y at lower Y/L levels indicates a stronger intention to realize faster technology absorption in developing economies as compared to that of the historical process in currently developed nations.

The degree to which a country succeeds in achieving this purpose cannot be precisely measured, but a first-hand rough indicator can be provided by the incremental output-capital ratio, ΔY/I. The value of this ratio for the currently developed nations showed a very modest rise during the pre-1960 period (15–17 percent). Developing countries, on the other hand, showed on average a greater magnitude of ΔY/I of close to 20 percent. The significance of this difference should not be ignored. As pointed out earlier in Table 1.1, ΔY/I is greater for developing than developed economies and this is again shown in Table 2.3. The high level of ΔY/I is not the result of a drastic decline of ΔY/I for developed nations. In fact, its decline in period (4), excluding 1980–1985, was very modest (17.1 in period 3 to 16.6 in period 4). The variance of ΔY/I is wide between subgroups A and B of developing nations. Although the performance of subgroup A is superior, the performance of subgroup B is not necessarily inferior as it is close to the historical level of developed nations. In other words, even subgroup B was able to maintain investment efficiency to the extent that was realized in developed nations. The advantage of latecomers in absorbing advanced technological knowledge from a richer backlog seems to be an explanatory factor.

How about the realization of a higher residual related to the performance of conventional inputs? On average, there is a notable difference in the labor term, $w\Delta L/I$, between developing and developed economies, despite a wide variance in the term among the latter group. This is a significant characteristic which differentiates the pattern of growth between developing and developed nations. Capital's contribution per investment, $r\Delta K/I$, for developed nations is 5.0 on average, ranging from 3.2 to 7.2 (see Table 2.2). A modest change over time is exhibited. The magnitude of $r\Delta K/I$ in developing countries is definitely greater, and even subgroup B shows a value greater than that of developed countries. The value of $r\Delta K/I$ for subgroup A is much higher and is almost twice that of developed nations. Although the magnitude of $r\Delta K/I$ for developed nations is conditionally estimated and needs to be qualified, this difference can be safely confirmed. An important conclusion is thus reached. The sum of conventional contributions is definitely greater in developing than developed economies. This has significant implications for elucidating the rate and type of technological advance in the development process.

There is a notable difference in the magnitude of $\Delta R/I$ between developing and developed nations. The term is 5.6 in period (1) for developing economies and 11.1 in period (4), excluding 1980–1985, and 7.0 in period (4') for the developed economies. This is to be expected as the higher $\Delta Y/I$ of the developing countries is canceled by the greater contribution of conventional inputs per investment. This finding can be broadly valid, even with the conditional value for developed nations and a wider variation of $w\Delta L/I$ for individual developed countries. It is our view that this is one of the most substantive characteristics which distinguish the performances of the developing nations and the historical path of developed nations. The sizeable difference in the residual performance of subgroups A and B of the developing countries is notable. On the one hand, in the case of vigorous innovational development, as represented by subgroup A in panel (b), $\Delta R/I$ surpasses the average historical performances of the developed nations. On the other hand, however, $\Delta R/I$ in subgroup B is extremely low and is much lower than the lowest level of the developed nations' historical performances.

Viewed from the perspective of investment performance, this significant differential between the developing economies is emphasized. We contend that for subgroup A the investment motive is supported by achieving high rates of residual creation through the realization of a growth mechanism similar to that of developed nations on the one hand; on the other hand, for subgroup B such a mechanism cannot be recognized because the intention of augumenting capital formation cannot realize expected residuals. The discussion in Chapter 1 is further substantiated by the comparison of the performances of developing countries with the historical records of developed nations. In terms of the relationship between PC and SC, sustenance of a balance between the two is broadly identified for the nations in subgroup A. Given the different components of production capacity as illustrated by $r\Delta K/I$ and $w\Delta L/I$, these nations created virtually similar rates of residual growth as those realized by the developed nations. It is our view that this is due to the upgraded social capability of these nations and it is possible to identify the "balance vs. imbalance" between countries in subgroup A and subgroup B in a historical evaluation.

A possibility of trend acceleration of increasing $\Delta R/I$ can be supposed for the historical path of developed economies in general, although its realization was often interrupted. We believe this supposition presents a significant suggestion. As discussed earlier, the performance of $\Delta R/I$ varies among the individual latecomer nations. Nevertheless, the performance of $\Delta R/I$ on average suggests a potential for realizing TA. The average value of $\Delta R/I$ in the table (Panel B) actually shows TA and this may be the result of cancelling out the effects of interrupting factors. As was touched upon earlier, the most relevant factor is the narrowing of the gap in technological level. The inverse U-shape of $\Delta R/I$ performance that was identified earlier for developing economies implies a decrease of $\Delta R/I$ from group III to group IV, instead of from group IV to group V. Why is this so? Finding an answer to this question is a challenge for further analysis. Tentatively, we can say that the potential of TA is hampered by certain interrupting factors at this increased Y/L level. (We will come back to discuss this important issue in depth.) At this juncture, the significance of TA as a potentiality for developing economies needs further elaboration.

So far the growth performance has been analyzed from the viewpoint of capital formation. Its dimension viewed from labor's standpoint requires additional clarification. The rate of productivity increase is conventionally discussed in terms of labor input and its indicator is approximated by $G_y = (G_Y - G_L)$ which is shown in Table 2.3. This term, G_y, is important in actual development planning. It may be a surprise to find that the average annual rate of labor productivity increase of the postwar developing economies is higher than that of the developed countries in periods

(1) and (2), i.e., 1901–1940. Despite a higher G_L, the much higher G_Y of the developing economies is the cause of the higher labor productivity increase. The effect of the higher G_Y viewed from the investment standpoint was discussed earlier in terms of $\Delta Y/I$. In terms of labor input, it can be similarly expressed by $\Delta Y/w\Delta L$ or the incremental output-labor ratio. (The incremental output-labor ratio was touched upon in Chapter 1, Section III.) The estimated magnitude of $\Delta Y/w\Delta L$ is shown in the last column of the table.

The value of $\Delta Y/w\Delta L$ for developed economies in past periods is much higher than that of developing economies and shows a decreasing trend from period (2). The average level of $\Delta Y/w\Delta L$ of developing economies, which is markedly lower, stands in sharp contrast to the higher level of $\Delta Y/I$ as compared to the historical performances of developed nations. The difference in $\Delta Y/w\Delta L$ between the two groups appears even more impressive than that of $\Delta Y/I$ which was discussed earlier. The difference between subgroups of developing economies is also considerable. From this additional information (which was also shown earlier in Table 1.6), we know that the magnitude of the incremental output-labor ratio shows an increasing trend through all the groups.

The term $\Delta Y/w\Delta L$ appears to be an incremental indicator of partial labor productivity, Y/Lw; the latter term, Y/Lw, is the reciprocal of Lw/Y, which is treated as β, labor's income share, in our analysis in Chapter 1. The value of β was assumed for each group and was used to calculate $w\Delta L/Y$ using $G_L\beta$, so that $\Delta Y/w\Delta L = 1/\beta \cdot G_Y/G_L$; that is, the variation of the ratio G_Y/G_L is significant. The value of G_Y/G_L is much lower for developing economies than the developed economies: the average value of this term is 1.9 for the developing economies and 2.6, 3.2, and 5.7 for the developed economies in periods (1), (2), and (3), respectively. Large differences in $\Delta Y/w\Delta L$ can be influenced by divergences in the ratio G_Y/G_L. We are convinced that the difference in the incremental ratio, $\Delta Y/w\Delta L$, is of great significance in characterizing the postwar developing economies because the pressure of the employment problem is adequately reflected by the ratio in our analytical framework.

II. The Historical Pattern of Japan: The Case of a Latecomer

In comparing developing economies with the historical performance of currently developed countries, an examination of Japan's case is particularly useful as it presents records of both processes, developing and developed. There is no doubt that Japan today belongs to the group of "industrial economies," but her historical growth performance presents a rich record of the long-term development process starting from a low Y/L level. Japan began its modern economic growth around 1885 following the Meiji Restoration in 1867. Eventually, the economy shifted to the fully-industrial phase at around the end of the 1960s. In the country groups presented in Table 1.1, where the countries were categorized in terms of Y/L in 1970, Japan is classified as a group V country.

Compared to the "follower" countries of Europe, Japan may be characterized as a "late" latecomer because of its low Y/L level in the initial phase. It is important to evaluate the initial economic levels of Japan and Europe on the one hand, and Japan and the contemporary developing countries on the other. Some assert that Japan's initial economic level must have been closer to that of the European followers rather than that of contemporary developing countries. This view supposes that if conditions were otherwise, Japan's success in development would have been impossible. However, this view is not substantiated by empirical evidence, either in terms of the Y/L level or the industrial structure. Another view asserts that Japan's case is unique, and some would say "exceptional," because in the initial phase both the level and structure of Japan's economy were not very different from those of other Asian countries (see earlier citation from Kuznets). If this is so, then Japan's uniqueness is a challenge for our analysis.

The initial level of per capita income or product is certainly an important factor. However, in our view, per capita income or product is not as crucial in determining growth performance as has often been asserted. Our framework emphasizes the significance of the interaction mechanism of growth, rather than the "given conditions." Yet, reliable measurement of the initial conditions is indispensable to shed light on our empirical analysis, and this topic will be discussed separately in Appendix I. The topics below will focus on using Japan's long-term record to enrich our knowledge and to place in historical perspective the growth pattern and mechanism. The characteristics of this Asian latecomer can be elucidated distinctly in view of the fact that in the 1870s, industrialization had already proceeded to a considerable extent in a number of Western countries. Similarities and dissimilarities with Western nations will be understood better by viewing and analyzing Japan's case as a bridge between the Western nations and the postwar developing nations which are much later followers.

Swings and Investment Spurts

An application of the analytical framework and procedure used in the previous chapter to Japan will place in international perspective the issues mentioned above. We have no intention of identifying the unique aspects of Japan's case. Instead, we believe this case is worthy of a detailed treatment in testing our hypothesis. A summary of the measurements is presented in Table 2.4.

Before discussing the measurements in the table, however, a few notes need to be discussed. First, the periods are demarcated by long-swings, which are composed of upswings (U) and downswings (D), in the growth rate of output. Periods denoted by Roman numerals refer to one full swing. This device is used to avoid a mechanical periodization, which was unavoidable in our previous approach to Western nations. The long-swings used here were derived from our previous analysis.[38] We do not necessarily claim a regular occurrence of repeated swings due to the sole operation of the endogenous factors. Fluctuations in the international environment were actually influential, and similar situations can be assumed for contemporary developing nations. Second, the years affected by World War II are excluded from the table, because these years were deemed to be abnormal. For these years the influence of the war on the Japanese economy was extraordinary. Japan's Y/L level immediately after the war, for example, went back to levels as low as those of the latter part of period II. An amazingly high rate of postwar growth in period IV is affected by the quick recovery process. The year 1953 marks the approximate starting point for a normal postwar growth performance.

Some facile analysis argues that Japan represents a typical successful case of an economy catching up quickly with its Western predecessors. This view is strongly influenced by the "miracle" high of postwar growth, without due knowledge of the country's long-term prewar performance which was rather moderate growth that was somewhat higher than that of Western nations. In particular, G_Y is 2.8, 3.5, and 5.5[39] for Japan in periods I, II, and III, respectively, while G_Y is 2.3, 1.9, and 3.4 for the Western countries in periods (1), (2), and (3), respectively (Table 2.3). However, Japan's growth of

---

38 Ohkawa and Rosovsky, *Japanese Economic Growth.*

39 G_Y is 1.8 percent for 1938–1953 and G_Y for period (5) is a rough average.

Table 2.4
The Historical Growth Performance of Japan [a] (percentage)

Period				G_Y	G_L	I/Y	ΔY/I	wΔL/I	rΔK/I	ΔR/I
I		1897–1919	(22)	2.8	0.6	17.6	16.5	1.6	5.9	9.0
	1 (D)	1897–1904	(7)	1.9	0.6	16.5	11.2	1.6	4.2	5.4
	2 (U)	1904–1919	(15)	3.4	0.6	18.1	18.9	1.6	6.7	10.6
II		1919–1938	(19)	3.5	0.9	19.6	18.4	1.9	6.9	9.6
	3 (D)	1919–1930	(11)	2.3	0.8	19.4	12.1	1.8	6.6	3.0
	4 (U)	1930–1938	(8)	5.2	1.0	19.7	26.5	2.1	7.7	17.0
III	5	1953–1969	(16)	10.0	1.6	30.2	33.0	2.4	11.6	19.0
IV		1969–1985	(20)	7.1	0.8	31.5	22.9	1.3	7.8	13.8
	6	1969–1979	(14)	8.5	0.8	31.3	27.2	1.2	9.1	16.9
	7	1979–1985	(6)	3.8	0.9	32.0	12.7	1.4	4.8	6.5

Period	ΔR/Y	rΔK/Y	wΔL/Y	ΔR/ΔY	wΔL/ΔY	rΔK/ΔY
I	1.6	1.0	0.3	54.6	9.7	35.7
II	1.8	1.9	0.4	52.2	10.3	37.5
III	5.7	3.5	0.7	62.7	7.3	30.0
IV	4.4	2.5	0.4	60.3	5.7	34.0

Note: a. The rates are based on seven-year (postwar, five-year) moving averages. U refers to upswings and D to downswings. Period 7 is specifically demarcated to show recent performance. The number of years that were converted by each period is noted in parenthesis. No figure is presented for the interval of 1939–1952, because this period was affected by the war. For 1945, no estimates of GNP and components are available. A rough estimate of G_Y, bridging 1938 and 1953, is 1.8 percent. For the prewar years, labor is used in terms of "gainfully occupied." For postwar years, L refers to the labor force employed.

Sources: Ohkawa et al., *Long Term Economic Statistics,* Vol. 1, *National Income;* Vol. 3, *Capital Stock;* and M. Umemura et al., Vol. 2, *Labor Force;* Postwar official statistics of EPA, Japan Government.

output was much smaller than that of contemporary developing countries. In the case of upswings, Japan had higher rates ranging from 3.4–5.2 percent which are comparable to the growth rates of developing countries in subgroup A. However, in the case of downswings, Japan's G_Y ranges between 1.9–2.3 which is close to the output growth of the Western nations in the interwar period but which is much lower than output growth of subgroup B. Japan's growth appears to have attained an intermediate level between the historical growth rates of Western nations and the postwar growth rates of developing economies.

A similar pattern is witnessed for G_L. The value of G_L was smaller than 1 percent in all periods and attained a minimum of 0.6 percent. Doubts may be cast on the statistical reliability of such a low rate; however, the low rate may be explained by Japan's peculiar demographic behavior of severe birth control which occurred towards the end of the premodern Tokugawa epoch. While this explanation cannot be ignored, the reliability of the low rate cannot be blindly accepted. At any rate, Japan's demographic pattern in a broad context shows a pattern similar to that of Europe. The magnitude of G_y as seen in the full swing is 2.2, 2.6, and 2.7 for Japan in periods I, II, and III, including 1935–1953, respectively; this is in contrast to 1.4, 1.3, and 2.8 for Western countries in periods (1), (2), and (3), respectively. During the prewar interval, Japan's pace of rising labor productivity was higher and was closer to that of developing nations on average.

Performance in the longer term including the postwar period requires special examination. The trend acceleration (TA) of G_Y that was observed earlier in Sweden, for example, is also observed for Japan through long swings. Two points merit particular attention. First, in the case of Japan the trend acceleration continues to the post-1960 period, then makes a turn and begins to decline in the 1970s. In the case of Sweden, the trend began to decelerate towards the 1960s. Second, the rate of postwar acceleration is specifically impressive in Japan's case, which followed the postwar upsurge immediately after normalization. Even in the 1970s, Japan's rate of output growth (and that of G_y as well) attained the highest level among the industrial market economies. To make comprehensive comparisons of these two cases of the trend acceleration is beyond the scope of our analysis. Nevertheless, some insights can be gained. In addition to the differences between Japan and Sweden in terms of the country size and the context of Europe vs. Asia, a sharp contrast is noted with regard to their economic histories: Sweden was able to follow a perfectly peaceful path, while Japan suffered greatly from war influences. Yet, the similar TA and the turn toward deceleration from a certain point in time is important for a historical understanding about the potentiality of growth for latecomers in general. Japan's sustained growth acceleration in the 1960s was certainly a result of its later start in industrialization. This time dimension is a crucial factor in grasping the growth pattern of economic development.

The performance of capital formation differs between Japan and the Western developed nations: Japan's postwar high I/Y of over 30 percent is similar to the high investment proportion of contemporary developing countries. According to the well-known Gerschenkron thesis of "a big spurt," the investment spurt (through producers' goods in his notion) is a necessary condition for achieving modern economic growth.[40] This proposition was derived from the European experience. In Japan, however, the investment spurts took place three times instead of just once and these investment spurt were the core of the upswings. Although the swing-average of I/Y in the table obscures this fact, the spurt pattern manifests itself for 7–10 years. First, during 1913–1918 in the upswing of period 2, investment was influenced by World War I. Second, in 1932–1937 in the upswing of period 4, the investment spurts was brought forth artificially by military mobilization. Third, in 1956–1962, in the upswing of period 5, which was the most remarkable investment spurt of all, the investment boom was the result of a vigorous move in the private sector.[41]

Japan's experience presents a link in understanding the investment performance of currently developing countries and the performance of European nations in the past. The big spurt, as demarcated by Gerschenkron, occurred years before the turn of the century. In the pre-1960 interval, I/Y was rather stable and was at a modest level in the developing countries. In 1960–1970, higher levels of investment proportion were witnessed for most of the developing economies. This contrasts with the pre-1960 period, when investment levels must have been very low (available data are, however, scanty). At a certain point, often in the 1950s, I/Y began to increase significantly. We call this point the "initial"

---

40 Alexander Gerschenkron, *Economic Backwardness in Historical Perspective,* Chapter 2 and Appendix (Cambridge: Harvard University Press, 1962). The average annual rate of increase in the manufacturing production index was 5.5 percent in Germany (1888–1896), 12.0 percent in Sweden (1888–1896), and 6.7 percent in Italy (1896–1908).

41 Kazushi Ohkawa, "Investment Spurts and Resource Allocation," Chapter 5 in *Nihon Keizai no Kozo* [Structure of the Japanese economy] (Tokyo: Keiso Shobo, 1974).

investment spurt.[42] The pattern appears analogous to "a big spurt" in European latecomers. What differs in the case of the developing economies is that the initial investment spurt was followed by subsequent spurts in the post-1960 interval.[43] The pattern of spurt repetition is analogous to the case of Japan. In realizing the vigorous path of innovational growth in latecomers, the subsequent repetition of investment spurts seems indispensable.

The relevance of Japan's pattern to the developing economies cannot be dealt with comprehensively.[44] Thus, we will focus on selected points. First, the incremental output-capital ratio, $\Delta Y/I$, shows a pronounced difference between downswings (D) and upswings (U) in which the investment spurt interval forms the core: that is, $\Delta Y/I$ is consistently lower in D and higher in U. The expected "efficiency" of capital formation was eventually fulfilled through the swings. Recovery from the downswing, including the most severe war-affected period, was difficult but possible. This brings us to the second point which is why was this recovery possible? A comprehensive answer cannot be given, but in our view, the major reason for the recovery was an increase in the pace of technological and/or organizational advance that was realized during the upswings in capital formation which depended upon flexible labor supplies but which also gave rise to a more inflationary situation on the demand-expenditure side. Third, investment spurts could carry out the function of lowering the "ceiling" set by aggravating deficits of foreign payments. A short-term view emphasizes export expansion in such cases, but viewed from a longer-term standpoint, the importance of the effects of

---

42 For example, I/Y in South Korea in 1911–1920 is estimated to have been 4.0–5.0 percent. Following a gradual increase in the 1920s, the level rose to 15 percent in the 1930s. Taiwan's case is similar in the estimates since 1903, but the tendency to increase started earlier (Toshiyuki Mizoguchi and Mataji Umemura, eds., *Kyu Nihon Shokuminchi Keizai Tokei 1895–1938* [Basic economic statistics of former Japanese colonies, 1895–1938] (Tokyo: Toyokeizai Shimposha, 1988).

43 Illustrative examples of a rapid rise in I/Y from 1960 to 1980 include: Sri Lanka (14 to 36), Thailand (16 to 27), the Philippines (16 to 30), the Dominican Republic (12 to 24), Turkey (16 to 27), Republic of Korea (11 to 31), Malaysia (14 to 29), Tunisia (17 to 28), and Egypt (13 to 31).

44 For more details, see Ohkawa and Kohama, *Lectures on Developing Economies*.

import substitution should not be ignored; the so-called secondary import substitution, in particular, producers' goods and/or investment goods was promoted during the first investment spurt due to the inability to obtain imports because of World War I; in the second investment spurt, the artificial acceleration of industrialization by military mobilization was the stimulus; and in the third spurt, it was the shift toward the phase of secondary export substitution.

Earlier, we paid attention to the unexpectedly large investment ratio in the contemporary developing economies. The governments of these economies often had an intention or a plan to stimulate investment spurts. Some of these efforts were successful, while others did not succeed in sustaining the spurt because of deficits in foreign payments which were criticized as "over-investment." The crucial issue is the possibility of enhancing technological and/or organizational advance in carrying out secondary import substitution.

Average values of $\Delta Y/I$ for the full swing, i.e., periods I, II, and III, show a broad tendency to increase towards period III. Subsequently, however, $\Delta Y/I$ declines. The pace of the decline towards the 1980s is particularly sharp, while the investment proportion is maintained at a high level. The first three intervals correspond to the TA and the latter intervals to the process of deceleration.

The small magnitude of $w\Delta L/I$ for Japan draws attention. The value of $w\Delta L/I$ is even smaller than the average European level in earlier periods, and, in addition, in Japan the term tends to increase rather than decrease as it did for the European nations, before making a sharp downturn in the 1970s. As mentioned earlier, this stems from the small value of G_L, combined with a somewhat higher level of I/Y. In terms of changes over time, a long-term increasing trend in G_L followed by a sharp decline, combined with the pattern of I/Y described earlier is the major determinant of the trend in $w\Delta L/I$. In this context, Japan's case seems to be irrelevant to the contemporary developing economies. On the contrary, the performance of the capital term, $r\Delta K/I$, appears to be relevant to the experiences of contemporary developing economies. It is true that the increasing trend of $r\Delta K/I$ is followed by a recent decline which resembles the experiences of the Western developed nations. But the trend of $r\Delta K/I$ is dissimilar to the Western nations as its magnitude tends to be much greater in Japan; in this sense, Japan is similar to the developing economies. For example, Japan's maximum value of 11.6 in period III is close to the value of 10.2 for developing countries in panel (b) of Table 2.3 which is a vigorous case of innovative growth.

Residual Growth: Trend Acceleration

The crucial term, ΔR/I, shows a TA from period I to period III and then decelerates towards the 1980s. This is broadly analogous to the shape found earlier in Sweden and conditionally for the other European latecomers. Two qualifications need to be mentioned. First, in Japan the wide variation in ΔR/I is associated with swings in the growth rates which were significantly affected by exogenous factors. The prolonged downswing of the interwar period, in which the unstable international environment was influential, negatively affected the value of ΔR/I in period II. Due to World War II, the recovery activities made for a large postwar value in period III. Taking into account these events, we would like to conclude that the pattern of ΔR/I broadly indicates the potentiality of trend acceleration of residual growth for latecomers in general and that the intervening factors, both internal and external, are influential in preventing realization of TA. Second, the magnitude of ΔR/I is greater in Japan than in the Western cases. This is to be expected as a result of the effects brought forth by differences in the time dimension: that is, compared to the European nations, Japan is a latecomer. In the output approach, this difference is particularly wide for later periods. The values for ΔR/Y can be derived from the data in Table 2.3, Panel (B). The value for ΔR/Y is 1.0 in period (1), 1.3 in period (2), 1.8 in period (3), and 2.5 in period (4). The magnitude of ΔR/Y is certainly greater for Japan than the Western average. Surely, a far greater investment proportion in Japan makes for such a phenomenon. The relevance of Japan's experience for developing economies is worthy of mention. The term ΔR/Y is fairly large for the developing nations. In subgroup (a), ΔR/Y is 1.2 and in subgroup (b) it is 1.6. The proportion is also greater. This illustrates the other side of the coin, i.e., the "telescoping phenomenon."[45] Because of the advantages of borrowing advanced technological knowledge, latecomers require much shorter intervals to achieve a certain level of technological innovation by subsector and by sector. Aggregation of these telescoping effects lead to a greater residual growth in the latecomers. Japan's case indicates that a greater investment proportion is required in order to realize this advantage.

45 For details, see Ohkawa and Kohama, Lecture 2, *Lectures on Developing Economies,* 82–84.

In this regard, the results of the "output approach" (i.e., Panel (B) in Table 2.3) are informative. The acceleration performance of the residual, $\Delta R/Y$, and its postwar deceleration, is similar to that of the investment approach. For the conventional inputs, $r\Delta K/Y$ and $w\Delta L/Y$, no additional discussion is needed. With respect to the ratios at the incremental level, our attention is drawn to two distinct patterns: (1) a very high $\Delta R/\Delta Y$ which follows a slow pace of increase (except for group I to group II), and (2) a very low $w\Delta L/\Delta Y$ which has a tendency to decrease, while $r\Delta K/\Delta Y$ shows a slight tendency to decrease. As pointed out earlier in Chapter 1, we have the simple relation $\Delta R/\Delta Y = \Delta R/I \cdot I/\Delta Y$. When larger $\Delta R/I$ is realized with greater ICOR, $\Delta R/\Delta Y$ would be greater. This is generally the case for the earlier phases in Japan and the Western nations when viewed in comparison with the contemporary developing countries. This can be illustrated by comparing the relevant figures in Table 2.3 with other data that has been recalculated. The values of $\Delta R/\Delta Y$ for the Western nations in the pre–1960 period is 44.1 in period (1), 51.9 in period (2), and 57.9 in period (3). These values of $\Delta R/\Delta Y$ are similar to that of Japan, though they are a bit smaller. For developing nations on average, $\Delta R/\Delta Y$ is much lower at 28.3 for subgroup (a) and 33.2 for subgroup (b). In terms of changes over time, both the Western nations and Japan show a similar pattern of $\Delta R/I$ change. The pattern is clear in the Western case and blurred in Japan's case, where only the postwar surge is conspicuous. This difference in the distinctness of the pattern is to be expected since the changes in the ICOR were minor for the Western nations, but distinctly decreased in the case of Japan. Why is the value of $\Delta R/\Delta Y$ so large in the earlier phase of Japan's development? Can we assert that the rate of technological and/or organizational advance in Japan was particularly large? The answers to these questions require sectoral analysis.

The issue of choosing between the output approach or the investment approach is a semantic problem. The appropriateness of the approach depends on one's analytical purpose, as was discussed earlier in Chapter 1. Actually, the terms $\Delta R/Y$ and $\Delta R/I$ appear to perform similarly and a large change or difference in I/Y would warrant particular attention. For our purpose of clarifying the characteristics of developing economies, however, we believe further evidence shows that the investment approach is indispensable. In the case of incremental ratios, special interpretation would be required. In Japan's earlier phase of development, $\Delta Y/I$ was relatively low, but because of the very small magnitude of $w\Delta L/I$, a fairly high level of the residual per investment was realized. Such a pattern may be peculiar to Japan.

Latecomers in the pre-1960 period, if characterized by a low rate of labor force increase, may follow this growth pattern insofar as the potential TA is realized; however, we do not witness such an example. Among the data in Table 2.3, let us focus our attention to $w\Delta L/\Delta Y$ which is the reciprocal of $\Delta Y/w\Delta L$, or the incremental output-labor ratio that was referred to earlier in our comparisons of developed and developing nations. Using Table 2.4, the value of $\Delta Y/w\Delta L$ in Japan is extremely high: $\Delta Y/w\Delta L$ is 10.3 in period (I), 9.7 in period (II), 13.7 in period (III), and 17.5 in period (IV). The tendency of $\Delta Y/w\Delta L$ to increase from period II is largely due to the performance of the ratio, G_Y/G_L, which was 4.7 in period (I), 3.9 in period (II), 6.3 in period (III), and 8.9 in period (IV) as compared to 2.6 in period (1), 3.2 in period (2), and 5.7 in period (3) for the Western nations. It goes without saying that a sharper difference is witnessed between Japan and developing nations.

Notes on the Estimates

1) Income shares

In Chapter 1, the conceptual and technical aspects of estimating income shares were discussed. In Japan's case, our approach is based on the same concepts, but somewhat different procedures are used. Estimates were made for the agriculture (A) and nonagriculture (N) sectors rather than manufacturing and nonmanufacturing because of the convenience of data availability. The β estimates used to compute the relevant terms in Table 2.4 are summarized in Table 2.5.

The aggregate value of β is extremely stable throughout the prewar period. Slight variations, downs in upswings vs. ups in downswings, are witnessed; this phenomenon is usually observed for developing economies with flexible supplies of labor. As expected, β increased distinctly in the postwar period, when the supply of labor became less flexible. The outcome seems broadly consistent with previous estimates of the magnitude and long-term changes of β derived from cross-sectional data.

The estimates depend upon the evaluation of labor earnings of the self-employed (proprietors and family workers) using prevailing wages for both the A and N sectors. In sector A, rent for land thus composes the substantive part of α, which is distinct from r, the rate of return on capital. If the marginal product approach is used to estimate the labor input of self-employed in agriculture, β would be smaller and α would be larger to a certain extent in the

Table 2.5
Income Share of Labor for Japan [a]

Periods	Income Shares(β)		Output Weights		Weighted Income Shares		
	A	N	A	N	A	N	Total (A+N)
(1) 1897–1903	55.8	55.8	39.7	60.3	22.3	33.5	55.8
(2) 1904–1918	55.5	55.7	33.9	66.1	18.9	36.7	55.6
(3) 1919–1930	58.5	57.1	25.1	75.0	14.6	42.8	57.4
(4) 1931–1938	57.3	51.7	19.3	81.3	11.7	42.0	53.7
(5) 1953–1969	62.2	60.4	15.4	84.6	9.6	51.1	60.7

Note: a. A = agriculture; N = nonagriculture (includes industry and services of private sectors). In applying these data to estimating $w\Delta L$ in Table 2.4, adjustment is made for NNP/GDP.

Sources: "Agriculture," Part 2, Chapter 4, and "Factor Incomes and Shares," Part 4, Chapter 11 by Ryoshin Minami and Akira Ono in Ohkawa and Shinohara, eds., *Patterns of Japanese Economic Development*.

prewar period when surplus labor was available.[46] This approach, however, cannot be applied in all cases. Calculations based on the marginal product approach assume that some part of the rent, which is measured as the land input of proprietors, is treated as wage earnings (assuming a sort of competitive equilibrium in the labor market). The procedure adopted in Chapter 1 also implicitly contains a similar assumption for the nonmanufacturing sector.

In general, the sectoral values of β do not show large differences. As will be analyzed in detail in Chapters 3 and 4, the average labor productivity is sustained at a much lower level in the agricultural sector than in the nonagricultural sector in developing economies. In Japan's case, if we take the macro-level of Y/L as 100, the ratio is 0.61 for A and 1.96 for N in period (1) and 0.49 and 1.55, respectively, in period (3). The difference in the marginal product of labor between the A and N sectors appears to parallel the

---

46 For example, output elasticity of rice cultivation in prewar Japan is estimated to have been 0.28 for labor, 0.29 for capital, and 0.43 for land. Kazushi Ohkawa, *Shokuryo Keizai no Riron to Keisoku* [Theory and measurement of food economy] (Tokyo: Nihon Hyoron-sha, 1945).

behavior of the differentials of Y/L, but in a narrower range. We used such a supposition in estimating β in Chapter 1.

2) Capital replacement

In principle, our procedure for estimating the capital term, $r\Delta K/I$, is to use αG_K / I/Y, that is, $rK/Y \cdot \Delta K/K \cdot Y/I$. Because the data for G_K are available for Japan, this procedure is applied directly.[47]

Continuous data on the postwar capital stock, however, are not readily available. The original data are available from the Japanese Government, Economic Planning Agency, Economic Research Institute, *Gross Capital Stock in Private Enterprises National Economic Accounts Quarterly.* Shigeru Ishiwata constructed a linked series of the data for the period 1955–1981 in 1980 prices by use of these original data available from various issues of the Quarterly. The term G_K is calculated using his worksheets.

The prewar and postwar data are in terms of gross capital stock but coverage is not consistent: residential buildings and public sector assets are included in the prewar series, but are excluded in the postwar series. Although adjustments cannot be made for this and though some distortions in the magnitude of G_K for the postwar years are unavoidable, for our purposes, the defects are not serious.

The value of G_K used in our estimates are a "bridge" calculation in that the values link the beginning and the ending years for each period with the use of smoothed values (more specifically, 5-year averages for the prewar period and 3-year averages for the postwar period). The values are in terms of percentage:

Period	G_K	Period	G_K
(1)	1.7	(5)	10.5
(2)	3.0	(6)	8.5
(3)	3.0	(7)	4.6
(4)	3.1		

47 *LTES*, Vol. 3, *Capital Stock,* for the prewar period and official data from the Economic Planning Agency, Japanese Government, for the postwar period.

III. Developing Nations: Changes Over Time

The purpose of this section is to observe again the postwar growth performance of developing nations in light of the insights gained from observing the historical growth path of developed nations. Although preliminary comparisons between the developing and developed nations were made earlier, a more systematic approach is used to focus on three aspects: (1) the initial level of product per worker, (2) the different features of growth performance over time in the post-1960 period, and (3) the problems raised by the postwar downswing, in particular, the influences of the unfavorable international environment. Finally, concluding remarks which incorporate what was discussed in Chapters 1 and 2 are made on the longer-term perspective of developing economies.

The Initial Level of Product Per Worker

It is difficult to define the "required level" of per capita income or per worker product that enables the inauguration of modern economic growth. Even from a very (often called "too") low level of per capita income or per worker product, economic development seems possible. The post-1960 experiences of countries in the Third World provide empirical evidence of this. However, the notion that the level of per capita income is the most basic factor in determining the growth possibility for economic development still prevails. We do not share this notion and refer to Japan's case as an illustration.

Let us begin by citing Kuznets' concluding remarks on this problem:

> With the single exception of Japan, the initial levels of per capita product in the developed countries, on the eve of their entry into modern economic growth, were fairly high . . . . Each currently developed countries, except Japan, began its modern growth with a per capita product of more than $200 (in 1965 prices) [see Table 2.6]. Japan is unique in that its initial per capita product was low enough to fall within the range of current per capita product levels of many of the populous underdeveloped countries in Asia and Africa.[48]

---

48 Kuznets, *Economic Growth of Nations,* 26.

We owe much to Kuznets' data in discussing this topic. However, we do not want to treat Japan as "an exceptional case."[49] Disagreement does not necessarily stem from the measurement results but rather from the way in which Japan is oriented into the general framework of the growth pattern of developing economies, including both the historical records of currently developed and developing economies. In our view, a high or low level of Y/L in the initial phase of each country's development is certainly one of the important factors stimulating economic development, but we do not believe it to be the sole determining factor. Rather, another important factor is the nation's social capability to undergo the process of what has been referred to as the "interaction mechanism" of growth which leads the economy towards a cumulative process that involves the operation of relevant components. These two factors, the Y/L level and the nation's social capability, have long been thought to be inseparable from the notion of per capita income level. This is perhaps true because based on the Western experience, the income level is thought to be the dominant factor determining capability. However, the possibility of borrowing advanced technological knowledge has been increasing over time. Beginning with the emergence of offshoots overseas in the nineteenth century, borrowing of advanced technology expanded under the lead of the United States in the twentieth century and finally in a drastic fashion, technology borrowing exploded in the postwar period. As a result, the relative importance of the two factors, level and capability, has changed. The second factor, a nation's capability, increased in importance as compared to the first factor, the Y/L level. Due to the richer backlog of technological knowledge, the level of a nation's social capability is much higher even for economies with a very low level of per worker product. Although there are a great number of "modern" versions of this in the postwar context of economic development, Japan represents a "classic" case of this in the Asian context.

The notion of the two factors, level and capability, can be found in Gerschenkron's well-known hypothesis of economic backwardness. The hypothesis suggests that the later the entry of a country into the modern economic growth process, the higher its growth rates are

---

49 This view was first asserted in Kazushi Ohkawa, "Initial Conditions: Economic Level and Structure," in *Japan's Historical Development Experience and the Contemporary Developing Countries: Issues for Comparative Analysis* (Tokyo: IDCJ, 1978). A major portion of this paper is reproduced in Appendix I.

likely to be, at least in the early phases.[50] A later entry is not exactly the same as a lower Y/L level, but it is nevertheless thought to be associated with faster rates of growth. To Gerschenkron, the advantages of greater "backwardness" are not necessarily the same as a nation's capability; yet, we are inclined to argue that the two factors were implicitly combined in his notion: that is, "a lower level tends to have greater capability."

In the European context from which Gerschenkron's hypothesis was developed, the TA witnessed for Sweden seems to exactly fit his hypothesis. The positive association of the two factors cannot be identified in contemporary developing economies, however. That is, one of the major reasons for demarcating two subgroups in Chapter 1 is to emphasize the significance of factors associated and not associated with the Y/L level.

In Table 2.6, the initial levels of product per capita and product per worker for currently developed and developing nations are listed in Panel (A). Estimates of the initial level of product per capita (and product per worker) and the demarcation of the dates for each country are issues of controversy among economic historians. For European nations, we depend solely on Kuznets' data as differences with other sources do not matter much for our purpose. For developing nations, the dates are not really "initial" because they did not start modern economic growth exactly at those dates. The dates are merely adopted depending on the data availability in order to get some idea of their initial development levels. The major indicator for developing nations is the level of product per worker, Y/L, instead of per capita product. For developed nations, per capita product is used as the major indicator due to lack of readily available data.

Panel (A) of the table provides an overview of the difference in the initial level of latecomers in economic development. The notion that the level of economic development of the Western countries was much higher than the level of non-Western countries is generally supported by the data. However, the data in the table require important qualification. The range of difference within Europe, even when not including Eastern Europe, is considerable, and among the nations in the non-Western world, the difference appears even wider. For example, between Asia and non-Asia, in particular Latin America, the difference is sizeable, notwithstanding the exceptionally high level of Argentina. An examination of the

---

50 Gerschenkron, *Economic Backwardness in Historical Perspective,* Essay I.

causes for these differences is beyond the scope of our analysis. However, what we are concerned with is that under such situations, it is hard to endorse the validity of the notion of a "required" initial level in general.

Let us specifically take up the case of Japan as an illustration in the Asian context. (This is discussed in more detail in Appendix I.) Kuznets' remark seems to be supported by the data despite the revision of Japan's level upwards in the *LTES* data insofar as the aggregate economy is concerned. However, we have two critical comments on the issue of Japan. First, Japan's initial level of product per worker is quite close to that of the European countries in terms of the nonagricultural sector and thus, Japan's level cannot be said to be "exceptional." Second, Japan's initial level of Y/L was distinctly higher than the level of currently developing nations in the early postwar years, and therefore, it is not correct to judge Japan's initial level as almost equivalent to the Asian low level. These two points involve conceptual and historical issues of judgment that are beyond the statistical estimates.

In the aggregate, Japan's level of per worker product, $274, is 42.0 percent that of the average European level, $645; even the low level of Sweden, $640, is distinctly higher than that of Japan. In comparing the nonagricultural sector (excluding the United Kingdom), the picture changes. Japan's level of $535 is 62.7 percent that of the European average of $850. Compared with the lowest level for the European countries, $746 for Italy, the difference is less than 30 percent. Exclusion of agriculture makes a substantial difference because Japan's Y/L level in agriculture is extremely low compared to that of European agriculture. This is due to Japan's poor natural resource endowments in relation to the population engaged in agriculture (this is usually expressed in terms of the man-land ratio). In our view, this places great pressure on the operation of a nation's capability. Exclusion of agriculture may be criticized as being "too mechanical," but comparison of nonagricultural activities, we believe, is much more appropriate when viewed from our concept of social capability.

A similar situation prevails in most of the Asian nations. For comparisons with these countries, the analysis of agricultural vs. the nonagricultural sectors is not required. In some Asian countries, the aggregate level of product per worker in the initial period of postwar economic development is considerably lower than that of Japan. In 1953–1957, for example, estimated product per worker was $202 for India, $258 for Pakistan, $186 for Thailand, $379 for the Philippines, $503 for South Korea, and $617 for Taiwan. The level

Table 2.6
Initial Levels of Product Per Capita and Product Per Worker and Rates of Growth [a]

Panel (A): (in US$ 1965)

	Date	Per Capita Product	Date	Per Worker Product	Date	Per Worker Product in NonAgriculture
Currently Developed Nations						
United Kingdom	1765-1785	227	1851	504	1801-1811	469
France	1831-1840	242	1856	618	1896	878
Netherlands	1831-1840	347		-		-
Germany	1850-1859	302	1882	788		906
Denmark	1865-1869	370	1882	794	1870-1869	1,008
Norway	1865-1869	287	1975	637	1865	1,057
Sweden	1861-1869	215	1870	640	1861-1870	886
Italy	1861-1869	261	1881	632	1861-1870	746
Japan [b]	1876-1887	154	1881-1882	274	1887	535

	Date	Per Capita Product	Per Worker Product	Per Worker Product in 1960
Currently Developing Nations [c]				
Argentina	1900-1904	443	692	1,568
Mexico	1895-1899	148	290	1,258
Jamaica (1)	1832	193	357	1,203
Jamaica (2)	1929-1931	194	359	-
Egypt	1895-1899	131	238	881
Philippines	1902	138	262	750
Ghana	1891	107	202	973
India (1)	1861-1869	47	85	177
India (2)	1881-1889	60	109	-

Panel (B): (percentage)

	Period	Average Annual Growth of Per Capita Product
Argentina (GDP)	1900/01 to 1925/59	1.6
	1925/29 to 1963/67	0.9
Mexico (GNP, GDP)	1895/99 to 1925/29	1.3
	1925/29 to 1963/67	2.0
Jamaica	1832 to 1930	0.0
(GDP, National income)	1929/1931 to 1951/51	0.9
Egypt (GDP, GNP, NDP)	1895/99 to 1945/49	-0.1
	1945/49 to 1965/66	2.6
Philippines	1902 to 1938	1.0
	1938 to 1950/54	-0.2

Ghana	1891 to 1911	1.9
	1911 to 1950/54	1.3
India	1811/69 to 1881/89	1.2
	1881/89 to 1901/09	-0.2
	1901/09 to 1950/58	0.6

Notes:
a. For details of the estimating procedure for product per worker in the aggregate and for nonagriculture in Panel (A), see Appendix I.
b. The per capita product estimates at the initial dates for Japan in Kuznets' data are revised from 74 to 154 using the new *LTES* estimates.
c. For developing nations, a simple conversion is made from per capita product to per worker product by using the proportion of the working age population to the total population from the *World DevelopmentReport 1982* for the year 1960.
d. - = data are not available.

Sources: Kuznets, *Economic Growth of Nations,* Table 2:24 and Table 3:30–31; Ohkawa, "Initial Conditions"; and sources of Table 1.1.

of product per worker is lower than Japan's initial level in the first three cases, but the levels in the latter three countries are fairly high due to their accelerated growth in recent decades.[51] This raises the controversial issue of the economic effects of colonization. Putting this question aside in view of the low rates of growth of the nations in other regions, we believe the levels of product per worker in the Asian countries were definitely lower than that of Japan in 1881–1882. Suppose, for example, a 1 percent rate of growth beginning at a level that is one-half of Japan's level at the time, i.e., $137. At this rate, the level of per worker product would just reach Japan's initial level in 1953–1957.

51 In the case of the Philippines, as is shown in Panel (B) of Table 2.6, the rate of growth was less than 1 percent in terms of product per worker. For Korea and Taiwan, similar growth rates were found. For the period 1913–1931, the rate of growth in per capita product in Korea was 1.5 percent, but accelerated to 5.2 percent during the 1930s. In Taiwan during 1908–1932, the growth of per capita product was also 1.5 percent on average but accelerated to 2.9 percent during the 1930s (Mizoguchi and Umemura, eds., *Kyu Nihon Shokuminchi Keizai Tokei,* Part 3, Tables 6, 8, and 23).

The level of per worker product is listed in Panel (B) for developing countries. Comparison is made between the level of per worker product in earlier periods and in 1960 (the data for 1960 are derived from the sources used in Table 1.1). Converted to 1965 prices, Argentina's initial level of per worker product is extremely high; and we can say that Argentina represents the case of a country with a high level and low capability. India, on the other hand, represents the case of a country with a low level and low capability of accelerating growth. Other countries fall between these two cases to varying degrees. The situation as a whole appears to be complex and permits no simple generalization; this poses a difficult problem in terms of explaining the wide range of difference in Y/L levels in 1960 among the group we call "developing countries." This issue, however, is beyond the scope of this work. Our concern is a comparison of the rates of growth that are suggested in Table 2.6, that are listed in Table 1.1, and that are summarized in Table 2.3 for the post-1960 period. An amazing finding is that the rate of growth, G_Y, is not very different between the two insofar as subgroup B is concerned. The value of G_Y is 3.3 for the total average of developing nations in group (a) and 3.7 for the vigorous group (b). Of course, G_Y of the developing nations is definitely greater on average than that of the developed countries in the prewar historical period, and in the case of subgroup A, the value of G_Y is much higher. Nevertheless, it is important to recognize that a drastic increase in the rate of growth in labor force supply is almost completely absorbed by the higher G_Y in subgroup B in the post-1960 period. Interpretation of these phenomena would require further examination of the postwar growth patterns.

Characteristics of Postwar Growth Patterns

In a historical perspective, the post-1960 growth performances are characterized by a wider and faster spread and diffusion of advanced technological knowledge from the developed to the developing nations, not only from the United States but from other industrial nations as well. The diffusion of the technology is unprecedented and epochal. The entire interval up to the 1980s has been eventful, despite the disruption caused by the two oil crises. In the analysis of growth performance, views may differ among economists, particularly with respect to the time dimension one may look at. We demarcate two broad intervals, 1960–1980 and the 1980s, and ignore variations in the shorter intervals. Within this time frame, our observations will focus on the growth performances of the developed and developing nations and its variance within the latter group.

One may wonder whether differences between the 1960s and the 1970s would be too large to be ignored in view of the effects of the oil crises. Let us thus begin by explaining why these two intervals were chosen. For some, the year 1973 is often taken as an important dividing line between the two periods. Using the GDP growth rate, G_Y, as a simple indicator, Table 2.7 presents our familiar groupings of nations (note that the major oil-exporting countries were excluded) broken down into four periods—1960–1970, 1970–1980, 1960–1980, and 1980–1985. There is not much difference between the first two periods. However, the average GDP growth rate in 1960–1980 declines drastically in the first half of the 1980s. We can simply say that the former periods represent an interval of upswing, while the latter period is one of downswing, where the term "swing" is used rather loosely. The underlying notion is analogous to the long-swings discussed earlier in the case of Japan. We will discuss this further in the following pages; at this point, however, we want to pay attention to the comparison between the developed and developing nations. A slowdown of the growth rate is conspicuous for group V from the 1960s to the 1970s. The average GDP growth declined from 5.0 to 3.0 percent. However, the slowdown, if any, is smaller for other groups of developing economies over the same period and in fact G_Y increases for group III. How can these variations in growth performance be interpreted? It is our view that these results are essentially the product of two different factors that operated during the intervals under consideration. One factor is the narrowing of the technological gap between the United States, the lead country, and other industrial countries in group V. This narrowing in the technological gap operated to slow down the growth tempo of developed nations by making a less favorable "international environment" for developing nations. In addition, the resulting changes in the external environment were aggravated by the oil crisis, which has often been considered as the major cause of the slowdown in the growth rate of developing economies.

We share the view which emphasizes the importance of such external factors. Nevertheless, the figures in the table induce us to think about the substantive operation of the other important factors, which can be called "internal" for each developing nation. Without taking the various operations of the internal factors into consideration, growth performance cannot be explained. This is most vividly illustrated by the variance between subgroups A and B. In subgroup B, G_Y slowed down distinctly in four groups without

Table 2.7
Changes in the Growth Rate of GDP Over Time: Comparison of Three Periods, 1960–1985 [a] (percentage)

Group		1960-1970	1970-1980	1960-1980	1980-1985
I		3.7	3.5	3.6	1.9
	A	3.7	4.4	4.1	2.2
	B	3.7	2.6	3.2	1.5
II		4.9	4.1	4.5	1.5
	A	6.2	6.0	6.1	2.4
	B	3.6	2.1	2.9	0.6
III		5.6	6.4	6.0	2.9
	A	5.7	7.9	6.8	4.0
	B	5.4	4.9	5.2	1.7
IV		5.6	4.8	5.2	0.6
	A	5.9	5.5	5.7	0.8
	B	5.2	4.1	4.7	0.4
V		5.0	3.0	4.1	2.1
	A	5.9	3.5	4.7	1.9
	B	4.2	2.5	3.4	2.2

Note: a. Growth refers to average annual growth.

Sources: See sources of Table 1.1.

exception from the 1960s to the 1970s. However, in subgroup A, G_Y accelerated in groups I and III and declined only slightly in groups II and IV. In view of the conceptual characterization of the subgroups which was given in Chapter 1, such a phenomenon is understandable. In short, in responding to the negative effects of the external factors, distinct discrepancies arise across countries due to the different ways in which the operation of what we call a nation's social capability takes place.

With regard to the first half of the 1980s, the drastic slowdown of G_Y is clear and contrasts markedly with G_Y of the period 1970–1980 and the period 1960–1980. In contrast to the other groups, a different interpretation seems appropriate for group V. In the industrial economies, the slowdown in the tempo of G_Y appears to be part of a process of deceleration from the 1960s onward. Therefore, to say that the 1970s represents a "downswing" for group V countries is not precisely correct. Rather, the lower rate of G_Y is a continuation of the path of deceleration to the 1980s. One may be tempted to define "trend deceleration (TD)" as a counter notion of the TA discussed earlier. Rather than follow this temptation, however, we would like to refer to this as a "deceleration path (DP)" instead of a trend deceleration. The reason for this is important. The TA refers to a long-term trend of economic development, but the "deceleration" witnessed for the group V countries does not involve such a long-term trend. Instead, the significance of the DP is in terms of the path which follows the terminal point of the TA. Conceptually, the deceleration path pertains to structural changes rather than downswings. Although, the movement of G_Y appears to be continuous, to call this a trend deceleration would imply that the slowdown in G_Y is a long-term trend. From historical data, we know this is not the case.

For the currently developed nations as a whole, the occurrence of a DP is unprecedented. The DP took place as a result of structural change: economic growth through industrialization ended or became closer to its terminal point somewhere in the 1970s (for some nations, this point was reached earlier while for others it was later). Almost all of the industrial countries show a typical DP in terms of G_Y throughout the periods 1960–1970, 1970–1980, and 1980–1985. Japan stands out as the most conspicuous case; G_Y declined from 10.9 to 5.0 to 3.8 percent in the successive periods, respectively. The United States, the lead country, also had a declining G_Y in the periods under review (4.0, 3.0, and 2.5, respectively). As a result of this situation, the narrowing of the technological gap continued. As a result, as shown in the table, the significance between the two subgroups practically disappeared for group V in the 1980s.[52]

In this regard, Abramovitz's view deserves particular attention. After having presented rich evidence from his empirical research, he concludes:

---

52 Our view may not be accepted by a number of economists. In particular, if one emphasizes the significance of the postwar high growth (i.e., the 1960s and 1970s taken together) in a historical perspective, the concept of the DP does not make sense. Furthermore, a few nations like the United Kingdom and France showed a revival in G_Y during the postwar period, and one may assert that no DP is witnessed for such cases.

> The record of postwar growth, therefore, appears to be consistent with the view that relative backwardness, [emphasis added] for countries with the proper apparatus of governmental and commercial institutions, with educated and skilled populations and advanced technological capabilities, is an important aspect of potential which shaped the postwar record.[53]

We share Abramovitz's view which emphasizes the significance of the postwar record of industrial economies. As was shown in the preceding tables, an accelerated G_Y in the post-1960 interval is compared to that of pre-1960 periods. This pre- and post-1960 comparison and the deceleration path (DP) since the 1970s onward does not, in our view, contradict each other.

For the developing nations, however, the notion of the DP cannot be applied in the same way as it was for developed countries. For all developing-country groups, i.e., groups I to IV, a sharper drop in the G_Y level occurs in the 1980s, particularly for subgroup B. In group IV, the drop is most conspicuous, even for its subgroup A. For the developing countries, these declines in G_Y are interpreted as downswings because, unlike the case of group V, the values of G_Y deviate distinctly downward from the past trend of growth. The decline in G_Y in the developing nations is not due to basic structural change which would imply the possibility of recovery to a "normalized" growth path. In this regard, the figures presented in Table 2.7 for the first half of the 1980s pose a problem: that is, how are we to understand the results of the interplay of the effects of external factors and the operation of the internal factors in this particular period? The duration of the period is too short to be able to answer the question. We will return to discuss the relevant issues further. Nevertheless, the essential point is clear. The problem developing economies face is how to make the basic adjustments needed which allow endogenous factors to respond more efficiently to the effects of exogenous factors.

The Downswing and Innovative Growth

Viewed from the perspective of innovative growth, domestic investment plays a central role. Thus, the following discussion focuses on investment and its efficiency, and the slowdown of the

---

53 Moses Abramovitz, "Rapid Growth Potential and Its Realization: The Experience of Capitalist Economies in the Postwar Period," in Edmond Malinvaud, ed., *Economic Growth and Resources*.

output growth rate is expected to manifest itself in a much more sensitive fashion.

As shown in Table 2.8, the investment proportion, I/Y, decreased slightly in the 1980s. As a result, the incremental output-investment ratio, $\Delta Y/I$, shows a drastic decline; note that the sharp decrease in $\Delta Y/I$ is the result of a much less sensitive decline in investment as opposed to changes in output. In other words, the ICOR increases to an abnormally high level. This pattern is typical in the observed downswings of currently industrial countries. However, an extremely sharp drop in the incremental increase of the residual per investment, $\Delta R/I$, for postwar developing economies does draw attention. In group V, the developed countries, the drop in $\Delta R/I$ is considerable, but is less drastic than that of developing countries for which the estimated $\Delta R/I$ is, in fact, negative for two groups (groups I and IV) in the downswing interval. Because the demarcation of the periods and the procedure for which the estimates are obtained are crude, we cannot claim that the values in the table are highly reliable. Nevertheless, conceptually it is possible to conceive of a "stagnation" in the sense of innovative growth, where no innovational activities take place: no contribution of the nonconventional factors can emanate in this growth path. Some nations, particularly those in groups I and IV, are in a crisis because of this stagnation.

The view of stagnation can essentially be accepted but would require qualification. Implicit incorporation of demand into our framework of the interaction mechanism of growth would make the meaning of the measured residual less clear. In the present framework, the effects of demand may in fact be included: its drastic decline from 1960–1980 to 1980–1985 may be due to the severe shrinkage of demand. Conventionally, estimation of deviations from the "normal" rates of utilization of production factors to measure residuals is used to adjust the effects of demand variation. However, we have not adopted this approach because while the results are generally acceptable insofar as the period of upswing is concerned, the situation differs in the downswing period. Because of the technical difficulty in applying such an adjustment, the results may tend to overestimate the stagnation and caution in interpreting the estimates is suggested.[54]

---

54 With respect to the stagnation in developing economies, particular attention is drawn to the labor term, $w\Delta L/I$, both technically and conceptually. Except for groups IV and V, $w\Delta L/I$ increases in the 1980s. Without repeating the explanation on the procedure of measuring the capital term, suffice it to say that it tends to decrease. Therefore, a

Table 2.8
Comparison of Investment Performance In Two Intervals, 1960–1980 and 1980–1985 [a] ***(percentage)***

(A):	I/Y		ΔY/I		wΔL/I		ΔR/I	
Group	1960-1980	1980-1985	1960-1980	1980-1985	1960-1980	1980-1985	1960-1980	1980-1985
I	17.3	19.2	17.5	10.1	7.0	7.5	1.3	-1.2
II	19.3	18.7	23.3	12.2	6.2	7.3	5.2	1.6
III	24.6	23.6	26.2	11.6	6.3	7.7	7.7	2.0
IV	23.8	21.0	20.9	0.9	4.2	4.3	5.7	-7.0
V	24.6	21.9	16.3	8.9	3.6	2.8	6.4	2.7

(B):	G_I		G_X	
Group	1960-1980	1980-1985	1960-1980	1980-1985
I	5.5	-4.0	3.1	-2.5
II	6.1	-8.1	3.7	3.2
III	8.9	1.1	6.5	3.9
IV	6.6	-6.6	5.4	2.3
V	3.9	1.7	3.9	4.4

(C):	I/Y				ΔY/I			
	Subgroup A		Subgroup B		Subgroup A		Subgroup B	
Group	1960-1980	1980-1985	1960-1980	1980-1985	1960-1980	1980-1985	1960-1980	1980-1985
I	17.7	19.6	16.7	18.9	23.0	11.2	19.2	8.0
II	19.8	21.0	18.8	16.3	30.3	11.4	15.4	3.7
III	21.8	25.3	25.3	21.9	31.2	15.8	20.6	7.8
IV	21.8	21.9	23.5	20.0	24.8	3.7	18.4	2.0
V	24.8	21.2	24.4	22.5	19.3	9.0	13.5	9.8

Note: a. The assumptions made and the procedures of estimating the measurements are the same as for Table 1.1.

Sources: See sources of Tables 1.1 and 6.8.

---

sustained large magnitude of the labor term plays a substantive role in reducing the residual. The economy depends heavily upon the

In terms of the downswing, clarification of its characteristics must be made within a confined scope: that is, within the framework of the interaction mechanism. The mechanism continued to operate through the preceding two decades of upswing for most of the developing economies, including the countries in subgroup B which were handicapped by what we call the nonassociated factors. The downswing made the operation of this mechanism much weaker and resulted in the stagnation which was referred to earlier. Today, demand shrinkage poses great pressure on the operation of the interaction mechanism. While no attempt is made to quantify the effect of shrinking demand on the operation of the interaction mechanism, an illustrative explanation that links demand with the operation of the mechanism is desirable. In Panel (B) of Table 2.8, the rate of change in domestic investment, G_I, is compared with the rate of export expansion, G_X. Contrary to the usual supposition, the decline in growth from 1960–1980 to 1980–1985 is greater for investment than for exports in developing economies. (This will be explained in detail beyond a comparison of simple averages in Chapter 6.) As is well known, the investment proportion, I/Y, is a supply-oriented term, while the rate of change of investment, G_I, is a demand-oriented term. The two terms behave differently. The moderate change in I/Y witnessed earlier is due to the relative decrease of Y. The endogenous factor, as represented by G_I, appears to play a more significant role in decreasing demand than does the external factor as represented by G_X.[55] A decrease in international trade, particularly in the external demand which stems from the demand of developed economies for exports from developing countries, is a well-known problem in the downswing interval. We wish to emphasize the significance of the interactive operation of endogenous factors with the exogenous factors.

Our conclusions can be summarized as follows. First, the interaction mechanism of growth is of a cumulative nature, in both a

---

contribution of the labor input. The term $w\Delta L/\Delta Y$ amounts to 0.60–0.70 for groups I, II, and III, which contrasts sharply to the 0.33 for group V. Group IV is an exceptional case.

55 The different performance between subgroups A and B in the downswing is not unexpected. The data in panel (C) of Table 2.8, which focus on investment performance, may help to understand this. The term I/Y decreases in subgroup B but increases in subgroup A in the 1980s. The drop of DY/I, however, is equally sharp for both subgroups.

positive and negative fashion. Its positive dimension is witnessed in accelerating innovative growth during the postwar upswings. In the downswing, its cumulative effects in the negative dimension operate for most of the developing economies. This may be the major cause of developing economies' stagnation of innovative growth and may provide an answer to the question posed earlier: why does a downswing occur for the developing economies while only a deceleration takes place in the developed economies?

Second, a few words need to be said on the possibility of technological and/or organizational advance. Following Schumpeter's classic view, one may assert that technological innovation is more possible during downswings. On the other hand, others contend that technological innovation is more promising during upswings due to the surge of effective demand. This is a controversial issue and we can apply either the former or the latter theory to the problem of developing economies today. Note, however, that the issue pertains to the possibility of developing economies borrowing advanced technological knowledge from developed countries rather than making technological innovations on their own initiative. Given the international backlog of technological knowledge, the issue is directly concerned with the level and composition of a nation's capability of absorption. In this context, we contend that the upswing is more favorable than the downswing for making technological progress in the development process.

Third, prolonged downswings would bring forth more differentiated effects in a cumulative mechanism for innovative growth. On the one hand, for a number of developing economies, recovery from a downswing has occurred since the latter part of the 1980s when vigorous growth began again. However, a period of prolonged stagnation seems to have occurred in a number of other developing economies. Given the decelerated path of developed economies, the differences in the endogenous capability of each nation will result in wider discrepancies in the growth performances of developing economies if the situation is not improved in the near future.

A Long-term Perspective

At this point, a methodological discussion on the relationship between the cross-sectional and time-series approaches is needed because the longer-term perspective integrates the results of both approaches. In view of the differences in the basic nature of these two approaches, a clear-cut answer to the question of whether there is conformity or nonconformity between the two approaches is considered to be neither possible nor useful. It is our view that the answer to this question depends on the methodology one adopts. We will look at this question within the confined scope of its relevance to the analytical framework and hypothesis that is being applied here.

First, the basic differences between the two approaches should be specified. The cross-sectional approach provides "averages" of multinational performance for comparison. The choice of what kind of average to use depends upon the analytical purpose. If conformity with time-series estimates is the goal, the data should be arranged to meet this purpose. In our case, product per worker, Y/L, is adopted as a possible common yardstick that links the cross-sectional and time-series estimates; to this end, we will be estimating two averages, one for each of the two subgroups A and B. This means that with respect to the factors responsible for the different performances between subgroup A and subgroup B, no attempt is made to search for conformity with the time-series data. The implication is that the results of the operation of these factors are involved in the historical series of individual nations and present varied growth performances. Unfortunately, any attempt to search for a regular pattern in these movements is beyond the scope of this study.

Second, with respect to the treatment of the time-series data, specific concepts will be required to ensure conformity with the cross-sectional analysis. The historical growth performance along the path of rising levels of product per worker varies among individual countries and depends on the operation of other factors. Therefore, many economists believe that no regularity can be established with regard to the historical path of growth, whether it be an acceleration or deceleration trend. Given this empirical finding, it may not be possible to achieve conformity with the cross-sectional findings. Therefore, it is useful to group nations in terms of such common characteristics, including factors that are directly relevant to the economy such as typology, country size, etc.[56] Our device of subgrouping A and B is for the purpose of obtaining an overview. As suggested earlier, with respect to the use of a time-series approach, greater significance was placed on the growth performances of the potential TA and DP rather than on empirical identification of

---

56 Affinity was asserted between Japan, Korea, and Taiwan in John C. H. Fei, Kazushi Ohkawa, and Gustav Ranis, "Economic Development in Historical Perspective: Japan, Korea and Taiwan," Part II, in Kazushi Ohkawa and Gustav Ranis, eds., *Japan and the Developing Countries* (Oxford: Basil Blackwell, 1985).

what actually took place. We believe this is one of the possible devices that link historical series with cross-sectional data in the search to obtain conceptual conformity.

Let us now move to more substantive discussion on the developing economies. In addition to the trend acceleration (TA), the deceleration path (DP) was discussed in terms of the growth path of currently developed nations. One may be tempted to raise the question: are these two concepts, DP and TA, totally irrelevant for the developing economies? The answer is "no."

It is of course impossible to find a regular or "normal" pattern or trend because reality is so complex. Classification by setting certain standards into several patterns may be one of the possible approaches. Because of our emphasis on the dynamic importance of the cumulative process, our focus is on a different dimension—the nation's potential which is implied from the long-term pattern of the nation's actual growth performance. The notion of a nation's potential is necessary in order to grasp the function of endogenous elements in the milieu of the varied effects of exogenous factors in a changing international environment.

Let us begin by looking at Table 2.7 again with respect to G_Y. In addition to the downswing, longer-term trends are implied for the developing economies, though the actual period is not long enough to be confirmed. First, let us compare G_Y in the periods 1960–1970 and 1970–1980. As was noted earlier, the general tendency is for G_Y to fall. Increases in G_Y do occur for subgroup A in groups I and III, but practically no change in G_Y occurs for subgroup A in group II. Furthermore, in the period 1980–1985, G_Y is maintained at a comparatively high level in these cases. The number of countries for which G_Y increases or remains unchanged in each group is 6 for group I, 5 for group II, 9 for group III, and 4 for group IV, out of a total of 12 for each group. We suggest that these numbers point to the possibility of trend acceleration.

The pattern is blurred for two reasons. The first reason, of course, is the effects of the downswing in 1980–1985. If a favorable international environment could be sustained, G_Y would be much higher, implying a possibility of the TA. Second, as described in Chapter 1, a number of developing countries inaugurated modern economic growth suddenly and discontinuously without having had a long period of moderate growth which preceded the path of TA of the currently developed nations. In other words, the rate of growth of developing countries today was "telescoped" in a shorter time interval and resulted in a modified TA. If the data were continuously available for the pre-1960 interval for individual nations, further evidence of this could be obtained. The Republic of Korea represents a good example. The average annual rates of G_Y in the pre-1960 period are:

1913–1921	1922–1931	1931–1936	1960–1970	1970–1980	1980–1985
3.8	1.3	7.0	8.6	9.6	7.9

Sources: See sources of Tables 1.1 and 2.6.

On the other hand, the possibility of a deceleration path is suggested by the nations in subgroup B. For these nations, a distinct decrease of G_Y occurs in the 1970s. This is followed by a sharper drop in G_Y in the first half of the 1980s: that is, the downswing deviated from the trend. For a number of developing nations, the deceleration path is recognizable for the post-1960 periods. The pattern of G_Y appears to be similar to the DP described earlier for the industrial economies. Can we then infer that the movement of G_Y in the developing nations is analogous to the case of industrial economies?

Second, the number of countries which indicate a possible TA is particularly small for group IV: a majority of the countries show a possible DP. In group V, almost all of the nations show a clear DP pattern. The growth performance of group IV appears to be similar to that of group V. Why is this so? The answer requires a more systematic interpretation, which can be pursued by incorporating the results of the cross-sectional analysis, in particular the inverse U-shape pattern. One may suppose a simple correspondence: that is, the earlier period exhibits a TA and the latter period exhibits a DP. This explanation implies perfect conformity between the historical and the cross-sectional data series. However, this can be possible only superficially and cannot be accepted without serious qualification.

The cross-sectional pattern is based on the average of various countries' experiences, demarcated at group III into two intervals. The historical pattern is of a selected nature as it is used to suggest a nation's potential. In this context, we can presume that the potential for TA works more intensively in the earlier interval, whereas the potential for DP operates more intensively for the latter interval. Thus, we may say that in general, the inverse U-shape pattern is not possible for all countries, but the pattern would be realized so long as the intervening factors do not disturb it. The second issue pertains to the Y/L level at which the turn from acceleration to deceleration takes place. The Y/L level of group III evaluated in 1970 cannot be expected to be the pivot point for all nations. Rather, the crucial issue for individual nations is the variation in the level at which the pivot takes place: for a certain country, the Y/L level at which the turn takes place may be lower or higher than others. In this regard, let us recall the performance of group IV which was discussed earlier: that is, a DP is witnessed for a number of countries in this group despite their having a much lower Y/L level than that of group V. This may illustrate well the limitations of directly linking the cross-sectional data with the historical series of individual nations even in our confined framework. Examination of the various situations of individual nations is beyond the scope of our study, but we believe it to be essentially relevant to a nation's capability to pursue innovative growth. The nation's social capability is not a "given" for each country and its level and components can be enhanced over time by the nation's willingness and efforts. To illustrate this using our subgroupings, we would say that countries in subgroup B have the potential to shift themselves to subgroup A, while subgroup A countries may move down to subgroup B. A longer sustenance of trend acceleration, whether it be telescoped or modified, to a higher Y/L increases the developing economies' chances of success in arriving at the developed state. This is in fact suggested by the recent performance of NIEs.

Thus, the possibility of trend acceleration is essentially open to all developing economies. The realization of the potential is hampered by intervening factors which operate in individual countries. The pattern of deceleration can emerge as Y/L rises and this is due to the weaker operation of endogenous factors in the path of innovation growth.

Third, more specific discussion on group IV is warranted. Referring back to Table 1.1, let us reconfirm the following: (1) the investment proportion, I/Y, does not change very much, dropping only slightly in a generally increasing trend from groups I to V; (2) investment efficiency, ΔY/I, begins to decrease and thus forms an inverse U-shape throughout the entire range of countries; (3) the rate of increase of labor, G_L, turns sharply and becomes smaller; (4) the incremental capital-labor ratio, I/wΔL (i.e., the reciprocal of wΔL/I), increases substantially from group IV to group V, after showing moderate changes through the lower Y/L groups. Finally, the relative magnitude of the residual per investment, ΔR/I, shows an increasing trend in groups I–III, begins to decrease at group IV, and increases again at group V. We believe that all of these phenomena can be interpreted consistently. At the higher Y/L level immediately before that of developed economies, developing economies face the difficult problem of speeding up the pace of capital intensification. This is essentially the problem of keeping the level of ΔY/I from falling by substituting capital for labor in sustaining the pace of technological and/or organizational advance. The level and components of a nation's capability differ in tackling this difficulty. The majority of the countries in group IV appear to have been less successful, though some of them seem to have overcome this problem. As will be discussed later in Chapter 4, a number of countries in group IV seem to be relatively highly industrialized. For example, the ratio of manufacturing output to GDP has already reached 40 percent or more in some of the countries in group IV. This ratio is almost the same as that of group V countries. This suggests that the turn or dip may not be necessarily caused by stagnation in the pace of industrialization. In other words, despite being industrialized in terms of output share, the group IV nations are behind the group V countries in terms of the Y/L level. We are tempted to refer to countries exhibiting this phenomenon as "early mature" in the sense that full industrialization in terms of share of output accompanies weaker operation of the interaction mechanism of growth at a comparatively lower Y/L level.

Qualifications to such a sweeping observation are necessary. More specifically,

1. Regional aspects. We have no intention of clarifying regional characteristics of growth performance. Nevertheless, a few words may not be out of order. The majority of countries in group IV are Latin American (LA) countries (LA 8 vs. non-LA 4). One may expect that LA countries would belong mostly to subgroup B, but the distribution of LA and non-LA (European) countries is in fact even (4:4 in subgroup A and 2:2 in subgroup B). Smaller values of wΔL/I is crucial for this group. Actually, the rate of labor force increase, G_L, is very small for the non-LA countries. However, the G_L of the LA group is 2.4 on average; this value is much smaller than 3.1 which is the average for group III. This is due to the inclusion of such countries as Argentina and Uruguay. The performance of group IV cannot necessarily be characterized regionally.

2. Capital intensification. We have often mentioned the difficulties of making the final shift to the phase of secondary export substitution. Most of the countries in group IV appear to have this problem as evidenced by their prolonged process of secondary import substitution. (This aspect will be discussed in-depth in

Chapter 6.) This problem cannot be clarified without knowing the internal mechanism: the speed-up of capital intensification could not accompany enhancement of the level of investment efficiency. The evidence for this is given in Tables 1.6 and 1.7. The term $\Delta R/\Delta K$ declines from 11.5 in group III to 7.2 in group IV, whereas $(G_K - G_L)$ shows a distinct increase from 2.7 in group III to 3.5 in group IV. Although the estimates are conditional, we are convinced that this is a basic problem for developing nations.

Looking at the terms which cause the inverse U-shape path of G_Y and G_L, namely, $\Delta Y/I$ and $w\Delta L/I$, what can we say about the performance of these terms for group V? The answer requires clarification of the so-called terminal point of development.

Group V is characterized by the sustenance of a high level residual per investment on average despite a low level of investment efficiency. This can be explained by the developing nations' successful realization of technological advance of the capital-intensive type in a situation of limited labor supply. While there are differences in the degree to which the countries succeeded—a comparison of subgroups A and B showed this to be the case—it is important to emphasize the significance of the narrow range in the differences. The difference is much narrower for group V than for other groups of developing economies. In the earlier segment of the inverse U-shape path, the differential in the intensity of the operation of the group mechanism tends to be wider, but as the Y/L level approaches that of the developed economy, the differential narrows. In general, at this level of Y/L, the gap in technological levels which continues to exist between these countries and the lead country (chiefly, the United States of America) is much narrower. As explained earlier, this is the basic cause of the slowdown in growth rates as the Y/L level rises, but the difference between the performances of subgroups A and B suggests that the operation of a nation's capability also tends to narrow the differentials. Even so, the factors responsible for the diversity in growth performances remain a matter of contention. One plausible explanation for the differences in performances is differences in the time dimension: that is, early attainment of the mature phase of industrialization may make the growth performance less vigorous as illustrated by the United Kingdom, while latecomer countries like Japan tend to be more vigorous.

The distinction thus presented between groups IV and V by the cross-sectional data cannot predict the final path of group IV to the terminal point of its development path, although the actual relevance is somehow closer in comparison between subgroup A in group IV in the 1970s and subgroup A in group V in the 1960s. However, further research is needed to deal with this issue and we hold off any in-depth discussion of this until after development phases have been more systematically demarcated in terms of structural changes in productivity growth and trade patterns (Chapters 4 and 6). At present, in discussing the "long-term perspective" for developing economies at the macro-level, we would like to define it simply as follows: the end of the development process can be conceptually demarcated by the point (or a range of years) at which the rate of residual growth stops its acceleration or begins to decrease. Empirically, this is clearly illustrated by such countries as Sweden and Japan which have realized TA. With regard to developing nations in the cross-sectional data, because of the dip (i.e., an early decline in the residual growth) in group IV, no real-life example can be presented. Yet, we are convinced that this notion is essentially applicable.

In light of our hypothesis, the problem of the balance between a nation's SC with its PC shifts to a different dimension: that is, the problem shifts from a nation's capability to absorb advanced technological knowledge to its capability of developing new technological knowledge on its own initiative. By sector and subsector, the shift may vary but both the SC and PC proceed concurrently to a certain extent. Yet, we can imagine a point of time at which the role played by SC in our original concept shifts from being a dominant role to a passive function. In our measurable terms, $\Delta Y/I$ begins to decrease and declines more than a decline in the magnitude of conventional inputs, thus bringing forth a decline in the residual growth. This is illustrated in Table 2.4 in the case of Japan.

Chapter 3

Production Structure and Employment:
Sectoral Inequality of Product Per Worker

I. Introduction: The Conceptual Framework and the Problem

Product per worker, Y/L, has been taken up as the yardstick in observing the macro-growth performances of developing nations. Also of interest are the sectoral aspects. Three major sectors—namely, agriculture (A), industry (I), and services (S)—are analyzed to obtain an overview of sectoral growth performance. Analysis of subsectors will not be attempted.[57]

Chapters 3 and 4 present our basic views on "production structure and employment." The overall design of the framework used in these chapters analyzes production structure from the aspect of labor employment and analyzes growth from the aspect of capital investment. The purpose of this study is to respond effectively to factual challenges rather than to follow logical requirements. Growth can be discussed in relation to employment (this was already done to some extent in the preceding chapters), and structural change can be dealt with in relation to investment. In this case the same approach that was used for the macro-analysis is required in the analysis of the three major sectors. While we recognize the usefulness of alternative approaches, we nevertheless believe that employment is of primary importance in analyzing structural changes. Thus, the sectoral inequality in product per worker employed is the basic yardstick used in measuring the direction of structural changes.

---

57 The grouping of subsectors into these three major sectors follows the usual international convention. Transportation and communication are not included in I but are included in S. Availability of data is the major reason for grouping the subsectors into the three major sectors.

The macro-level of product per worker, Y/L, is decomposed into Y_A/L_A, Y_I/L_I, and Y_S/L_S with $Y = Y_A+Y_I+Y_S$ and $L = L_A+L_I+L_S$. The term Y/L is a weighted sum of sectoral product per worker:

(1) $$Y/L = Y_A/L_A \cdot L_A/L + Y_I/L_I \cdot L_I/L + Y_S/L_S \cdot L_S/L$$

Equation (1) simply tells us that the level of Y/L varies first by sectoral output per worker (Y_A/L_A, etc.) and second by the weights of the sectoral allocation of labor (L_A/L, etc.).

In the cross-sectional approach used in Chapter 1, countries were classified into five groups according to their level of Y/L in 1970. For each of these same five groups, this sectoral decomposition will be applied. Corresponding to the preceding chapters, both cross-sectional and time-series observations are used. The problem covers wide areas pertaining to both demand and supply, and various approaches are possible and useful. We will begin with what we call the "ratio approach" where the patterns of change in the sectoral relationship of product per worker are examined. The "growth approach," which treats differences in the sectoral rates of productivity growth (G_{YA}, G_{YI}, and G_{YS}), is the subject of Chapter 4.

In Section I, the ratio of the sectoral product per worker to the total product per worker—that is, Y_A/L_A / Y/L or (θ_A), Y_I/L_I / Y/L or (θ_I), and Y_S/L_S or (θ_S)—is analyzed for the five country groups as a preliminary step. In terms of the sectoral relationships, specific ratios such as θ_N/θ_A where ($N = I+S$) are first investigated; then the ratios θ_I/θ_A, θ_S/θ_A, and θ_S/θ_I are discussed (these will be expressed as I/A, S/A, and S/I hereafter for simplicity). It is easy to see that these ratios are the combined result of variations in the product shares and labor shares as follows,

(2) $$\theta_A = Y_A/Y \;/\; L_A/L,\ \theta_I = Y_I/Y \;/\; L_I/L, \text{ and } \theta_S = Y_S/Y \;/\; L_S/L$$

In Section II, the ratio approach is applied to the historical performances of the developed nations, in particular, selected Western countries and Japan. The purpose of the analysis in this section is to further clarify the problems and characteristics of developing economies through comparative observation. Following the procedure taken in Sections I and II, the variations in the sectoral shares of product and labor are investigated in Section III in order to confirm and extend the preceding findings. This is the "share approach." In both the ratio and the share approaches, subgrouping is attempted from the viewpoint of industrialization.

As is widely known, research in this area was pioneered by Colin Clark and later developed more comprehensively and in greater depth by Simon Kuznets. We have learned a great deal from their findings. However, for the developing economies, we believe that questions remain and need to be clarified further. We are convinced that the best starting point for our discussion is Kuznets' concluding remarks on the results of his elaborate research:

> Over the long periods up to the recent decade or two, despite the sustained rise in per capita product, the (I+S)/A ratio [N/A in our terms] did not decline; in other words, the relative products per worker in the two major sectors did not converge, <u>as they did it in the recent cross-section</u> [emphasis added]; and in many countries they tended to diverge. It was probably only after the 1920s that convergence became evident in several countries . . . . given the economic depression of the 1930s, the convergence probably emerge during World War II. For any substantive analytical purposes, its concentration in the post-World War II decade or two is significant.
>
> Apparently through much of the period of modern economic growth, the percentage rise in product per worker in the A sector could not keep pace with that in product per worker in the I+S sector. That does not mean that per worker product in the A sector did not rise, but rather that it rose at a lower rate than in the rest of the economy. And it is in this crucial respect that the long-term past fails to conform to the current cross-section. The recent reversal of these older disparities between the A and the I+S sectors in growth rates in product per worker was striking, since the sharp rise in the relative product per worker in the A sector means a high rate of growth of labor productivity in the sector. There is a strong suggestion of changes in the A sector (possibly initiated elsewhere but affecting the A sector) that amount to <u>a recent agricultural revolution in the developed countries</u> [emphasis added].[58]

The problem presented for currently developed nations—that is, convergence vs. divergence—is, in fact, more important for developing economies. We are concerned here with the direction of change in the production structure and follow Kuznets' concept and method in using the ratio approach to examine the long-term trend of convergence or divergence. We will also be comparing the richer postwar cross-sectional data of recent years and the historical

58 Kuznets, *Economic Growth of Nations*, 294–95.

records of currently developed countries to confirm the pattern of performance in search of a consistent interpretation. The procedure and results will not necessarily be the same as Kuznets'. For example, in the latter part of his concluding remarks, he paid particular attention to the agricultural revolution in the context of sectoral productivity growth without linking the ratio approach to the growth approach. In our view, the latter approach should be developed further. The technological characteristics of productivity growth, particularly in relation to labor employment, should be analyzed for all sectors, not only agriculture but also industry and services, in order to obtain an integrated analysis of aggregate performance.

With regard to the ratio approach, what is the real significance of convergence or divergence? Our answer is that the shift of the production structure manifests itself in this form and that the direction of change in the sectoral ratios is of structural significance in developing economies. Thus, we define the "production structure" of the economy in terms of the sectoral relation between product and labor. In the growing economy, structural changes are defined as the relationship between productivity growth and reallocation of employed labor in the sectoral context. As mentioned earlier, this relationship can be directly measured using equation (1) and/or in terms of the sectoral composition of shares of product and shares of labor in equation (2). In this regard, some conceptual clarification is needed. Usually, the term "structure" is used in a different context. As is illustrated by conventional terms such as "industrial structure" or "structural change," the term structure is inseparable from economic growth. No doubt growth certainly brings forth structural changes in sectoral output. Similarly, growth accompanies structural change in labor employment. However, we do not share the view that structural change is merely a concomitant result of growth, but believe that structural change pertains to changes in the sectoral relationship between product and labor. Thus, according to our definition of structural change, the same aggregate output growth would be possible with different sectoral relationships between productivity and employment.

Further conceptual clarification may be needed in the treatment of developing economies, in particular, with respect to the structure of the wage determining mechanism. In our framework, a competitive equilibrium in wage determination is not necessarily presumed for all phases of development. Instead, differences in the sectoral marginal product of labor is assumed to arise in the earlier phases of development. As will be discussed later, the process of possible shifts in the supply and demand curves for labor is assumed to be substantially significant. At what point in terms of the change in production structure does the so-called turning point emerge? To answer this question, changes in the production structure can be analyzed for the individual countries. This approach is desirable, and if the turning point is demarcated exactly, this would contribute to clarifying the crucial time dimension involved in structural change. "Disequilibrium" in the early phases could be elucidated and one may discuss the validity of the classical vs. the neoclassical mechanism. However, as has already been mentioned in the interpretation of the macro-performance of $w\Delta L/\Delta K$ in the investment approach, we do not intend to apply *a priori* a certain theoretical framework to the actual situation of developing economies. Rather our intent is an empirical clarification that is flexible with regard to the varied performances of the so-called disequilibrium.

On the other hand, from the viewpoint of neoclassical thought, the ratio approach is often criticized for its use of an "average" instead of a "marginal" analysis, asserting that the average approach has nothing to do with equilibrium analysis. The concept of shifts in the production structure aims at setting a broad and flexible framework at the sacrifice of rigor. The following illustration may help clarify our intention.

Because of the practical difficulty in treating wage rates and/or earnings in the sectoral approach which arises due to the data limitation, we use the assumption made earlier concerning labor's income share, β, in Chapter 1, Technical Notes. That is, the relationship between wages (w) and product per worker (Y/L) for sector (1) which has a higher Y/L and sector (2) which has a lower Y/L is as follows:

$$(3) \quad L_1w_1/Y_1 < L_2w_2/Y_2 \text{ or } w_1/w_2 < Y_1/L_1 \; / \; Y_2/L_2$$

Equation (3) is derived from $\beta_M < \beta_N$, where β_M is labor's income share in manufacturing and β_N is labor's income share in nonmanufacturing. If equation (3) can be applied to our three-sector approach, the ratio approach would suggest an important implication: a trend of convergence (or divergence) in Y/L means a trend of convergence (or divergence) in the sectoral wage ratio, although the two trends may not exactly parallel one another.

Yet, in the extreme case where Y/L in the manufacturing and nonmanufacturing sectors are equal, i.e., at the terminal point of convergence between the two sectors, we can expect wage equalization in the two sectors. In a simple theoretical setting, we would say this is a state of equilibrium in which the marginal product of labor is equalized in the two sectors. In the real world, wages are composed of a wide range of varied elements (skill, sex, age, etc.) and equation (3) can only be used for conceptual clarification. Nevertheless, the equation can be used as the underlying hypothesis for the ratio approach.[59]

Finally, the substantive problems in this chapter are as follows. First, the hypothesis stated in Chapter 1 implies that the growth performances of developing economies and the historical path of currently developed economies have different features. In Chapter 2, empirical evidence was presented showing the different features. But what about the issue of convergence vs. divergence? It may be natural to hypothesize that different features may be consistently identified in the process of shifting production structure. If so, convergence (or divergence) does not necessarily emerge as a general linear path for developed and developing economies. This is one of the important issues of the analysis which follows. Second, unlike the macro-analysis in the preceding two chapters, the characteristics of sectors, historically and technologically, play a crucial role. Although this aspect will be discussed in depth in Chapter 4, it is an important issue here as well. In carrying out the ratio approach, the trend to converge or diverge will be examined by emphasizing the importance of the performance of the ratio between the two major sectors. We will want to observe the process of industrialization in this context.

---

59 In the development process, which involves the modern enterprises as well as the peasant economy, how can we conceive of equilibrium or disequilibrium in terms of labor-related behavior? Mixed income of peasant households makes it difficult to apply a simple concept merely in terms of wage rates. An equilibrium state in the neoclassical sense may not be realized to the extent that nonwage income plays an important role in their utility evaluation; if this is true, then equality in the marginal product may not be realized. One may say this is underemployment, but in terms of income, there is a sort of equilibrium. This may illustrate the complex background that is implied in equation (3) and is relevant to the reason for adopting the "imputation" procedure in estimating β in Chapters 1 and 2.

II. A Sectoral Comparison of Product Per Worker Using Cross-sectional Data

Kuznets' study linked historical series data with recent cross-sectional data. Let us begin with the latter.

Agriculture vs. Nonagriculture

An overview of sectoral performance of product per worker can best be commenced, because of its historical significance, with a discussion of N/A, that is, the ratio of the level of the nonagricultural sector (N = I+S) to the level of agriculture. In the contemporary developing economies, the agricultural sector dominated the economy in terms of product and particularly in terms of labor. Development starts with economic modernization, and the relationship between industry, I, and services, S, is an important issue; however, I and S will be examined later.[60] At this point, a two-sector approach, i.e., A vs. N, is taken up. What will its performance look like through the process of industrialization?

The pace and style of industrialization are not uniform among the nations but vary as a result of the different responses of each country, even those countries with a similar level of Y/L. The ratio N/A is observed by subgroup (in addition to observing the average observations) in order to see the effects of industrialization. The yardstick for the subgrouping is the product share in the sector I (i.e., Y_I/Y). Each group of countries is divided into two subgroups: subgroup C includes nations with higher Y_I/Y and subgroup D includes nations with lower Y_I/Y.

The subgroups will be used in all cases that follow in this chapter. An alternative approach is to use all ten groups to see more intensively the relationship between the ratio N/A and the Y/L level. We have no reason to reject this approach, but will (as was stated in the Introduction) define industrialization in terms of shifts in the production structure. This assumes that shifts in the production structure will not be simply associated with rises in Y/L, but will also be affected by factors not associated with the Y/L level.

---

60 Statistically weighted measures of sectoral inequality, taking the three sectors together, is the conventional approach. However, we do not follow this because it presents less significant results.

Selected Countries

(the total number of countries is shown in parenthesis)

Group I (10)	C:	(5)	Malawi, Madagascar, India, Somalia, and Tanzania
	D:	(5)	Ethiopia, Nepal, Niger, Benin and Burma
Group II	C:	(6)	Pakistan, Bolivia, Philippines, Thailand, Kenya, and Ivory Coast
(12)	D:	(6)	Sierra Leone, Senegal, Sudan, Sri Lanka, People's Rep. of Congo, and Mozambique
Group III (12)	C:	(6)	Rep. of Korea, Turkey, Nicaragua, Egypt, Rep. Arab, Morocco, and Rep. Syrian Arab
	D:	(6)	Paraguay, Tunisia, El Salvador, Honduras, Colombia, and Malaysia
Group IV	C:	(6)	Brazil, Peru, Yugoslavia, Jamaica, Portugal, and Argentina
(12)	D:	(6)	Rep. Dominica, Mexico, Greece, Costa Rica, Uruguay, and Singapore
Group V	C:	(6)	Japan, France, Netherlands, Belgium, United Kingdom, and Fed. Rep. of Germany
(12)	D:	(6)	Denmark, Canada, Finland, Norway, Sweden, and United States of America

Table 3.1 presents a summary of the measured ratios. The cross-sectional series is of primary interest, but the changes over time are also important. Though we need periods of longer duration, the information provided by the three years are informative.

First, the magnitude of the ratio in the lower groups is strikingly high. This indicates an abrupt start to the industrialization path which characterizes the postwar developing economies (which was pointed out earlier). This will be discussed further in comparison with the much lower ratio of presently developed nations. For now, let us focus our attention on the pattern of changes. Broadly speaking, the ratio N/A appears to follow "a tendency of convergence" but notable variations along the path of changes in the level of Y/L also occurs.

From all groups, the ratio of product per worker in the nonagricultural sector to that in agriculture, i.e., N/A, shows a general tendency to decline as Y/L rises (Panel A). However, a careful look reveals two deviations. First, from group I to II, N/A in fact increases to a surprisingly high value for group II, suggesting the possibility of divergence. From that point, however, the ratio declines significantly without remarkable exception for all three

Table 3.1
Comparisons of Product Per Worker Between Nonagriculture and Agriculture (N/A) [a] in 1960, 1970, and 1980

(A) Group by Y/L level	1960	1970	1980	Average
I	5.01	5.21	4.42	4.88
II	5.44	5.56	5.04	5.35
III	3.44	2.96	2.55	2.98
IV	3.14	3.08	2.98	3.07
V	1.86	2.21	1.79	1.95

(B)	Subgroup C				Subgroup D			
	1960	1970	1980	Average	1960	1970	1980	Average
I	5.03	5.85	4.85	5.24	4.95	4.48	3.95	4.46
II	5.10	5.21	4.90	5.07	5.84	6.12	5.28	5.75
III	4.44	3.47	3.11	3.67	2.75	2.54	2.31	2.53
IV	3.13	3.86	4.22	3.74	3.15	2.30	2.10	2.37
V	2.08	2.59	2.09	2.35	1.65	2.04	1.49	1.73

Note: a. The ratio N/A is expressed by assuming the Y/L of agriculture or A is equal to unity.

Sources: See sources of Table 1.1; World Bank, *World Tables*, various issues.

years under consideration, thus confirming the convergence tendency. Another deviation is witnessed between groups III and IV in 1970 and 1980. In these two years, N/A is higher for group IV than for group III in 1970 and 1980, though on average there is almost no change. Thus, we face a problem of judgment: how do we interpret these patterns? We could assume the validity of the convergence trend as the normal path of growth and explain the distortions either statistically or as exceptions. Or, rather than make sweeping generalizations, we could leave the significance of these deviations open to further explanation. We choose to take the latter approach and regard these cases as factual challenges.

The variation in the magnitude of the ratio across groups is significant. The tempo of change does not appear to be uniform. Between groups I and II, the difference in the ratio's magnitude is not large, while between groups II and III, a larger difference is witnessed for all three years. Between groups III and IV, the difference is negligible. Finally, between groups IV and V, the difference is much wider. Is there any reason for such a zigzag performance? Two factors are relevant: the factors associated with Y/L and the nonassociated factors. If Kuznets' proposition of a general trend of cross-sectional convergence is valid, the Y/L-associated elements should be dominant. A smooth path of convergence is then naturally expected. If the zigzag pattern is confirmed, however, the nonassociated factors deserve closer attention.

The N/A ratios in Panel (B) of Table 3.1 confirm what was found from the average data across groups, and furthermore, helps us to present new aspects, if any, with respect to the different performances of the subgroups which were classified by the degree of industrialization. As is widely known, the variation in the share of industrial output (Y_I/Y) has primarily been explained by the increasing trend in the income level (usually GNP per capita). This explanation, however, is only concerned with the demand side. If an analogous view can be applied to the supply-production side, we would expect that Y_I/Y would be associated with Y/L. However, we do not assume such an association for the countries within the same Y/L-level group since the nonassociated factors may operate. This requires further examination.

On average the ratios for subgroups C and D exhibit a similar zigzag pattern (Figure 3.1). In general, differences in the rates of industrialization appear to be of minor relevance. However, a significant difference in the magnitude of the ratio is witnessed between subgroups C and D: the ratio N/A is high for C vs. low for D in the higher Y/L groups (i.e., groups III, IV, and V). Industrialization does have considerable effects. At the same Y/L level, more (less) industrialized economies are accompanied by larger (smaller) ratios of N/A. This pattern, however, is not witnessed for lower Y/L groups (i.e., groups I and II). The zigzag shape pointed out earlier now becomes clearer, and an important suggestion can be made. To be realistic in identifying a trend of convergence or divergence, an assumption of linearity is not appropriate. At least two intervals should be demarcated for the shifts in the production structure: the primary interval, PI, and the secondary interval, SI. For the developing economies, the cross-sectional data suggest that the PI is represented by groups I and II and that the SI is represented by groups III and IV. The basic shift in the production structure (from PI to SI) appears to emerge between groups II and III.[61] Of course its possible conformity with the development path over time remains to be examined.

Although the table does not cover an interval long enough to analyze over-time changes, certain suggestions can be drawn. The broad trend is toward convergence, but is often interrupted by divergence during the 1960s. Differences across groups are noteworthy. Groups of higher Y/L (i.e., groups III and IV) show a convergence path (stronger in III and weaker in IV) throughout the two decades without interruption. This is also witnessed by the subgroupings, except for subgroup C of group IV in the 1960s. As for groups of lower Y/L levels, the over-time performance is mixed: there is divergence in the 1960s, followed by convergence in the 1970s. By subgrouping, the trend is the same except for subgroup D of group II in the 1970s.

From such a seemingly mixed performance for the three years, it appears that no suggestions can be drawn. Yet, the different performances witnessed between the two groups of lower vs. higher Y/L level deserves attention because it corresponds to PI and SI, respectively.[62] With respect to the SI, the trend of convergence can

---

61 Kuznets was aware of the different performance observed over time between PI and SI in the 1950s:

> In the lower per capita product countries, despite substantial rises in per capita product, relative product per worker in the A sector did not rise, and the inter-sectoral inequality ratio, (I+S)/A, did not decline. In the higher per capita countries . . . . relative product per worker in the A sector tended to rise and the inter-sectoral inequality ratio tended to drop. . . . This additional case of differential non-conformity between cross-section estimates and observed time trends suggests an important conclusion: in the post-World War II period, the industrialization growth process in the less developed countries, marked in the share in product, was limited in the case of labor force, widening the already wide contrast between An and I+S sectors (Kuznets, *Economic Growth of Nations,* 314).

In Table 3.1, a similar time-series performance is exhibited by groups I and II in the 1970s, but we do not claim nonconformity because a divergence trend is witnessed in the cross-sectional estimates. Therefore, the limitation of industrialization Kuznets observed is identified more regularly for demarcating the PI.

62 Their relevance to the phase of development will be discussed later in Chapter 4.

Figure 3.1
The Ratio of Nonagricultural and Agricultural Product Per Worker (N/A)

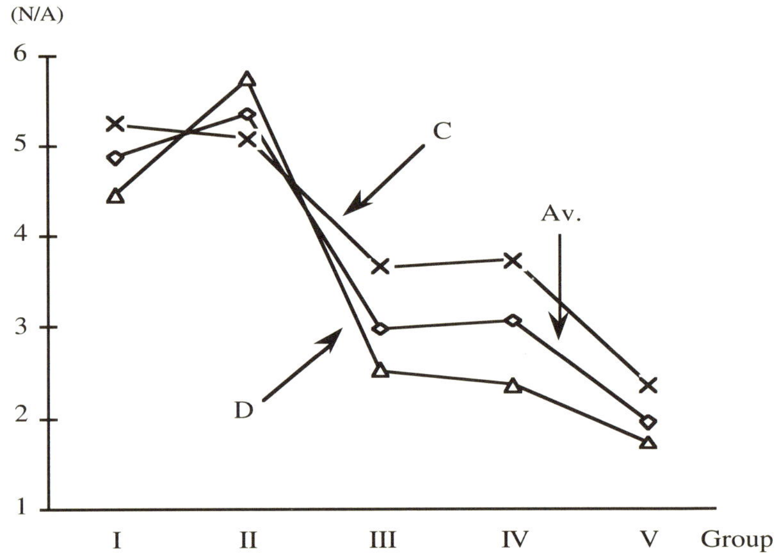

Sources: See sources of Table 3.1.

generally be expected and the mixed performance of the PI may not be inconsistent with the property of PI. A few words may be required regarding group V. In our framework, the nations in group V are at the state of a terminal trend of convergence. Its further possible change is beyond the scope of our analysis. How can we interpret the mixed performances of N/A in subgroups C and D—that is, divergence in the 1960s and convergence in the 1970s? An answer would require an examination of the factors that are relevant to the rates of productivity growth (this is the topic of the following chapter).

Industry vs. Services

The ratio approach in terms of N/A is a preliminary attempt to give an overview of the shifts in the production structure. Further decomposition of this ratio is the focus of this section. The characteristics of the three major sectors—A, I, and S—should be clarified even within the confined scope of the ratio approach. We expect that some, if not all, of the issues raised by the preceding analysis will be solved in so doing. As suggested earlier, the changing relationship between the two sectors, I and S, is important. The performance of I/S will be examined, together with I/A and S/A.[63]

Table 3.2 which corresponds to Table 3.1, presents the average observations by group (Panel A) and by subgroup (Panel B). The ratios of product per worker of the individual sector to the national average (i.e., θ_A, θ_I, θ_S) are shown in Panel (A). These ratios are preliminary information and are informative for grasping the aggregate position of each sector. For example, θ_A shows that for agriculture, a broad decreasing trend from around 0.60 in group I to around 0.40 in group IV turns to increase to 0.60 in group V. Another example is a broad declining trend in the industrial sector, which finally comes close to unity in group V. However, for our present purposes, these ratios are useful only in a relative sense, that is, in comparison with the corresponding ratios of the other two sectors. Nevertheless, such ratios have often been used as indicators for revealing the pattern of convergence or divergence. This is misleading as the aggregate average includes the value of the individual sector for which the comparison is being made.

With respect to the sectoral ratios, let us first consider the relationship between industry and services (I/S, the last column in the table). With respect to changes over time, a distinct and regular pattern is witnessed over the two decades. Without exception, I/S increases. This is significant in that it indicates a sustained effect of industrialization through economic growth. The product per worker of sector I in 1960 was distinctly lower than that of sector S ($I/S < 1$) for groups IV and V. Since then, because I tended to increase, the ratio became close to or exceeded unity in 1980, that is $Y_I/L_I > Y_S/L_S$.[64] The significance of this will be discussed in greater detail

---

63 For sector I, at least two subsectors, manufacturing (M) and nonmanufacturing (NM) are needed, but labor employment data are not available for international comparisons. Sector S is much more mixed and its decomposition into several subsectors is also desirable. However, this can only be done for selected cases. It is beyond our ability to provide comprehensive data for these categories.

64 The performance of the ratio across subgroups (Panel B) generally shows the same pattern, but an additional aspect deserves attention. For group I, the performance differs significantly between C and D; this is the reason for the larger value of I/S in 1980 in the group observation. Inclusion of mining in sector I is the cause of this. A relatively larger ratio for group IV in 1980 may be similarly explained, but to a less degree. Otherwise, the

Table 3.2
Comparisons of Product Per Worker Between the Major Sectors in 1960, 1970, and 1980

(A) By Group

Group		Agriculture	Industry	Services	I/A	S/A	I/S
I	1960	0.63	3.52	4.43	5.59	7.03	0.79
	1970	0.61	4.38	3.72	7.97	5.98	1.33
	1980	0.59	5.69	2.99	9.64	5.07	1.90
II	1960	0.49	3.41	3.45	4.92	7.04	0.70
	1970	0.48	2.79	3.43	5.81	7.15	0.81
	1980	0.49	2.72	2.64	5.55	5.39	1.03
III	1960	0.47	1.41	2.15	3.00	4.57	0.66
	1970	0.52	1.45	1.72	2.79	3.31	0.84
	1980	0.53	1.53	1.46	2.89	2.75	1.05
IV	1960	0.47	1.40	1.67	2.98	3.56	0.83
	1970	0.46	1.33	1.38	2.89	3.00	0.96
	1980	0.42	1.51	1.21	3.60	2.88	1.25
V	1960	0.61	1.02	1.18	1.67	1.93	0.86
	1970	0.57	1.00	1.12	1.75	1.96	0.89
	1980	0.66	1.07	1.03	1.62	1.56	1.04

(B) By Subgroup

Subgroup		I/A			S/A			I/S		
		1960	1970	1980	1960	1970	1980	1960	1970	1980
I	C	4.59	6.31	4.25	7.34	7.31	5.29	0.63	0.86	0.80
	D	6.85	7.97	15.48	6.61	5.78	4.87	1.04	1.38	3.18
II	C	4.64	5.79	5.00	6.24	6.62	4.82	0.74	0.87	1.04
	D	5.21	5.81	6.12	7.88	7.56	5.96	0.66	0.77	1.03
III	C	3.19	3.71	3.26	4.22	4.04	3.26	0.76	0.92	1.00
	D	2.36	2.24	2.58	3.54	2.84	2.35	0.66	0.79	1.10
IV	C	3.65	3.88	5.17	3.65	3.21	3.47	0.96	1.21	1.47
	D	2.46	2.50	2.50	3.20	3.20	2.46	0.77	0.78	1.02
V	C	1.66	1.65	1.67	1.66	1.74	1.68	0.96	0.94	0.98
	D	1.67	1.75	1.60	2.14	2.17	1.46	0.78	0.81	1.10

Sources: See sources of Table 3.1.

---

pattern found above presents the effects of industrialization fairly well in relation to the services sector activities.

in Chapter 4 after examining the performance of the relevant terms, but a preliminary suggestion can be made here. The process of industrialization has often been treated in comparison with the other goods-producing sector, agriculture. A theoretical model of development that is composed of two sectors, industry and agriculture, is a good example. We do not deny the importance of the relationship between industrialization and the agricultural sector, and I/A will be discussed later. But the factual challenge we currently face is that the relationship, in particular the competitive relationship between I and S, is an important aspect that requires further clarification. Subsectors of an infrastructural nature, such as the transportation and communication subsectors, and modern facilitating subsectors, such as banking and financial activities sectors, may have developed in a complementary fashion with industries. But, viewed from investment, the competitive relationship is important. Moreover, subsectors of a traditional nature which are often found in the commerce, trade, and personal services sectors, may have become an area of underemployment and to that extent Y_S/L_S may be relatively smaller. (This leads us to a later discussion of S/A.) Recognizing this, we are keenly aware that the faster rate of technological progress is a salient characteristic of sector I and is much less so for sector S as a whole. This essentially explains the pattern of the increase in I/S shown in Table 3.2. How to explain the faster rate of increase in product per worker as a result of technological advance in industry is the topic of Chapter 4.

From the cross-sectional performance data, clear regularity is recognized in two different patterns: the magnitude of I/S declines in lower Y/L-level country groups (from group I to group II) and rises in the higher Y/L-level country groups (from group III to group IV). Only a very minor change is witnessed between groups II and III. As this pattern seems to generally correspond with the changes over time, a fairly strong suggestion may be made that the data support the interval demarcation made earlier: that is, a PI for the former and a SI for the latter. Confirmation of consistency between the two cases is of substantial significance. This does not mean, however, that a uniform structure can be supposed to prevail in both country subgroups. Instead, as is indicated by the ratio performance shown in Panel (B), the pattern differs distinctly between subgroups C and D. Subgroup C tends to sustain a "normal" pattern of a rising ratio through all groups of developing economies, whereas subgroup D does not. The declining ratio in the latter group is particularly influential in forming the pattern of the lower Y/L groups. For the higher Y/L groups, its role tends to be weak and even negative in raising the ratio I/S. Hence it is suggested that, for development phasing in Chapter 4, we must look carefully at the structural relationship between industry and services. In the relationship between the two sectors, PI is characterized as prenormalized while SI is characterized as normalized.

Finally, a few words need to be said for group V. On average, the change in the ratio I/S over time is the lowest for group V and in 1980, it is close to unity. This conforms with our common sense expectation. By subgroup, C shows the least over-time change, and the magnitude of the ratio is sustained at a value close to, but under, unity. Subgroup D presents a somewhat different pattern: the ratio I/S tends to increase from a relatively low level to a high level and exceeds unity. The reason for such a difference can only be explained on the basis of sectoral productivity analysis. Nevertheless, even without sectoral productivity analysis, the essential point can be made that this occurs towards the end of the industrialization drive with the emergence of new service activities. The performance of subgroup C represents this normal pattern well, whereas subgroup D may represent countries which are somewhat behind. What is important is the identification of the pivot point for group V, as indicated by the ratio, in the normal course of industrialization.

Industry and Services vs. Agriculture

Coming back to the ratio N/A, its decomposition into I/A and S/A allows us to examine further the issue of convergence or divergence. Beginning with changes over time, different tendencies are witnessed in the I/A and S/A ratios: for the former, there appears to be an increasing trend while for the latter, the trend is a declining one. This is an important finding. Instead of a uniform trend, two directions of change are suggested: a trend of divergence in the industry-agriculture relationship and a trend of convergence in the services-agriculture relationship. The trend performance of the ratio between agriculture and nonagriculture that was discussed earlier can now be interpreted as a composite outcome of the trends of the two terms which counteract one another.[65] For group V, neither

---

65 By group (Panel A), S/A shows a tendency to decline in all groups of developing economies with the exception of group II. An increasing tendency for I/A is seen for these groups with the exception of groups II and III. In a direct comparison between 1960 and 1980, no exception is seen for the former and only a slight exception for group II is witnessed for the latter. It may be easy to confirm these tendencies by subgroup (Panel B). Accounting for the exceptional cases, subgroup C, instead of subgroup D, is responsible for the trend relationship in both the industry-agriculture and services-agriculture ratios.

convergence nor divergence is distinctly identified, although a decrease in both S/A and I/A from 1970 to 1980 is noted.

Do the performances of the over-time ratios conform with the pattern of the cross-sectional data? Broadly the answer is "yes" for the services-agriculture relationship, but "no" for the industry-agriculture relationship. Figure 3.2 presents a simple picture of the average ratios at the three points in time shown in Panel (B) of Table 3.2. Taking a broad overview, for both S/A and I/A, one's first impression is that there may be a trend of convergence for the entire range from groups I to V. For S/A this is expected, but for I/A it is not when we recall the observation of over-time changes. The reason for the appearance of nonconformity in the latter case cannot be fully explained, but we suggest two distinct shifts—the first shift occurs between groups II and III and the second shift occurs between groups IV and V—may be the major factor responsible for the pattern of the ratio.

The shifts imply a distinctly greater decline in the level of the ratio. On average for the three years, the magnitude of I/A by group first declines from 5.43 in group II to 2.89 in group III and from 3.16 in group IV to 1.67 in group V. As shown in Figure 3.2, deviations between subgroups C and D do not matter much. A careful look at the performance of S/A shows similar shifts in the range of groups II–III and groups IV–V in terms of an accelerating pattern of a trend decline: first, from 6.54 in group II to 3.54 in group III, and second, from 3.15 in group IV to 1.82 in group V. Naturally, one may suppose a common factor to be responsible for these concurrent shifts. Changes in the relative productivity of agriculture must be one of the major common factors. The importance of the "agricultural revolution," which was stressed in quoting Kuznets' proposition earlier, is discussed in depth in the chapter that follows.

At this point, we focus our discussion on a summary of the findings. In the case of I/A, divergence is witnessed for the two intervals: first, from group I to group II in subgroup C (note that subgroup D is not normal), and second, from group II to group IV for subgroups C and D and the group average. In this sense, we are reminded of the zigzag pattern found earlier for N/A. With conceptual conformity, we would assert that industrialization sustains its potential effect of producing a trend of divergence throughout the entire path of development. The realization of this trend depends upon the other sectors' performances in terms of productivity and employment. The two intervals, the intervals between groups I and II and groups III and IV, are characterized by a strong operation of forces which produce diversion. In group IV, the greater tendency is toward diversion in subgroup C.

Figure 3.2
Ratios of Product Per Worker Between Industry and Agriculture (I/A) and Between Services and Agriculture (S/A), 1960–1980

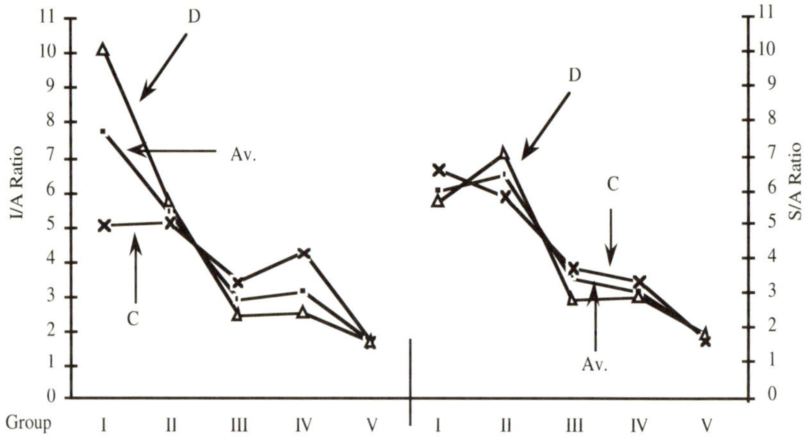

Sources: Tables 3.2 and 3.3.

The pattern difference is also identified for the case of the services-agriculture relationship, although it is less distinct in the interval from group III to group IV. This is clearly shown in Figure 3.2. These pattern differences cannot be solely attributed to industrialization. At the conceptual level, we are reminded of the two-interval demarcation proposed earlier. The decomposition of the nonagricultural sector into industry and services does not seem to bring forth any insights that would alter the notion and we believe that the explanation given is substantive.

By way of concluding remarks, let us draw attention to the following points.

1. There exists a wide variation in the magnitude of the ratios. A direct comparison of two groups, I and V, indicates an enormous difference in the sectoral differentials of product per worker. On average, the ratio of sectoral product per worker between nonagriculture and agriculture is 4.9 for group I vs. 2.0 for group V; between industry and agriculture, the ratio is 5.9 for group I vs. 1.7 for group V; and between services and agriculture, the ratio is 5.9 for group I vs. 1.8 for group V. These ratios eloquently suggest the significance of the problem of inducing the shift in production structure by industrialization, and at the same time the serious difficulty in realizing its possibility. According to our presumption

Table 3.3
Ratios of Product Per Worker Using Time-series and Cross-sectional Observations, 1960–1980

	N/A		I/A		S/A	
	1980	1960	1980	1960	1980	1960
(1)	4.4 (I)	5.4 (II)	9.6 (I)	4.9 (II)	5.1 (I)	7.0 (II)
(2)	5.0 (II)	3.4 (III)	5.6 (II)	3.0 (III)	5.4 (II)	4.6 (III)
(3)	2.6 (III)	3.1 (IV)	2.9 (III)	3.0 (IV)	2.8 (III)	3.6 (IV)
(4)	3.0 (IV)	1.9 (V)	3.6 (IV)	1.5 (V)	2.9 (IV)	1.9 (V)

Sources: See sources of Tables 3.1 and 3.2.

on the effects of "abrupt" industrialization, the problem must be interpreted as one that is characteristic of the postwar developing economies, rather than a common phenomenon for all paths of development in general. The essential issues involved concern the concurrent existence of enormous differences in production structure in the world, ranging from an extremely diverged state on one hand to an extremely converged state on the other.

2. There exists the possibility of phase shifts occurring. Although phasing has not been dealt with in this text thus far, it would nevertheless be useful to discuss the possibility of shifting from the present phase to the next one by comparing the two intervals, PI and SI, by examining the cross-sectional and over-time performances. The over-time data series is not long enough, but the direction and the magnitude of changes may be informative if adequately compared with the cross-sectional data. The technique used here simply compares the ratio of each group in 1980 with the ratio in 1960 of the next group according to the order of the Y/L level. In the case of near conformity between the over-time and cross-sectional data, these two ratios would be nearly equal. Otherwise, we may find some discrepancy in either direction or magnitude. For simplicity, let us ignore 1970 and subgroup differences; the comparisons between the group averages are summarized in Table 3.3.

Viewed from a broad trend of convergence, the most important point is that a great difference in the magnitude of the three ratios is witnessed in cases (2) and (4), and minor exceptions occur for (1). Over-time changes are largely on a convergence trend, but its pace lags too much to be able to make a shift from group II to group III and from group IV to group V, which were the situations presented in the cross-sectional analysis. The backwardness is particularly strong in case (4). This feature may appear in sharper contrast if comparison is made with case (3), from group III to IV. Here, the difference in the magnitude of the ratio is minor and even exhibits a tendency to diverge. Case (1) shows prenormal performance as is expected from the earlier discussion. Thus, in sum, two points attract our attention. If the cross-sectional data are assumed to present largely the phase situation, realization of the shifts from group II to group III and from group IV to group V would need a surprisingly longer time than expected. This conforms with the earlier discussion that referred to Figure 3.2. To fit our analysis with reality is a great challenge. The situation represented by the cross-sectional data thus requires much more qualification to link it with over-time changes with regard to the phasing of developing economies. In particular, the near-perfect converged state of group V is the accumulated result of the long historical path of economic development and cannot be simply compared with group IV. The factors which affected the historical path is the topic of the section that follows.

III. Sectoral Comparisons of Product Per Worker: Historical Patterns of Presently Developed Nations

This section examines the historical performances of the sectoral product per worker for currently developed nations which were treated as group V in the preceding sections. A comprehensive analysis is beyond our reach, but the discussion focuses on the aspects closely relevant to elucidating the problem of contemporary developing economies, the clarification of which is our major purpose. Stated simply, Kuznets' proposition was that the historical performance of developed nations shows a long-sustained trend of divergence; and this trend has turned to convergence in recent years. If this is really so, the performance of developing economies elucidated in the preceding section appears to differ: the major trend that was found was one of convergence, and divergence was limited to the lower Y/L level groups. Why is the trend of divergence not sustained in the higher Y/L postwar developing economies? Or, why do we see a trend of convergence taking place at Y/L levels that are lower than expected from the Kuznets' viewpoint? More concretely, we face two issues. The first issue is confirmation of the fact itself: that is, do these differences really exist? Second, if confirmed, we must then deal with the interpretation. That is, is it possible to interpret the cases of both the developed and developing economies consistently in a general framework of development?

The case of Japan is taken up first in detail, in the reverse order of treatment of Chapter 2, because we believe this case is most directly informative in answering these questions. Second, a sketchy observation, which basically depends upon data arranged by Kuznets, is made of Western nations. Some qualifications of his proposal are voiced. Third, for the postwar interval, i.e., 1960–1980, which was considered in the preceding section, further discussions will be developed in comparison with Japan and Western developed nations. This discussion hopes to gain insights into the internal differences between the countries in group V. We believe this may further help to clarify the characteristics of latecomers.

Japan: An Illustrative Case

In describing Japan's historical experience, the methodology and procedure used are the same as those used in the preceding sections. That is: (1) the relative sectoral product per worker (θ) is used as the major term. The pattern shown since the mid-1960s will be discussed separately in subsequent sections, but for now we focus on clarifying the pattern of the developing process; (2) the data series that is expressed in constant prices is also used. In the preceding section this was not done because of the lack of available data. As a result, there was a defect in our analysis as the influence of the variation in relative prices is often of considerable magnitude (in Chapter 5 an attempt is made to supplement the analysis of the postwar developing economies); (3) the major sector classification is essentially the same except for one point: transportation and communication are included in industry, instead of services, in the *LTES* which is the source of the data used in our analysis. This is also true for Kuznets' data for Western nations. Thus, interpretation of the results will need some qualification.

The data for Japan are summarized in Table 3.4. Let us begin with the series in current prices (i.e., Panel A), and first take up the ratio of product per worker in the nonagricultural sector, N, to that of agriculture, A (i.e., N/A). The ratio appears to vary rather little, but a closer look reveals certain trends: in the periods 1927–1933 and 1936–1940, which are periods of fast industrialization, the ratio N/A increased from its level in previous years; but towards the early postwar years, the value of N/A declined. During the earlier period of development, i.e., prior to the 1930s, however, the ratio shows no distinct trend (either divergence or convergence) of

Table 3.4
Sectoral Comparisons of Product Per Worker for Historical Japan,[a] 1885–1965

(A) In Current Prices

Periods Selected	N/A	Decomposed into: Industry	Decomposed into: Services	I/A	S/A	I/S
1885-1889	3.26	1.10	2.16	2.30	4.19	0.55
1894-1899	2.82	1.13	1.69	2.31	3.28	0.70
1901-1907	3.12	1.29	1.83	2.80	3.39	0.83
1908-1914	3.26	1.69	1.57	2.94	3.03	0.97
1915-1922	2.93	1.58	1.35	3.13	2.75	1.14
1927-1933	3.80	2.06	1.74	4.03	3.72	1.08
1936-1940	3.51	2.24	1.27	4.24	2.69	1.58
1937-1955 [b]	3.04	1.64	1.40	3.33	2.54	1.31
1955-1960	3.08	1.72	1.36	3.46	2.69	1.29
1960-1965	3.26	1.76	1.50	3.48	3.03	1.15

(B) In Constant Prices [c]

Periods Selected	N/A	Decomposed into: Indusry	Decomposed into: Services	I/A	S/A	I/S	Relative Prices of Output: I/A	Relative Prices of Output: S/A	Relative Prices of Output: I/S
1885-1889	3.46	0.69	2.77	1.44	5.34	0.27	160	0.78	205
1894-1899	3.80	0.89	2.91	1.82	5.68	0.32	127	0.58	219
1901-1907	3.73	1.04	2.69	2.25	5.02	0.45	124	0.68	182
1908-1914	3.61	1.30	2.31	2.26	4.42	0.51	130	0.69	188
1915-1922	4.01	1.55	2.46	3.08	4.96	0.62	102	0.55	185
1927-1933	3.56	1.87	1.69	3.65	3.61	1.01	110	1.03	107
1936-1940	4.28	2.54	1.74	4.81	3.70	1.30	88	0.73	121
1937-1955 [b]	4.18	2.19	1.99	4.42	3.58	1.23	75	0.71	106
1955-1960	3.76	2.20	1.56	4.44	3.10	1.43	78	0.87	90
1960-1965	3.82	2.63	1.19	5.23	2.40	2.18	67	1.26	53

Notes:
a. See Table 3.1; Y stands for NDP. The weights used for the decomposition are labor's share in the two sectors, I and S.
b. The figures for 1937--1955 are roughly estimated to link pre- and postwar series. However, less reliable data were used.
c. Base year of output prices is 1934–1936.

Sources: Ohkawa et al., *LTES* , Vol. 1, *National Income,* revised version of Table 3.5:56 and Vol. 2, *Labor Force;* Table 8.3:192, in Kazushi Ohkawa, "Keizai Hatten to Kozo" [Economic development and structure], in Kazushi Ohkawa and Ryoshin Minami, eds., *Kindai Nihon no Keizai Hatten* [Economic development of modern Japan] (Tokyo: Toyokeizai Shimposha, 1975).

change.[66] This differs clearly from what was identified earlier from the cross-sectional evidence. How can we account for this surprising result?

Japan's initial magnitude of N/A (3.3) is much smaller than that of the groups with the lowest Y/L level (4.9 for group I and 5.5 for group II) and is closer to the ratio of the higher groups (3.0 for group III and 3.1 for group IV; See Table 3.1). Japan's genuine modern economic growth started around 1885, so that the phase indicated by groups I and II, which were demarcated earlier as PI, cannot be identified in Japan's recorded historical experience. This naturally leads us to the following reasonable interpretation of the basic reason for the nonexistence of a sharp decline in the ratio for Japan. (We will return to the issue of the effects of abrupt industrialization.)

The decomposition of N/A into industry and services in the table shows a sharp contrast between the two. The industry component, I/A, shows a distinct increasing trend, while a declining trend is exhibited by the services component, S/A, throughout the prewar period. These two terms operated to cancel each other out and no distinct trend of structural change is found in terms of N/A. These broad performances, however, actually tend to have deviations. In terms of the industry component, a slight exception is seen for the period 1915–1922, which was an interval abnormally affected by the World War I boom in Japan. The magnitude of I/A becomes higher in 1927–1933 and 1936–1940 than the longer trend; this jump in I/A was brought forth by faster industrialization resulting from the military mobilization. With respect to the services component, noticeable deviations are mainly due to long swings in demand which were discussed in Chapter 2. The ratios between the two sectors thus help our interpretation of N/A. The ratio I/A exhibits a trend of sustained increase throughout the entire prewar period and begins to decrease toward the postwar 1960s. The ratio, S/A, on the other hand, shows a trend of decline.

---

66 This statement is made in the context of the international situation. In phasing Japan's development path, changes in the production structure between the primary phase before 1915–1922 and the secondary phase, since 1915–1922, is the major demarcation (Ohkawa and Kohama, *Lectures on Developing Economies,* Lecture 1).

The pivot and increasing trend towards the postwar 1960s is expected in light of our earlier observation of group V.

The magnitude and direction of changes in I/A and S/A differ between Japan and the developing economies. The value of I/A starts at a very low level in Japan (2.3 for Japan vs. 5.9 for developing economies in group I and 5.4 for developing economies in group II) and clearly shows an increasing trend throughout the prewar period. This is in sharp contrast to the decreasing trend for the postwar developing economies (though there is a deviation for group IV where the value of I/A is 3.2, which is close to Japan's postwar values). If taken together with the pattern of S/A, this seems to answer the query that was left unanswered earlier. The ratio S/A shows a generally decreasing trend similar to that of the postwar developing economies. However, the level of the ratio differs: in Japan, S/A is 4.2 in 1885–1889, while it is 5.9 for group I and 6.5 for group II. In other words, in the initial phases for the postwar developing nations, not only services but also industry plays a significant role in creating an enormously diverging structure. In Japan's case, this role was played chiefly by services. The abrupt effects of industrialization are not identified and we believe this is the major reason for such a difference.

The contrasting trends of I/A between Japan and the developing nations, however, needs careful interpretation, because the role played by agriculture is seriously involved. In Japan, the depressed state of agriculture contributed to a considerable extent in forming the strong trend of divergence. This was pointed out earlier for postwar developing economies. Actually, I/A does show deviations from the generally declining trend. Nevertheless, Japan's case should be interpreted as illustrating the stronger effects of divergence due to rapid industrialization.

Finally, the relationship between industry and services warrants special attention. The pattern of I/S shown in Table 3.4 shows an increasing trend throughout the entire prewar period, with the exception of 1927–1933 which was a period of severe downswing. The ratio then begins to decline in the postwar period up to 1960–1965. The pattern of I/S for Japan illustrates well the case of the stronger force of divergence due to industrialization, which was witnessed earlier in a milder form for the postwar developing economies. We can confirm this assertion. In addition to the over-time changes shown in Table 3.1, using the cross-sectional data, the average ratio at the three points in time can be calculated from Table 3.2: I/S is 1.00 for group I, 0.83 for group II, 0.87 for group III, 1.00 for group IV, and 0.92 for group V.

We know the ratio of I is not normal (for subgroup D, the ratio is 1.76). Japan's postwar pattern corresponds to the declining trend from group IV to group V. For the interval between groups II and IV, the ratio does follow a rising trend but does so at a mild pace. In the case of subgroup C, the changes appear to be faster: 0.76 for group I, 0.83 for group II, 0.87 for group III, 1.23 for group IV, and 0.98 for group V.

Yet, compared to the pattern of I/S for Japan, the change in I/S for the developing economies is milder. The difference in the classification of transportation and communication should be mentioned, particularly with respect to the magnitude of the difference in the ratio, I/S. However, in terms of its relative change, the effects are perhaps much less for industry than services in the postwar developing economies in general. We cannot deal with this trend in terms of the efficiency and sectoral allocation of investment due to data constraints. Nevertheless, we contend that the degree of the difference between the two sectors is substantial in the competitive situation. If sector S, particularly its traditional segments, continues to provide better opportunities for profit-making, investment for modernization in industry will face a more unfavorable situation than it would otherwise.

Conventionally, the different output growth performances of the two sectors have been discussed from the aspect of demand, especially in terms of differences in income elasticities. With respect to the supply-production side, we adopt the classic assumption of "increasing returns" in industry, but no clear notion has yet been given for services. Viewed from the historical shifts of production structure, we think it realistic to pay more attention to the operation of the competitive mechanism between the two sectors. Japan's case illustrates the importance of this competitive mechanism between the two sectors, which is also implied in the postwar records of developing nations. Our use of subgroupings helps to clarify this. The performance of the ratio, I/S, on average for subgroup D is 0.80 for group II, 0.85 for group III, and 0.84 for group IV of developing economies; this pattern corresponds to the pattern of I/S for subgroup C. No distinct trend of increase is witnessed for this subgroup, and the competitive situation never turned out to be distinctly in favor of industry.

Significance of Relative Output Prices

Indispensable in discussing production structure is the difference of and changes in the relative sectoral output prices. Each developing nation has a domestic price system that is indigenous to its economy. A well-known illustration is the effects of rich or poor endowments of natural resources which reflect upon the prices of agricultural products. By opening the economy to international competition, international prices begin to affect local markets. Yet the indigenous price system continues to play a substantive role. (This process will be discussed in detail in Chapter 5.) The significance of the over-time changes in relative output prices among the major sectors is the focus of this section.

Any discussion or analysis of production structure involves two aspects: income and productivity. The common view is that the sectoral ratio of product per worker in current prices involves both income and productivity aspects, while the ratio in constant prices involves only the productivity aspect. In the columns on the right-hand side of Panel (B) in Table 3.4, changes in the relative output prices are shown. Relative to the price of agricultural products, the price of industrial goods (I/A) show a distinct long-term declining trend throughout the entire period. In contrast, the relative price of services (S/A) changes only slightly with short-term variations. Towards the postwar 1960s, the S/A price index tends to increase, while the I/A price index continues to decrease at an even faster pace. The contrasting movements of the two indexes are even more clearly shown by direct comparison, that is, by looking at I/S. The index I/S declines dramatically especially towards the 1960s. Note, however, that statistical estimates of these prices over such a long-term range are not highly reliable. This is especially so for prices of the "output" of the services sector. Thus, the patterns of output prices mentioned above should be carefully interpreted with reservations for possible biases. Nevertheless, we believe the impressive contrast presented between the two major sectors broadly describes the actual trends.

Such patterns of change in sectoral relative prices of course cannot be representative for developing economies in general. Instead, differences in the relative sectoral prices across nations present a problem (this is discussed further in Chapter 5).

A full explanation of the factors responsible for bringing forth such different performances in output prices is beyond the scope of this work. However, we believe that the essential factor is the differential rates of productivity enhancement among sectors insofar as the long-term trend is concerned. The industrial sector has the possibility of being superior in this regard as compared to other sectors. A faster shift of the supply curve to the right makes possible a relative decrease in output prices so long as the demand situation permits. Much has been talked about a slower rate of productivity increase in agriculture, but changes in productivity for the services sector need to be emphasized more. We would argue that productivity in the services sector is demand-determined. In the case of Japan, the rate of total factor productivity increase in sector S during the prewar period is estimated to be close to zero,[67] while the swing variation of productivity in the services sector is caused by demand. On the other hand, as suggested earlier, shifts in the supply function in agriculture seem to be more noticeable in recent years as compared to the classic notion of "decreasing returns" in the postwar economies.

Japan's records seem to illustrate the case of a rapid shift of the supply function in the industrial sector, combined with the typological feature of poor natural resource endowments for agriculture. As a matter of fact, patterns that are both similar and dissimilar with that of Japan appear to exist among developing nations. This aspect cannot be ignored in discussing a trend of convergence or divergence for all developing economies.

The estimates of sectoral product per worker in constant prices are shown in Panel (B). Although the patterns are rather clear, the following points are noted.

First, the pattern of the overall indicator, N/A, generally mirrors the pattern found in the current price series if the rather high ratio for 1885–1889 is judged to be abnormal. Similarly, what has been called the effects of speedy industrialization are distinctly found in both 1915–1922 and 1927–1933. This is understandable as these periods are characterized as "spurts" in Japan's prewar growth.

Second, with respect to the components, I/A and S/A, the period at which their magnitudes almost equal one another differs between the current-price series and the constant-price series. In the case of current prices, the ratios are almost equal to one another (near 3.0) in 1908–1914. In the case of constant prices, this occurs much later in 1927–1933 (near 3.6). Because of the sharp decline in the relative prices of industrial goods, the effect of changes in the industrial components and hence the increase in I/A is apparent much later. To this extent, the divergence effects are diluted. In the postwar developing economies, a similar type of dilution effect may take place, and, in fact, such effects of dilution as a result of industrialization appears to have actually occurred in a number of developing nations (Chapter 5).

---

67 Kazushi Ohkawa and Nobukiyo Takamatsu, "Shogyo to Sabisu no Seisansei" [Productivity of commerce and services], Chapter 12, in Ohkawa and Minami, eds., *Kindai Nihon no Keizai Hatten*.

Third, a few further remarks about I/A should be made from the aspect of agriculture. In constant prices, the trend of divergence that was recognized earlier in the current-price series becomes much sharper, especially in the postwar years. One of the reasons for this is the elimination of the effects of nonmarket factors which became especially influential in the postwar period, especially the effects of domestic protection policy on agriculture. This basically reflects the differences in the rate of sectoral productivity growth between the two goods-producing sectors. The trend of sectoral divergence has been "diluted" by policy intervention which have kept the relative prices of farm products higher than a market-determined level. Through the entire period under review (some seventy-five years), the index, A/I (the reciprocal of I/A which is shown in the table) decreased from 62.5 in 1885–1889 to 15.0 in 1960–1965. Such a big difference in I/A between the current-price and the constant-price series illustrates well the discrepancy which emanates from the nominal income approach and the productivity-focused approach, although the magnitude of the discrepancy varies depending upon the different situations of the developed and developing nations. Equalization of sectoral income per worker has been an important policy subject, especially from the standpoint of agriculture. Analytically, this implies that the production structure brought forth by competitive power is judged to be undesirable from the social welfare standpoint. Of course, a rise in relative prices enhances output growth and modifies the production structure; thus, the problem is complex. In the case of industry, protected domestic output prices are rather conventional for carrying out import-substitution policies, whether they be primary or secondary, in almost all the development paths of developing economies. This affects the performance of I/A.

Despite differences in the nature of the two major sectors, the production structure and its possible shifts are affected to a considerable extent by nonmarket factors in the latecomer countries. Evaluation of these policies for the individual countries is beyond the scope of this work. Viewed from the long-term perspective of shifts in the production structure, these nonmarket factors operate positively or negatively on the convergence or divergence which is identified in terms of product per worker in current prices. While we have no intention of evaluating the policy option, we would like to emphasize the importance of distinguishing the productivity effects and the price effects brought forth by policy intervention in the operation of the market mechanism.

The Western Experience

Let us now turn to the historical experience of the Western nations. As was presumed earlier, the trend of convergence or divergence cannot be dealt with as a linear trend that links the historical path of developed nations and the postwar performance of developing nations. In clarifying this issue, historical records are discussed first and the recent performance of Western nations are discussed thereafter. Depending heavily upon Kuznets, we select and recalculate the data to obtain the sectoral product per worker ratios in terms of current and constant prices. The limitation of the data does not allow us to discuss the changes over time precisely. Nevertheless, we use these data to (1) focus on the initial magnitudes of the ratios which are comparable to those of Japan and developing nations, and to (2) roughly grasp the long-term changes insofar as the data permit.

From Table 3.5, Panel (A), it is clear that the initial magnitude factor which determines the subsequent trend of convergence. This assertion can be further backed up by comparisons of forerunners with latecomers among the Western nations. For latecomers such as Sweden and Norway, the initial ratio is relatively large (2.1 and 2.7). These can be linked to the ratio of 3.2 for Japan, which is a more lagging latecomer. The initial position of agriculture relative to industry and traditional services was much higher in the Western nations' historical records compared to that of developing economies. This is shown by the ratio of product per worker in agriculture to that of sectors I and S, i.e., I/A and S/A. Even for latecomers such as Sweden and Norway, I/A is 2.0 and 2.9, while S/A is 2.2 and 2.6, respectively. In Table 3.2, Panel (A), we have shown that in 1960 even for developing nations in groups III and IV, I/A is 3.0 and S/A is much greater. No doubt definite differences are manifest there.

The ratio between agriculture and services, S/A, is of particular importance in evaluating the initial conditions as it shows the relationship between the two traditional major sectors at the time of initiation of industrialization. In the 1880s, S/A varies within a narrow range for most of the countries (around 2.0–3.0), except for the atypically low value of the United Kingdom. This suggests that the initial production structure did not vary widely across these nations. The ratio S/A is certainly smaller for the Western nations than the ratio for Japan which is closer to that of developing countries ingroup III from the cross-sectional data. The Western ratios fall

Table 3.5
Sectoral Comparisons of Product Per Worker: The Historical Record of Selected Western Nations [a]

(A) In Current Prices		N/A	I/A	S/A	I/S
United Kingdom	1801/11	0.80	0.54	1.10	0.49
	1907 (1901/11)	1.09	0.87	1.35	0.64
	1907 (1901/11)	1.04	0.91	1.32	0.68
	1963-1967 (1961)	1.05	1.07	1.03	1.04
Netherlands	1913 (1909)	1.73	1.39	2.17	0.64
	1950 (1947)	1.39	1.41	1.38	1.02
	1950 (1947)	1.38	1.51	1.21	1.25
	1960 (1960)	1.28	1.33	1.23	1.08
Denmark	1950-1951 (1950)	1.22	1.38	1.02	1.35
	1963-1967 (1960)	1.62	1.83	1.38	1.33
Norway	1865	2.72	2.92	2.46	1.19
	1950	2.11	2.15	2.04	1.05
	1950	1.90	2.07	1.62	1.28
	1963-1967 (1960)	2.51	2.71	2.22	1.22
Sweden	1861-1870 (1860/70)	2.11	1.97	2.24	0.88
	1963-1967 (1960)	2.08	2.19	1.88	1.16
Italy	1861-1870 (1861/71)	1.34	0.92	2.15	0.43
	1881-1900 (1881/1901)	1.60	1.04	2.83	0.37
	1963-1967 (1964)	1.97	1.95	2.02	0.97
United States	1839	1.93	2.40	1.55	1.55
	1899	2.71	2.96	2.41	1.23
	1929	1.98	2.27	1.71	1.32
	1963-1965	1.52	1.91	1.25	1.52

(B) In Constant Prices		N/A	I/A	S/A	I/S
France	(1954 prices)				
	1896	2.25	2.56	1.81	1.41
	1963 (1960)	2.59	2.77	2.39	1.16
Belgium	(1963 prices)				
	1910	2.95	2.10	4.65	0.45
	1963-1967 (1964)	0.95	0.90	1.01	0.90
Germany	(1913 prices)				
	1850-1859 (1952/55/58)	1.32	0.82	2.53	0.32
	1935-1938 (1938)	1.72	1.91	1.40	1.36

Sweden	(1959 prices)				
	1861-1870 (1860/70)	2.35	1.80	2.89	0.62
	1963-1967 (1960)	2.48	2.59	2.30	1.13
Italy	(1963 prices)				
	1861-1870 (1861/71)	1.74	0.88	3.35	0.26
	1881-1900 (1881/1901)	1.85	1.04	3.64	0.29
	1963-1967 (1964)	1.90	1.90	1.88	1.01
United States	(1929 prices)				
	1899	1.47	1.49	1.46	1.02
	1929	2.04	2.44	1.67	1.46

Note: a. National income excludes income from dwellings except for Belgium, the Netherlands, and Denmark; and finance is excluded for the Netherlands, Denmark, Norway, Sweden, Italy, and the United States. Data for years connected by a slash (/) are for the single years indicated; when connected by a dash (–), the data refer to all years in the interval. Years and periods in parenthesis refer to dates for the labor force data when these differ from those for products.

Source: Kuznets, *Economic Growth of Nations*, Table 45:290–92.

somewhere between group IV and group V.[68]

How can we interpret the data presented in the table in terms of long-term performance? Finding an answer to this question is not easy because the data are not uniform in terms of dates, length of coverage, etc. And yet, can we confirm a trend of divergence in contrast to the trend of convergence found from the postwar cross-sectional data? The long-term performance of N/A does not appear to be similar but is rather mixed across the countries under review.

68 For the sake of simplicity, the original data for θ_A, θ_I, and θ_S (the aggregate ratios) are not shown in the table. In order to demonstrate the significance of the difference in the initial structure of production, the higher initial levels of θ_A for the Western countries are shown, including the constant-price series in Panel (B) (these are marked by *). They are 0.59* for France, 0.87 for Germany, 0.80 and 0.73* for Denmark, 0.61 for Norway, 0.75 and 0.71* for Sweden, 0.88 and 0.77* for Italy, 0.75 and 0.78* for the United States. Comparable data for currently developing nations, which are given in Table 3.2, Panel (A), are much smaller. Even the largest ratio for group I in 1960 is 0.63 (For Japan, see Appendix I).

In the current-price series, convergence is seen for the Netherlands, Sweden, the United Kingdom, and Norway (although there is divergence in the 1960s); divergence is exhibited for Italy; and first divergence and then convergence for the United States. In the constant-price series, divergence is evident for France, Germany, Italy, Sweden, and United States; convergence is only seen for Belgium.

A speculative interpretation of such a mixed performance may be as follows. Given the broad trend of convergence identified from the cross-sectional analysis in the preceding section, it is natural to not expect an analogous historical performance of the Western economies, primarily because of the large differences in the initial production structures. Suppose that a distinct trend of convergence had long been sustained in these nations; if so, then these nations would have been in a situation similar to that of group V much earlier in the prewar period. In this hypothetical example, it may not be a surprise to witness "no convergence" insofar as a general broad trend is concerned. On the other hand, there is no evidence that a trend of divergence occurs in these countries, although it is interesting to find such a pattern for not a small number of cases which Kuznets did not ignore. Thus, with regard to the ratio N/A in the current-price series, we cannot definitely assert either convergence nor divergence. However, in terms of constant prices, the ratio shows a strong trend of divergence. If Belgium is assumed to be a questionable case, then the ratio for all remaining five countries exhibit a divergent trend.

The different performances between the current- and constant-price series were also found for Japan. In Japan's case, the trend of the ratio in the current-price series is somewhat blurred, while in the constant-price series, the trend is quite distinct. Thus, it is our view that in comparing the historical records of currently developed nations and developing nations, this distinction in terms of current- vs. constant-price series is indispensable in order to avoid possible confusion. Insofar as the comparison of the ratios in the current-price series is concerned, we would conclude that (1) a tendency of convergence is more distinct for developing economies in the cross-sectional data, while it is blurred in the historical path of developed nations, and (2) this difference is essentially caused by differences in the initial production structure between the two.[69]

---

69 To avoid too much statistics, the trend in I/S is not discussed in the main text despite its importance in light of preceding discussions. As is clear from the table, the performance of I/S shows an increasing trend, though with some exceptions. This trend is particularly distinct in the

A Recent Turn: Towards Convergence?

We want to follow upon Kuznets' idea of the analytical significance for further scrutiny of recent decades. For Kuznets, the problem at issue was clear: when did the turn from divergence to convergence take place for developed countries? In our view, the question is now modified to: when did the turn from a blurred to a distinct pattern of convergence occur? Although the second question is more difficult, let us nevertheless try to obtain an answer. The scrutiny will be carried out chiefly in terms of the current-price series, though a brief suggestion will be made for the constant-price series. Our concern with this topic is based on the belief that a better understanding of it is useful for our understanding of the characteristics of developing economies.

The data in Table 3.5 lack information on the possible changes that took place during the 1950s. Thus, we must make speculative judgments in order to link the data for the earlier postwar years with those of the 1960s. First, in looking at the changes in the 1950s and the 1960s, the data on N/A in current prices are only given for three countries in Table 3.5: the Netherlands (convergence, 1.38 to 1.28), Denmark (divergence, 1.22 to 1.60), and Norway (divergence, 1.82 to 2.65). That is, the performance seems mixed. Second, for the seven countries listed in Panel (A), the ratio of relative product per worker in the nonagricultural sector to the agricultural sector, N/A, is 1.74 for the 1960s. This is fairly close to the average value of N/A for group V in 1960 in the cross-sectional data of Table 3.1 (1.85), and the two can be roughly linked. Third, from the cross-sectional data, the corresponding ratio is 2.21 in 1970 and 1.79 in 1980, and when compared with group IV (the ratio for group IV is 3.14 in 1960, 3.80 in 1970, and 2.98 in 1980), a distinct convergence to group V is observed.

Taken together, is it possible to interpret these trends in a fairly acceptable way? We believe that a turn from a mixed to a less-mixed convergence trend took place in the postwar decades. It goes without saying that a distinct convergence from group IV to group V in the cross-sectional data cannot represent the historical path of the nations of group V. No conformity can be asserted in this case between the cross-sectional and time-series data. However, the trends for Japan, which were discussed earlier for the period to the 1960s, seem to support the view of a recent turn to convergence.

---

constant-price series and together with a recent decline in I/S, the trend is analogous to that of Japan. Changes in this ratio over time of all groups in the cross-sectional data perform in a similar fashion.

Our hypothesis can be investigated further by looking back at the data given in Tables 3.1 and 3.2 in more detail and examine the over-time changes of the relevant terms for group V in the two decades. The movement of N/A is summarized for group V below.

	1960	1970	1980	Average
Group V	1.9	2.2	1.8	2.0
Subgroup C	2.1	2.6	2.1	2.4
Subgroup D	1.7	2.0	1.5	1.7

As seen in the figures, N/A increases slightly in the 1960s and decreases slightly in the 1970s in all three cases. However, a direct comparison between the years 1960 and 1980 shows almost no change. A breakdown of the ratio by sectors is summarized below.

	I/A			S/A		
	1960	1970	1980	1960	1970	1980
Group V	1.7	1.8	1.6	1.9	2.0	1.6
Subgroup C	1.7	1.7	1.7	1.7	1.7	1.7
Subgroup D	1.7	1.8	1.6	2.1	2.2	1.5

In subgroup C, both I/A and S/A are perfectly stable during the entire period, whereas in subgroup D, I/A is relatively stable in the 1960s rising from 1.7 in 1960 to 1.8 in 1970; but a decrease is witnessed in the 1970s. Broadly speaking, however, despite some difference in the degree of industrialization, no sizeable difference is seen between the two subgroups for the two decades.

Our interpretation of the factors responsible for bringing forth the trends of convergence for groups III and IV, which are the relative increase in product per worker in agriculture and the relative decrease in product per worker in services, can also be applied to the over-time changes of the ratios for group V. The situation of the countries in group V in 1960, i.e., the very low level of both I/A and S/A, is the outcome of the operation of these factors in the pre-1960 period. In the post-1960 interval, that outcome had generally been sustained. If this view is accepted, then why do we refer to the convergence as being only "less-mixed"? The reason for using this term is suggested by the increase in these ratios during the 1960s in contrast to their decrease during the 1970s. This pattern is chiefly brought forth by the different pace of industrialization. When the pace of industrialization is faster, there is a tendency toward divergence; when the pace is slower, the tendency is toward convergence. These "mixed" patterns are still observed for the post-1960 decades, though in a much narrower range as compared to the developing economies.

In the case of constant prices, we proposed that in the historical records of Western nations a trend of divergence is more clearly observed than in the case in current prices. Examination of a "recent turn" in the trend encounters greater difficulty due to the data limitation. Though examination is postponed to the growth analysis in Chapter 4, we contend that a turn to a trend of convergence may be a broad possibility for group V nations. However, we call this a "mixed" pattern because the different degrees of industrialization may also have substantive implications.

A brief description of Japan's case may be suggestive in illustrating this. Japan is the most recent latecomer in the nations of group V, as its shift to the developed-economy phase took place toward the end of the 1960s. The magnitude of the ratios pertaining to the production structure is extremely high among this group and are rather similar to those of subgroup C of group IV in 1960. However, its change towards 1980 differs a great deal because of the faster rate of industrialization during that period. And yet, even in 1980, the magnitude of the ratios are sustained at the highest level in group V. Let us draw our attention to the ratio I/S. It is especially informative to look at the effects of industrialization on this ratio since the services sector occupies a dominant share of the nonindustrial sector. The ratio I/S decreases from 1960 to 1980 in Japan, whereas it increases in subgroup C of group IV:

	I/A		S/A		I/S	
	1960	1980	1960	1980	1960	1980
Japan	3.5	2.8	2.5	2.5	1.4	1.2
Subgroup C, Group IV	3.7	5.2	3.7	3.5	1.0	1.5

Except for a few countries such as Korea and Brazil, most of the nations in groups III and IV show a similar pattern of an increase in I/S. It is interesting to find that this pattern of an increasing I/S is also witnessed for the majority of group V nations (for three nations—the United Kingdom, France, and Finland—it remains unchanged). Thus, one may argue that Japan's performance is atypical. However, Japan cannot be treated merely as an exception.

Instead, in our view, the Japanese case represents a noticeable phenomenon—a distinct tendency of decrease in the relative prices of industrial products.

	N/A		M/A[a]	
	In Current Prices	In 1980 Prices	In Current Prices	In 1980 Prices
1970–1975	1.35	1.21	3.71	3.47
1976–1980	1.26	1.26	3.67	3.87
1981–1985	1.25	1.34	3.53	4.33

Note: a. M, manufacturing, is used in place of industry because comparable data are not readily available.

Sources: Official statistics of Japanese national accounts and labor force surveys.

For the most recent period, 1981–1985, a trend of convergence is seen both in N/A and M/A in current prices, but a trend of divergence is witnessed in constant prices for both terms. From the sources of Tables 3.1 and 3.2, N/A is 1.24 in 1960, 1.17 in 1970, and 1.07 in 1980. This tendency of convergence certainly turns to a tendency of divergence in the constant-price series. In the earlier discussion, a trend of divergence was witnessed in the historical records of Japan since around 1920, and this trend has been sustained throughout recent years. The major factor responsible for this continued trend of divergence is the sustained decreasing trend in the relative price of manufactures, which was essentially brought forth by a faster shift in the supply curve which resulted from the faster rate of productivity increase in this sector.

This factor is supposed to work in the developing nations in groups III and IV, in particular in subgroup C nations including the Asian NIEs, and the trend would be sustained in the future growth path towards the phase of a developed economy. We believe this observation may be valid insofar as these nations succeed in realizing vigorous industrialization. This is not a forecast but a logical outcome induced from our factual analysis. Thus, we are keenly concerned with distinguishing two kinds of trends—the trend in current prices and the trend in constant prices—in identifying convergence or divergence in the recent changes in trend directions.

IV. Discrepancy Between Product and Labor Shares

Production structure can be analyzed by looking at the discrepancies between the shares of product and the shares of labor by sector. This share approach has a long history, and a most comprehensive treatment is the one presented by Kuznets both for cross-sectional and long-term trends.[70] The product per worker approach in the preceding section does not ignore the importance and usefulness of this share approach. In responding to the factual challenge of the complex process of shifts in the production structure, the two approaches have substantively the same analytical objective: that is, clarification of convergence or divergence. Sectoral product per worker was first defined relative to Y/L or the aggregate product per worker, that is, $\theta_A = Y_A/L_A$ / Y/L, etc. In terms of product (Y_A/Y, etc.) vs. in terms of labor (L_A/L, etc.), it will be Y_A/Y / L_A/L, etc. The methodological difference between the two approaches is that the share approach is a decomposition into output (product) and input (labor). In this sense, the approach is informative in providing additional knowledge on the product and labor performances, respectively.

If one prefers to investigate the hypothesis that product and labor sectoral shares are determined by independent factors, then the share approach should be the major approach used. Such a hypothesis has never been proposed explicitly in its full context. However, partially and/or implicitly, such a notion often appears to be involved in the share approach. For example, among the factors responsible for determining product shares, the level of income per capita is thought to be most influential. In this kind of demand approach, the notion of "derived demand" is applied to labor. As discussed earlier in Chapter 1, we do not follow this approach but choose to emphasize the importance of nonassociated factors. On the other hand, as for the labor shares, an example of an increasing trend in the labor share in the services sector is found a lá Colin Clark.

Sectoral discrepancy between the product and labor shares has hitherto been recognized as the basic phenomenon. However, no a priori hypothesis that specifies the determining factors has been proposed. Additional discussion in this section will accordingly be focused on an analysis of the difference (cross-section) and changes (over-time) in the relation between the two shares. In the section

---

70 Kuznets, *Economic Growth of Nations*.

Table 3.6
Product and Labor Shares in the Agricultural vs. the Nonagricultural Sectors by Country Group, 1960–1980 (percentage)

Group	Y_A/Y		$L_A/L(2)$		(Y_A/Y)-(L_A/L)	
	1960	1980	1960	1980	1960	1980
(A) Level						
I	55.2	46.1	86.1	77.0	-30.9	-30.9
II	35.3	27.8	74.8	66.0	-39.5	-38.2
III	32.4	23.8	63.6	43.8	-29.9	-20.0
IV	18.5	12.2	41.5	29.3	-20.0	-17.1
V	9.4	3.8	16.2	6.6	-6.8	-2.8
(B) Changes						
I - II	-19.9	-18.3	-11.3	-11.0	8.8	7.3
II - III	-2.9	-4.0	-12.5	-22.2	-9.6	-8.2
III - IV	-13.9	-11.6	-20.8	-14.5	-7.9	-2.9
IV - V	-9.1	-8.4	-25.8	-22.7	-15.2	-19.9

Sources: See sources of Table 3.1.

that follows, a detailed discussion on the postwar cross-sectional patterns is given. A comparative observation of the historical records of currently developed nations will be tried only briefly due to data limitations.

Agriculture vs. Nonagriculture

Let us begin the discussion with a comparison of Y_A/Y and L_A/L. The ratios are shown in Table 3.6 for 1960 and 1980, the beginning and the end years of our analysis. In the cross-sectional data, both the product and labor shares tend to decrease in a straight line, with the labor shares always being greater than the product shares (Panel A). This is a well-known phenomenon. What we are concerned with is the changing relative magnitudes of these discrepancies, that is $(Y_A/Y)–(L_A/L)$. The magnitude of the difference between the two ratios does not follow a smooth line of decline. In fact, $(Y_A/Y)–(L_A/L)$ goes up from group I to group II, and drastically declines from group IV to group V. Panel (B) also presents the changes in terms of product and labor, respectively. In the case of labor, the tempo of decline generally appears to be linearly accelerated. In the case of product, the pattern is different and follows a zigzag shape. Two points draw our attention. First, in the initial interval the discrepancy between the two ratios tends to increase and then turns to decrease, forming a normal path. Second, developing economies keep to the normal path, arriving at another point of turn from group IV to group V, resulting in a considerably large magnitude for the discrepancy. These findings correspond perfectly with what was found earlier in the zigzag path of the product per worker approach. It is now clear that the pattern of shifting production structure is chiefly caused by changes in the product share rather than in the labor share.

What about changes over time? Even a short interval of two decades can tell us something useful in terms of checking conformity with the cross-sectional data. To this end, a comparison of changes in the product and labor shares over time between 1960 and 1980 are shown below (the values are negative for the agricultural and positive for the nonagricultural sector).

	I	II	III	IV	V
Output	9.1	7.5	8.6	6.3	5.6
Labor	9.1	8.8	18.5	12.2	9.6

No uniform pattern is seen for the changes in the discrepancy. There seem to be no changes for group I, moderate changes for groups II and V, and remarkable changes for groups III and IV. These appear to broadly conform with the earlier suggestion that shifts in the production structure are conceivable between groups II and III, and groups IV and V.

Table 3.7 is arranged in terms of the nonagricultural sector by subgroups. The performance of the discrepancy, i.e., $(Y_N/Y)-(L_N/L)$, for subgroups C and D at first glance appears mixed (see Panel A). However, on average, the discrepancies are distinctly larger for subgroup C than for subgroup D insofar as groups III and IV are concerned. This indicates that more industrialization tends to be associated with a greater discrepancy between product and labor, which thus contributes to the divergent trend. Since this is found for the majority of developing economies, it is of substantive importance. As for the mixed pattern in the initial interval (i.e., groups I and II), the reason is not necessarily clear. The influence of the different initial conditions may be relevant to the abnormal pattern in the earlier phases of development. As for group V, the difference becomes minor, as can be expected. A closer look at the performances in 1960 and 1980 reveals some distortions from the pattern pointed out above. Some reservation is certainly needed as

Table 3.7
Product and Labor Shares of Labor and Output by Subgroup and by Sector (Nonagriculture vs. Agriculture), 1960 and 1980 (percentage)

	Subgroup C						Subgroup D					
	Y_N/Y		L_N/L		$(Y_N/Y)-(L_N/L)$		Y_N/Y		L_N/L		$(Y_N/Y)-(L_N/L)$	
	1960	1980	1960	1980	1960	1980	1960	1980	1960	1980	1960	1980
(A) Level												
I	45.2	54.0	14.1	19.5	31.1	34.5	44.2	53.9	13.9	22.6	30.3	31.3
I	67.6	74.8	28.8	37.7	34.6	34.5	61.8	69.6	21.7	30.3	40.1	39.3
III	68.7	80.0	33.1	56.2	35.6	37.1	66.7	73.8	40.4	56.1	26.3	17.7
IV	81.7	89.8	58.8	67.6	22.9	23.8	81.4	86.2	58.2	73.1	23.2	13.1
V	92.0	97.0	84.7	93.9	7.3	3.1	91.6	96.2	83.8	92.7	7.8	3.5
(B) Changes												
I–II	22.4	20.8	14.7	18.2	3.5	-5.0	17.6	5.7	7.8	7.7	9.8	8.0
II–III	1.1	5.2	4.3	18.5	1.0	2.6	4.9	4.2	18.7	25.8	-13.8	-21.6
III–IV	13.0	9.8	25.7	11.7	-12.7	-13.3	14.7	12.4	17.8	17.0	-3.1	-4.6
IV–V	10.3	7.2	25.9	26.3	15.6	-20.7	7.2	10.8	25.6	19.6	-15.4	-9.6

Sources: See sources of Table 3.1.

the interval is not long enough. Yet, at least evidence for nonconformity is not given. We are convinced that the suggestion made earlier on the positive association between industrialization and the discrepancy between the shares can be generally confirmed.

In Panel (B) in the table, changes in the shares are shown for each successive group in the cross-sectional data. An additional point found is that the magnitude of the discrepancy is distinctly different between groups II–III and III–IV. In groups II–III, the discrepancy increases for subgroup C, while for subgroup D, a remarkable decline is witnessed. In groups III–IV exactly the reverse takes place: a much greater decrease is seen for subgroup C, while for subgroup D, the decline is moderate. These patterns are similar for 1960 and 1980 and may be of a sustained nature. As stated earlier, in our aggregate observation the performance of the discrepancy is found to be relatively stable for these intervals. Now it becomes clear that the stable shape is the result of a cancelling out which has occurred by the diverse behavior of the two subgroups. The reason for this phenomenon is unknown and requires further scrutiny.

On the one hand, the discussion using the N–A two-sector share approach can present further factual confirmation of the findings of the ratio approach including the zigzag pattern and the two-step shifts in the production structure. On the other hand, it suggests that such problems as the performance discrepancies in the demarcated intervals as well as the relationship between the pace of industrialization and the magnitude of these discrepancies requires further scrutiny. Yet, at this stage the evidence suggests that the supply-production aspect is important in view of the conventional approach, which merely emphasizes the demand side in terms of the magnitude and changes in income elasticities.

Industry vs. Services

The nonagricultural sector is decomposed into industry and services. In the earlier discussion, we have often pointed out noticeable differences in the behavior of industry and services. This aspect is further clarified by the share approach: in particular, the converging function of services and the diverging function of industry is supported by the findings of this approach.

First, changes in the discrepancies, (Y_I/Y)–(L_I/L) and (Y_S/Y)–(L_S/L) over time by group show notable features (Table 3.8, Panel A). The magnitude of the discrepancies tends to decrease over time in services for all groups of developing nations, except group I. However, an increasing tendency is witnessed for all groups of developing nations, including group I, in industry. Thus, the services sector contributes to a trend of convergence, whereas industry contributes to divergence. The aggregate result is a tendency of convergence, because the magnitude of the discrepancy decline is larger and its trend is stronger in services as compared to the increase of the discrepancy in industry. A closer look, however, would find differences in the degree of change in the discrepancy between groups. With respect to industry, changes in the discrepancy are greater for groups I and II, and smaller for groups III and IV. In contrast to the behavior of developing nations, group V shows a decrease which is a movement in the reverse direction. Although two decades is not a sufficiently long enough period from which to make hard conclusions, these patterns are relevant to the discussion of the two-step shift in the production structure.

A summary presentation of the average discrepancies, including that of 1970, is shown in Table 3.8, Panel (B). For services, a tendency of convergence is clearly seen except for groups I–II. Regarding industry, however, no distinct tendency can be seen. Rather, the discrepancy for industry follows a zigzag pattern: first the discrepancy increases from group III to group IV and then sharply decreases from group IV to group V. This is similar to the zigzag pattern suggested earlier in the N–A two-sector approach.

Table 3.8
Product and Labor Shares of the Industry and Services Sector by Country Group, 1960 and 1980 (percentage)

(A):

Group		1960			1980		
		Product	Labor	Discrepancy	Product	Labor	Discrepancy
I	Industry	11.1	4.5	6.6	17.7	7.1	10.6
	Services	32.7	9.5	21.7	36.4	9.5	26.9
II	Industry	17.0	9.4	7.6	24.6	10.4	14.2
	Services	45.7	15.8	29.9	46.0	23.7	22.3
III	Industry	21.8	14.5	6.3	30.8	23.7	7.1
	Services	46.6	25.8	21.8	46.2	33.4	12.8
IV	Industry	28.8	22.6	6.2	37.3	27.4	9.9
	Services	48.8	29.6	14.9	49.2	44.7	4.5
V	Industry	40.9	39.0	1.9	37.1	36.5	0.6
	Services	51.4	44.2	7.2	60.1	58.9	1.2

(B): Average Shares, including data on 1970

	I	II	III	IV	V
Discrepancy					
Industry	5.3	14.8	6.8	8.9	-1.7
Services	26.1	26.2	17.2	11.6	3.6
Share of Product					
Industry	11.2	22.2	25.8	33.2	38.5
Services	35.4	45.6	46.8	49.5	56.4
Share of Labor					
Industry	5.9	12.6	19.0	24.3	40.2
Services	9.3	19.4	29.6	37.9	52.8

Sources: See sources of Table 3.1.

More detailed observation of the changes in the product and labor shares will tell us more about the features of the two shifts in the production structure, i.e., shifts between groups II and III and shifts between groups IV and V. In the case of industry, in both intervals the share of labor increases noticeably, while the share of product changes only moderately. These features are not seen in the other intervals where both the product and labor shares changed parallel with one another. One may be tempted to suppose that these patterns are the result of the type of industrialization that occurred. In our view, however, this is only half the story, because the shift in the production structure in the interval from group II to group III seems to emanate from the performance of agriculture. The reason for such an interpretation is explained as follows by considering the performance of services together with industry.

The sum of the product shares of industry and services (I+S) increases from group II to group III much less as compared to the noticeable increase in the same share from group I to group II (67.8–72.6 vs. 45.5–67.8). In other words, the product share of agriculture decreases very moderately in the interval from group II to group III as compared to a drastic decrease from group I to group II. This implies that the pace of product growth in agriculture is remarkably high in a relative sense. On the other hand, similar to industry, the labor share of services increases distinctly in this interval. This implies a faster decrease in the share of labor in agriculture. This was known in the N-A two-sector approach, but here we are concerned with the relative performance of industry and services. Thus, the important finding is that the interval from group II to group III, which represents the first shift in the production structure, is basically composed of a remarkable increase in productivity growth in agriculture.

The story differs in the case of the interval from group IV to group V. A much larger increase in the labor share vis-à-vis the product share is also recognized for services in this interval. The two sectors, i.e., industry and services, taken together show a drastic increase in the labor share, although the growth in the labor share differs between the two. Although new activities arise in services, industry is close to the end of the time where it acts as the driving force. However, our attention is focused on the noticeable increase in the share of labor in this interval which is much more impressive when compared to the case of the first interval from group II to group III (note the magnitude of the discrepancy of industry in group V where the value is negative). One may suppose that because of the increased demand for labor from new services activities, the share of labor in agriculture must show a trend of decline. This phenomenon has, in fact, been recognized recently in the nations of group IV. However, we are concerned with the possible shift of production structure from group IV toward group V. In this context, we would argue that the effect of the acceleration of industrialization, in terms of an increase in labor employment, is an indispensable factor in realizing the shift toward the developed phase.

As was mentioned earlier, changes in the production structure in terms of employment have often been discussed in terms of the type of industry and/or technology (particularly, in manufacturing). Our discussion suggests that an overall observation is more desirable

Table 3.9
Shares of Product and Labor and Discrepancies by Subgroup, Averages of 1960, 1970, and 1980 (percentage)

		I	II	III	IV	V
Product Share						
Industry	C	14.5	24.1	27.8	38.2	41.4
	D	12.2	20.2	23.8	28.3	35.9
Services	C	34.5	45.3	47.2	43.2	51.9
	D	35.9	45.8	46.4	55.4	59.0
Labor Share						
Industry	C	6.0	12.5	19.2	26.0	40.2
	D	4.3	8.2	18.7	24.0	38.6
Services	C	9.5	20.3	29.8	39.5	49.6
	D	9.3	18.8	29.1	40.4	55.8
Discrepancy						
Industry	C	8.5	11.6	8.6	12.2	1.2
	D	7.9	12.0	5.1	4.3	-2.7
Services	C	25.0	25.0	17.4	3.7	2.3
	D	26.6	27.0	17.3	15.0	3.2

Sources: See sources of Table 3.1.

than a narrower assessment based on industrial and/or manufacturing performances to obtain a more accurate understanding of reality.

Using observations by subgroup, we can examine in greater detail what was found earlier from the cross-sectional data. The changes by subgroup over time tell us virtually nothing more than what was found earlier in the discussion by group. However, we now know the different functions of the two sectors: industry with a divergence vs. services with a convergence. How do these different activities operate in subgroups C and D? The average figures of the relevant data are summarized in Table 3.9 for three points in time—1960, 1970, and 1980. Let us first look at the discrepancies. The pattern of industry's discrepancy appears to be rather mixed. Some regularity in the changes is suggested. In the primary interval (i.e., the interval between groups I and II), the discrepancy increases, but from group II to group III the discrepancy declines sharply. However, a distinct difference emerges between the subgroups in the interval from III to IV: the discrepancy of industry increases for subgroup C but decreases for subgroup D. In the case of services, a declining tendency appears for both subgroups, with a minor exception in the primary interval (I–II), where it remained relatively unchanged for both subgroups. However, a specific case is noted for group IV, in which the discrepancy in subgroup C turns out to be much smaller than for subgroup D. Thus, the issue is how to interpret this specific distortion.

The discrepancies in industry are greater for subgroup C than for subgroup D with the minor exception of group II. As shown in the table, this is due to the higher share of product in subgroup C and to the smaller differences in the shares of labor between the two subgroups. We have suggested that the more industrialized an economy is, the larger the discrepancy tends to be; this is confirmed by the discrepancies of the industrial sector. The specific case of group IV requires special attention because of the wide difference between the discrepancy for subgroups C and D. Our interpretation of this difference is that the industrialization drive is particularly stronger in subgroup C, and is accompanied by a sharp increase in the discrepancy.

A corresponding phenomenon is seen in the case of services. Namely, the much smaller discrepancies in subgroup C than in subgroup D can be explained by the smaller share of services in product of subgroup C as compared to subgroup D. If this view is accepted, we speculate that the sharp difference between subgroups C and D for subgroup IV emerges because of the competitive activities of industry and services.

Measures of Sectoral Inequality of Product Per Worker

Now, let us make some general remarks on the problem of sectoral inequality. The so-called trade-off relationship—that is, more (less) industrialization is associated with larger (smaller) discrepancies—appears to be valid for countries at more or less the same Y/L level. However, this relationship requires qualification. The proposition is deduced using the share approach, so that it differs from the macro-standpoint of measuring sectoral inequality of per worker product. The greater discrepancy that is brought forth by faster industrialization actually emerges in relation to the varied performance of other sectors, agriculture and services. In this regard, we need some device which integrates the measurements. To this end, a simple formula is adopted where ε stands for the coefficient of sectoral inequality of per worker product and

$$(4) \quad \varepsilon = \{(Y_A/L_A - Y/L) + (Y_I/L_I - Y/L) + (Y_S/L_S - Y/L)\}Y/L \text{ or}$$

$$\varepsilon = ((Y_A/L_A \;/\; Y/L)-1)+((Y_I/L_I \;/\; Y/L)-1)+((Y_S/L_S \;/\; Y/L)-1)$$

For each sector these terms are nothing more than θ_A-1, θ_I-1, θ_S-1, respectively, the numerical values of which were given earlier in Table 3.2. We know that the value of θ_A-1 is negative through all groups and that the coefficient is defined as the sum of absolute values, following the conventional way of measuring the coefficient of variation. Taking the average of 1960 and 1980, the magnitude of ε with its sectoral components is shown for subgroups C and D in Table 3.10, Panel (A).

For both subgroups C and D, the coefficient ε declines as the Y/L level increases. This is to be expected given our preceding findings. In this section, however, we are concerned with possible differences between subgroups. In general, the value of ε is smaller for subgroup C than for subgroup D suggesting that faster industrialization is accompanied by smaller sectoral inequality. An exception is witnessed for group IV, however, and this deserves further attention.

The share of output not only acts as a yardstick for our observations but is the basis of Panel (B) of the table which presents the results of another subgrouping E vs. F. The yardstick is the rate of macro-productivity growth, G_y. The group pattern by Y/L is the same as for the case of subgroups C and D. Through the major portion of developing economies, the magnitude of ε is greater for subgroup F than for subgroup E: this is basically the same pattern that was observed before, including group IV. However, a reverse relationship is witnessed in this case for group V.

The notion of a trade-off between growth and equity has often been discussed. Our analysis aims to clarify the problem by analyzing sectoral performances rather than by assessing income distribution. Measurement of ε is its concomitant, no information being included as for intrainequality. As far as the pace of industrialization or macro-productivity growth is concerned, the notion of a trade-off between growth and equity for countries in the initial range of the inverse U-shape development path is not validated. The possibility of a trade-off is suggested, however, for countries in the latter range, as represented by group IV in the case of industrialization pace and by group V in the case of macro-productivity growth.

Our conclusion is that for the developing economies, the more vigorous path of growth and/or industrialization can accompany less inequality of sectoral product per worker, but we have to

Table 3.10
Coefficients of Sectoral Inequality of Per Worker Product by Subgroup, 1960–1980 Average (percentage)

		$\theta_A - 1$	$\theta_I - 1$	$\theta_S - 1$	ε
(A) C vs. D[a]					
C:	I	-0.40	1.67	2.76	4.83
	II	-0.57	1.07	1.38	3.02
	III	-0.54	0.51	0.67	1.72
	IV	-0.61	0.72	0.38	1.71
	V	-0.31	0.05	0.04	0.40
D:	I	-0.38	5.95	3.30	9.63
	II	-0.54	1.62	2.19	4.35
	III	-0.42	0.73	0.71	1.86
	IV	-0.53	0.17	0.33	1.03
	V	-0.38	0.06	0.13	0.57
(B) E vs. F[b]					
E:	II	-0.57	-0.98	1.32	2.87
	III	-0.47	0.41	0.63	1.51
	IV	-0.48	0.33	0.22	1.03
	V	-0.34	-0.18	0.26	0.78
F:	II	-0.57	1.45	1.69	3.71
	III	-0.49	1.18	0.65	2.32
	IV	-0.54	0.39	0.19	1.12
	V	-0.12	-0.38	0.04	0.54

Notes: a. θ_A was obtained using the N–A, two-sector approach. θ_I and θ_S were then calculated using I/A and S/A. Due to the crude indirect approach, some discrepancies remain.

b. Subgroup by the magnitude of G_y is adopted for E (higher) and F (lower) nations. For details, see Chapter 4, Section I, by subgrouping.

Sources: See sources of Tables 3.1, 4.4, and 4.10, assuming $w_A/v_A = \theta_A$, $w_I/v_I = \theta_I$, and $w_S/v_S = \theta_S$. No reliable data are available for I.

recognize a certain limitation in pursuing this favorable path. If the notion of a trade-off is deduced from the knowledge about presently developed nations, it may not be applicable to the developing economies.

Historical Patterns of Presently Developed Nations

As was pointed out earlier, the levels of Y/L in the initial periods of the Western nations' development were much higher than those of currently developing nations. This is an important factor in shaping the subsequent growth performance. In the share approach, Y_A/Y and L_A/L and the discrepancy between the two shares in the two-sector (A vs. N) approach should be examined in this regard. However, the available data are not rich enough and is no more informative than the ratio approach.[71] Therefore, the discussion will focus on the decomposition of sector N into industry (I) and services (S), and will try to clarify further the combined effects of the two forces: that is, the divergence in industry vs. the convergence in services.

The case of Japan is first taken up because long-term, continuous data are available (Table 3.11). For convenience of direct comparison, the preceding cross-sectional data by group average are rearranged and shown in Panel (B).

First, the product share of services $(100-(Y_I/Y_N))$ is much greater than that of industry in the initial phases of development; this is similar to the results of the cross-sectional data which showed a higher product share of services in groups at lower Y/L levels. However, during the second phase from 1916–1936, this trend reverses, $Y_I/Y_N > Y_S/Y_N$ though there are some swing variations. During the postwar years, this trend continued until the recent re-reversal. The cross-sectional data did not reveal such a pattern. Is Japan's performance atypical? Or do the product shares of industry for groups III and IV lag?

Second, the magnitude of the difference between the labor share of the two sectors in Japan is much more moderate in the initial phase and becomes even smaller in subsequent years.[72] Except for the initial phase, we believe that the labor allocation between the I and S sectors changed little over the long-term trend

71 In the initial phase, L_A/L is 51.4 percent on average for the seven countries, France, Belgium, Denmark, Norway, Sweden, Finland, and Italy. Y_A/Y (in current prices) is 41.1 percent on average for the four countries, Denmark, Norway, Sweden, and Italy (Kuznets, *Economic Growth of Nations,* Table 38:250–53).

72 For Japan, the trend is somewhat mixed, perhaps being affected by swing variations. Average values are 8.6 for 1892–1914, -1.5 for 1916–1936, and 0.9 for 1952–1969.

Table 3.11
Structural Changes in the Nonagricultural Sector of Historical Japan: Industry vs. Services[a] (percentage)

			Share Discrepancy		
Selected years	Y_I/Y_N	L_I/L_N	$(Y_I/Y_N)-(L_I/L_N)$	$(Y_I/Y_N)-(Y_S/Y_N)$	$(L_I/L_N)-(L_S/L_N)$
(A) Japan:					
1892	33.7	50.7	-17.0	-32.6	1.4
1904	36.8	57.9	-21.1	-26.4	15.8
1916	55.3	45.2	10.1	10.6	-9.6
1925	51.9	52.8	-0.9	3.8	5.6
1936	59.0	49.7	9.3	18.0	-0.6
1952	51.4	50.9	0.5	2.8	1.8
1961	53.4	50.8	2.6	6.8	1.6
1969	50.0	50.4	-0.4	0.0	0.8
(B) International Cross-section by Group:					
I	30.8	40.0	-9.2	-38.4	-20.0
II	33.3	36.8	-3.5	-33.4	-26.4
III	34.8	38.4	-3.6	-30.4	-23.2
IV	40.4	39.6	0.8	-19.2	-20.8
V	40.3	42.3	-2.0	-19.4	-15.4

Note: a. The product and labor shares are both calculated from five-year averages centered on the year shown in the table. The shares are presented as a percentage of the nonagricultural total, so that $Y_S/Y_N = 1-Y_I/Y_N$ and $L_S/L_N = 1-L_I/L_N$. In comparing (A) and (B), some adjustment should be made to take account of differences in coverage.

Sources: Ohkawa and Shinohara, eds., *Patterns of Japanese Economic Development,* Tables A10 and A11:273–77; EPA, national accounts statistics; and *LTES, Labor Force*, Vol. 2.

until the recent turn toward an increase in the labor share of services. Japan's performance can be interpreted as being analogous to the pattern identified earlier for groups of different Y/L levels in the cross-sectional data. However, an important question arises: why is the magnitude of the difference between the two sectors so small in Japan's case? Or alternatively, why is the magnitude of the difference so large for the contemporary developing economies?

We will try to answer these questions in the following paragraphs. The basic reason for the different performances of industry and services that was observed for both Japan and the developing nations is that the product share in industry tends to increase while in services it tends to decrease. At the same time, the labor share tends to be much more stable in both industry and services. The term Y_I/Y_N in Panel (A) shows a long-term rising trend of increase from the earliest years to the 1960s despite the effect of long-swings, and then declines. In the cross-sectional data, a similar trend is seen throughout the four groups of developing economies with a decrease in group V. In both cases, Y_S/Y_N performs in a reversed manner. Thus, conformity is witnessed between the time-series and cross-sectional data. One may be tempted to point out the differences in the magnitude for industry where the discrepancies are greater in Japan than in the developing economies. However, because of the inconsistency between the two data sets, we choose to be conservative.[73] A near-constancy of the labor share is evident in Panel (A). In the cross-sectional data in Panel (B), one may see some increase from group II to group IV, but we consider the labor share to be near-constant as well. The share discrepancies shown in the table confirm the combined results mentioned above.

Despite the similarities in the share performances, there is an important difference between the shares of the historical records of Japan and the shares obtained from the cross-sectional data for the developing economies. More specifically, the product share of services is greater than that of industry for all developing countries in groups I–IV for the cross-sectional data, whereas in Japan, the product share of services tends to be smaller than that of industry.[74] Essentially the same can be said with respect to the labor share. The different performance patterns suggest that in the developing nations, the greater shares of product and labor of services in the initial period of industrialization appear to have been sustained in the development process. However, in Japan's historical records, the initial pattern is altered quickly toward the end of the initial phase of industrialization.

---

73 According to Japan's postwar data, the effects of shifting transportation and communication from sector I to sector S are estimated to be 12–15 percent for Y_I/Y_N and Y_S/Y_N (Ohkawa et al., *LTES , National Income,* Vol. 1, Table 12). For the labor share, no data are available for adjustment.

74 These estimates, of course, need to be adjusted. However, even the use of a 15 percent conversion, would keep the pattern virtually intact.

Table 3.12
Coefficient of Sectoral Inequality of Product Per Worker for Historical Japan [a]

Selected Intervals	In Current Prices	In Constant Prices
1885-1899	2.08	2.41
1900-1914	1.88	2.05
1915-1922	1.47	2.41
1927-1933	2.13	1.71
1936-1940	1.64	1.84
1955-1965	1.60	1.73

Note: a. The procedure used is the same as that used for deriving values in Table 3.10.

Sources: See sources of Table 3.4.

Finally, all these findings are integrated by simply using the coefficient of inequality of product per worker, i.e., ε. The results are summarized in Table 3.12, including the case of a constant-price series.

The value of ε in current prices for Japan is comparable with those of developing economies (Table 3.10). The coefficient of inequality of product per worker generally follows a long-term trend of decline, but in Japan's case, there is no drastic change which was found in the lower groups of the cross-sectional series. Although Japan's magnitude of ε may seem small compared to those of developing nations in the semi-industrialized phase, this perception is not altogether correct. The magnitude of ε for Japan is rather close to the magnitude of ε for nations of faster industrialization (subgroup C), and its postwar value is greater than the magnitude of ε for group V. These are all understandable in light of our analysis of the features of Japan's pattern of development.

Lastly, we cannot miss an important point, which is the long-term declining trend of ε that accompanies swing-wise deviations where ε is smaller in upswings (1915–1922) and greater in downswings (1922–1933). We can examine this point further by looking at the change in the magnitude of ε in the constant-price series. The data indicate that the change in the magnitudes of ε are brought forth by drastic swing-wise changes in the relative sectoral prices. Such deviations take place in all developing economies and

Table 3.13
Product Share and Labor Share in the Nonagricultural Sector of the Historical Records of Selected Western Nations: Industry and Services [a, b] ***(percentage)***

Country	Year [c]	Product Share in Industry [b]	Year [c]	Labor Share in Industry
Great Britain, United Kingdom				
	1851/61	45.1	1851/61	54.1
	1907	52.2	1851/61	72.6
	1924	55.5	1921	64.7
	1955	59.0	1961	57.1
	1963-1967	56.6		
France (1954 prices)				
	1895	61.6	1856	59.0
	1963	55.7	1962	54.5
Belgium (1953 & 1963 prices)				
	1910	50.4	1846	75.6
	1953-1955	51.3	1964	55.7
	1963-1967	53.1		
Netherlands				
	1913	45.3	1849	53.8
	1950	56.3	1960	56.7
	1963-1967	54.2		
Norway				
	1865	48.2	1865	54.8
(1963 prices)	1960	61.6	1890	58.9
(1963 prices)	1963-1967	61.0	1920	59.7
			1960	60.4
Sweden				
	1861-1870	31.5	1860	46.7
	1951-1955	64.6	1891	62.3
(1959 prices)	1963-1967	58.3	1960	61.1
Italy				
	1861-1870	44.4	1861/71	60.7
	1891-1900	41.8	1880/1901	68.9
	1950-1952	61.3	1964	62.0
	1963-1967	54.2		
United States				
	1839	45.1	1839	48.1
	1889/99	53.7	1869/79	56.4
	1953	49.4	1929	65.3
	1963-1967	45.0	1965	40.3

Notes: a. Unfortunately, product and labor shares could be found for the same year or band of years in only a few cases. Therefore, the two shares are presented separately. Although data on the product share are available in both current and constant prices, only the former is used so as to be comparable with the figures in the preceding table.

b. Product is GDP in current prices, except for the following: Great Britain, 1851/61 and 1907 (NDP); France (1954 prices); Belgium (1953 and 1963 prices); Netherlands (national income); Norway, 1960 and 1963–1967 (1963 prices); and Sweden, 1963–1967 (1959 prices).

c. The coverage and classification of the two major sectors, I and S, is the same as that used for Japan, except for Great Britain 1851/61.

Source: Kuznets, *Economic Growth of Nations,* Table 21:144–47, and Table 38:250–51.

we must therefore pay particular attention to the problem of sectoral inequality, especially in the downswing.[75]

Turning to the historical records of Western nations, the data based on Kuznets' estimates are arranged in Table 3.13 so as to be comparable with those in the preceding tables. Not all of Kuznets' relevant data are used; rather nations and dates that are directly useful for our purpose were selected.

First, the trend increase in Y_I/Y_N and the trend decrease in Y_S/Y_N means that the pattern of the Western nations' historical records is virtually similar to that of Japan and the developing nations.[76] The performance of the labor share, L_I/L_N, appears

75 It is beyond the scope of this study to discuss the significance of sectoral inequality in the general framework of income distribution analysis. Yet, we believe its significance is greater than is usually thought. With regard to historical Japan, the analysis was more carefully and thoroughly carried out in this regard by applying the coefficient of variation by Akira Ono and Tsunehiko Watanabe, "Changes in Income Inequality in the Japanese Economy," in Hugh Patrick, ed., *Japanese Industrialization and Its Social Consequences* (Berkeley and Los Angeles: University of California Press, 1976), 363–89.

76 The average of Y_I/Y_N for the 1960s is roughly 50 percent, which is 10 percentage points higher than that of group V in the cross-sectional data. The major reason for this is classification differences, as mentioned earlier.

mixed among nations: there are increasing, decreasing, and unchanged trends. However, the magnitude of Y_I/Y_N and L_I/L_N seems larger in the Western nations' historical records than in Japan and is often a percentage greater than 60 percent. Earlier, we found that for groups of lower Y/L in the cross-sectional data and for Japan's earlier years, the shares are very small. Nevertheless, although the Western case appears to show a relatively higher magnitude, the shares imply a similar phenomenon.

Second, with respect to the labor share, however, the magnitude is greater than 50 percent in the majority of cases. Values slightly smaller than 50 percent are found only for Sweden and the United States. The magnitude of $L_I/L_N>L_S/L_N$ is the dominant case in the early years of industrialization. This phenomenon, together with a similar pattern found earlier for the initial phase in Japan's industrialization, contrasts with the pattern observed.

Third, because of the poor correspondence of the dates between the share estimates of product and labor, we cannot derive the exact magnitude of the discrepancy between the two shares. Nevertheless, on average the labor share can be assumed to be greater than the product share to a considerable extent for the industrial sector in the early industrialization period: the simple average is 64.0 for industry and 36.0 for services for the countries in the table. Broadly speaking, Japan's pattern which was mentioned earlier can be said to belong to this type. This is dissimilar to the pattern shown by contemporary developing economies in the cross-sectional data.

In concluding this chapter, we emphasize again the importance of different initial conditions and their effects on the performance of production structure in the subsequent growth path and its significance in characterizing postwar developing economies. This also answers the question raised earlier regarding Japan's case. In view of all the discussions in this chapter, we would argue that the employment problem of developing nations should be understood in light of the characteristics of its production structure thus clarified.

Chapter 4

Production Structure and Employment:
Sectoral Productivity Growth and Development Phases

The preceding discussions on convergence vs. divergence lead us to the topic of this chapter, the problems of growth performance in terms of productivity and employment. In Chapter 3, the ratio method was applied in analyzing product per worker using cross-sectional and/or time-series data. In this chapter, the direction and pace with which these differences or changes were formed are observed as growth phenomena. The factual confirmation of a trend of convergence or divergence is a preparatory task. Further analysis is required in an aggregate framework of economic growth: for example, the significance of the same trend of convergence may differ if the rate of growth of output and/or productivity differs. In Section I of this chapter, treatment of convergence or divergence in terms of growth rates, that is, a "growth approach," is attempted. The cross-sectional data are arranged to correspond with those in preceding chapters—that is, in five country groups according to the Y/L level—and are supplemented by time-series data for 1960–1980.

Two subgroups, E and F, are defined in terms of the macro-growth rate of product per worker (or more simply, the productivity growth rate), where subgroup E refers to the countries with higher productivity growth and subgroup F refers to countries with lower productivity growth in each group. The use of subgroups E and F is analogous to the subgrouping employed in Chapter 1. Similarly, the factors responsible for determining the relationship between productivity growth and employment performance are factors associated with the Y/L level and factors that are not associated with the Y/L level. It is important to make the distinction between the level and the rate of growth.

The insights to be gained from the macro-approach is limited and at least in this area, the macro-approach is less fruitful than the sectoral approach. The significance of sectoral characteristics is particularly emphasized in terms of technological property. The relationship between productivity growth and employment performance for the three major sectors—namely, agriculture, industry, and services—is examined in Section II. Following the classical school, several well-known theories have been developed on this subject, but most of these theories are based on the historical path of currently developed nations. While there is no doubt that these theories provide us with the basic line of thought, they cannot be fully applicable to the postwar developing nations because of the different characteristics of the developing countries.

Finally, all of the observed results are integrated into an aggregate performance pattern in Section III. The integration is essentially intended to distinguish the phases of development. We believe the phase concept can be appropriately used for this purpose in terms of basic changes in the production structure according to our definition. Three major phases will be demarcated—the preprimary phase, the primary phase, and the secondary phase. These phases are deduced primarily from observing the postwar performances of developing economies, and the historical path of currently developed economies will be observed next to finalize the integrated phasing. This procedure may help to further clarify the characteristics of contemporary developing economies.

I. Productivity and Employment by Sector: Growth Relationships

Enhancement of labor productivity would certainly reduce demand for employment. This trade-off notion in the static sense often confuses the interpretation of the actual problem of the relationship between productivity and employment. A legitimate answer would be given by the growth approach which takes into account the effects of technological and/or organizational change in terms of speed and type. One can say that changes in the relationship depend on the pattern of the shift of the production function. Because of data limitations, particularly for the sectoral allocation of capital investment, measurement of the shift cannot be done widely enough for a sectoral analysis of this type. Nevertheless, we intuitively believe that the relationship would change over time not in one simple direction but in multiple ways somewhere between the extreme case of a trade-off and the other extreme case of a parallel change. Using the limited data that are available, we attempt in this section to find out how to identify actual performance.

By Y/L Level

In Table 4.1, simple averages of the growth rates of all relevant terms are presented for two periods, 1960–1970 and 1970–1980. The growth rates for the aggregate economy differ from the corresponding figures in Table 1.1 in that the growth rates in Table 4.1 exclude 1980–1985. For simplicity, output per worker will be called "productivity" (although some qualification is needed since the data for L are not adequate in representing "input").[77] With respect to the aggregate pattern, the inverse U-shape identified in Chapter 1 manifests itself for all terms, G_Y, G_L, and G_y. What about by sector? With respect to output, the same pattern is witnessed for all sectors, G_{YA}, G_{YI}, and G_{YS} without exception. Surely common factors must operate in determining the rates of output growth: accelerating factors for the former interval (PI), i.e., groups I–III, vs. decelerating factors for the latter intervals (SI), i.e., groups III–V. If both productivity and labor were to exhibit exactly the same shape, our problem would virtually disappear. However, this is not the case and they differ a great deal.

In agriculture, growth of both labor, G_{LA}, and productivity, G_{yA}, follow unidirectional trends throughout all groups: for labor the trend is decreasing, except for group II, and for productivity the trend is increasing. The significance of this is that productivity growth is substantively associated with reducing labor employment. The rate of productivity growth for industry (G_{yI}) exhibits a tendency to increase unidirectionally throughout all groups much like that of agriculture. However, the performance of employment in industry (G_{LI}) shows an inverse U-shape, except for group I. For the lower Y/L levels, the data suggest that productivity and employment are enhanced concurrently, but at higher Y/L levels, productivity would rise faster the more rapid the decrease in labor employment; this would seem to be analogous to the case of agriculture. One may be tempted to assert that a shift takes place in

---

77 We do not attempt to narrow the gap between the given labor data that were available and the actual labor input by distinguishing between male and female labor with different efficiency units and by counting changes or differences in labor hours and labor quality (including skill and age). In this sense, our treatment is very rough, but this is unavoidable for an overview approach as the data for such an approximation are not internationally available. In the growth approach, however, the effect of this treatment is relatively minor.

Table 4.1
Sectoral Growth Rates of Output, Labor, and Productivity [a]
by Group: 1960–1980 (percentage)

		Group				
Sector	Growth Rates	I	II	III	IV	V
Aggregate	G_Y	3.1	4.9	6.6	5.1	4.3
	G_L	2.0	2.4	2.9	2.0	1.2
	G_y	1.1	2.5	3.8	3.1	3.1
Agriculture	G_{YA}	1.7	3.2	4.5	2.8	1.1
	G_{LA}	1.3	1.8	1.1	-0.7	-3.0
	G_{yA}	0.4	1.4	3.4	3.5	4.1
Industry	G_{YI}	5.2	6.7	8.0	6.6	4.4
	G_{LI}	4.7	3.8	4.8	3.0	0.7
	G_{yI}	0.5	2.9	3.2	3.6	3.7
Services	G_{YS}	4.4	6.2	6.4	6.1	4.4
	G_{LS}	4.9	4.4	4.7	3.7	2.8
	G_{yS}	-0.5	1.6	1.7	2.4	1.6

Note: a. G_y is approximated by G_Y–G_L.

Sources: See sources of Table 1.1.

the type of industrial technology, from a labor-using technology in the lower Y/L groups to a labor-saving technology in the higher Y/L groups. Such a shift would be possible corresponding to changes in factor prices (i.e., a rise in the level of real wages). Under these conditions it is possible to see a movement from group III to group IV. This needs to be examined further. For the moment, we recognize that the productivity-employment relationship cannot be readily assumed to be unchanged in the long term, but in fact changes through differing development paths.

For services, the growth rates appear to be less regular. However, apart from group I for which the data are less reliable, the range of difference among groups is narrower as compared to the other sectors for both productivity growth (G_{yS}) and labor employment increase (G_{LS}). This corresponds to a narrower inverse U-shape of G_{YS}. A second feature is that the rate of employment change is much greater in services as compared to the same rate in the goods-producing sectors. This is even more true for the rate of productivity growth. One may be tempted to assert that the narrower range of the differences is caused by the minimum changes in technology and/or organization in this sector and that the second feature suggests significant characteristics of employment performance of this sector. We generally agree with these assertions and will discuss them later. The second feature is of substantial interest and a most noticeable problem for currently developing nations. Furthermore, we perceive the performance of services in general to be most dissimilar to that of the historical path of currently developed nations.

The significance of the inverse U-shape was repeatedly emphasized in Chapter 1. But what about its disaggregation into sectoral components? Evidence can be obtained using the conventional procedure of calculating weighted terms to see the changes in the contribution of sectors forming the aggregate pattern in terms of output and labor, in order to obtain weighted sectoral productivity growth. The results of such a procedure are shown in Table 4.2. If we look first at productivity, two different sectoral patterns are distinguished. The weighted rate of productivity growth for the goods-producing sectors, agriculture and industry, accelerates for all groups of developing economies, i.e., groups I–IV. A decrease is only seen in the final move from group IV to group V. Services appear to follow an inverse U-shape pattern similar to that of the aggregate. Since the percentage contribution of the services sector is larger than that of the goods-producing sectors, it may appear that the aggregate pattern is chiefly shaped by services. This is not what we would usually expect because we tend to believe that industry is the driving force of economic development and must play the leading role in shaping the aggregate pattern. Is this a contradiction? We do not think so. The peak in the value of services occurs at group II rather than group III, but the weighted rate of productivity growth in services decreases to group III. In the upper Y/L levels, two different movements begin to operate: services continues to decelerate while the goods-producing sectors sustain their acceleration until it finally decelerates with services. Taking these together, we conclude that the aggregate pattern is not dominated either by industry or services. Instead, the shape of the aggregate pattern is the combined result of the different sectoral performances, including agriculture. This is an important fact that will be incorporated into our integrated phasing analysis.

The negative values for the rate of productivity growth of agriculture in the weighted case (groups I and II) may need interpretation. The negative rate of productivity growth stems from

Table 4.2
The Inverse U-Shape Pattern of Productivity Growth: Sectoral Decomposition, 1960–1980 Average [a] (percentage)

Output

Group	w_A	$G_{YA}w_A$	w_I	$G_{YI}w_I$	w_S	$G_{YS}w_S$	Sum
I	49.7	0.84	14.7	0.77	35.6	1.57	3.2
II	31.6	0.96	22.9	1.53	45.5	2.82	5.3
III	23.3	1.04	25.9	2.10	50.8	3.25	6.4
IV	17.0	0.46	33.4	2.18	49.6	3.03	5.7
V	5.0	0.11	38.4	1.69	56.6	2.49	4.3

Labor

Group	v_A	$G_{LA}v_A$	v_I	$G_{LI}v_I$	v_S	$G_{LS}v_S$	Sum
I	83.3	1.08	5.9	0.24	10.8	0.51	1.8
II	72.8	1.31	11.4	0.42	16.8	0.73	2.5
III	51.5	0.57	18.9	0.90	29.6	1.33	2.8
IV	37.4	-0.03	25.0	0.75	37.6	1.39	2.1
V	6.4	-0.02	40.2	0.28	53.4	1.26	1.5

Productivity [b]

Group	G_{yA}	G_{yI}	G_{yS}	Sum
I	-0.24	-0.53	1.06	1.4
II	-0.35	1.11	2.09	2.8
III	0.47	1.21	1.92	3.6
IV	0.49	1.43	1.64	3.5
V	0.13	1.41	1.23	2.7

Notes:
a. The terms w and v are the weights (i.e., the proportion to aggregate) for output and labor, respectively. The weights are the averages of the sector proportion in 1960, 1970, and 1980. The weighted sectoral terms do not exactly add up to the aggregate terms in Table 4.1 due to rounding.
b. The weighted growth rates of productivity are calculated as follows: $G_{YA} \cdot w_A - G_{LA} \cdot v_A = G_{yA}^*$. Similar formulae were used to calculate G_{yI} and G_{yS}.

Sources: See sources of Table 1.1.

the higher proportion of labor (v_A) than output (w_A). The implication of this does not appear to be substantive, but rather statistical.[78] However, we believe it is, in fact, substantively important because the turn in productivity growth to a positive rate of considerable magnitude from group II to III draws attention. The meaning of this turn will be discussed later in the discussion of demarcation of phases.

Productivity Growth by Subgroup

The productivity-employment relationship in the dynamic context can be clarified in the Y/L level framework. The reason for this is that the productivity-employment relationship changes according to the differences in technology and/or organizational and industrial structure which, in turn, varies with changes in the Y/L level. This can be illustrated by the inverse U-shape pattern that was discussed earlier: at the lower Y/L levels, the trend of accelerating rates of productivity growth accompanies the tendency of increasing rates of employment growth, whereas at the higher Y/L levels, the rates of productivity and employment growth tend to decline. It is impossible to deduce anything directly about the productivity-employment relationship from this pattern alone.

Setting the same Y/L level is impossible, but an approximation is possible by classifying two subgroups in each Y/L group. The yardstick for subgrouping is the productivity growth rate, G_y: that is, subgroup E is composed of nations with higher G_y and subgroup F is composed of nations with lower G_y and the number of countries in each subgroup is equal. An alternative yardstick for creating the subgroups could have been used; for example, the performance of labor employment change (G_L) could have been the criteria for the formation of the subgroups but this criteria would not be as effective as its range of variation is much narrower. The subgrouping of countries are shown below (the number of selected countries is shown in parenthesis):

---

78 In order to avoid negative values, a procedure of weighting sectoral productivity by output proportion is possible ($G_{yA} \cdot w_A$, $G_{yI} \cdot w_I$, $G_{yS} \cdot w_I$). The value of $G_{yA} \cdot w_A$ is 0.26 for group I, 0.47 for group II, 0.79 for group III, 0.59 for group IV, and 0.21 for group V. However, because of what we call the "structure term," productivity of the aggregate cannot be fully decomposed using this method (See Appendix II).

Group I (10)			Malawi, Burma, Tanzania, Benin, India, Nepal, Ethiopia, Somalia, Madagascar, and Niger
Group II (12)	E	(6):	Pakistan, Bolivia, the Philippines, Thailand, Kenya, and Ivory Coast
	F	(6):	Sierra Leone, Senegal, Sudan, Sri Lanka, Congo (People's Rep.), and Mozambique
Group III (12)	E	(6):	Korea (Rep. of), Turkey, Egypt, Syrian Arab Rep.,Tunisia, and Malaysia
	F	(6):	Nicaragua, Morocco, Paraguay, El Salvador, Honduras, and Colombia
Group IV (12)	E	(6):	Brazil, Portugal, Yugoslavia, Dominica, Greece, and Singapore
	F	(6):	Peru, Jamaica, Mexico, Costa Rica, Uruguay, and Argentina
Group V (12)	E	(6):	Japan, France, Belgium, Germany (Fed. Rep.), Denmark, and Finland
	F	(6):	Netherlands, United Kingdom, Canada, Norway, Sweden, and the United States

In Table 4.3, the data are arranged to correspond with the data in Table 4.1. The difference in the rates of productivity growth for the aggregate (G_y) is remarkable between the two subgroups E and F. In fact, the difference between G_y for the two subgroups is often far greater than the difference between G_y of two successive country groups. The aggregate rates of labor employment, however, differ between the subgroups only within a very narrow range (except for group V). These two features, taken together, cast a preliminary but important suggestion. One could say that without changing the rate of employment increase, it might be possible to realize a considerably greater or smaller rate of productivity growth. Such an aggregate pattern may be conceivable, but is, in fact, not realistic. A realistic supposition would be that such a pattern will emerge as the combined result of various sectoral performances in the productivity-employment relationship among which two phenomenon that are mutually interrelated are substantive. One phenomenon is the difference in sectoral composition and the other is the different performance of technological and/or organizational advance.[79] For

---

79 The measurement of the production function and its shift in the two subgroups with regard to the three sectors is not possible. However, G_y

Table 4.3
Sectoral Growth Rate of Output, Employment, and Productivity by Subgroup [a]: 1960–1980 (percentage)

Growth Rate	Group	Aggregate E	Aggregate F	Agriculture E	Agriculture F	Industry E	Industry F	Services E	Services F
G_Y	II	6.3	3.4	4.2	2.4	7.8	5.7	6.9	5.6
G_L		2.6	2.5	1.8	1.9	4.5	3.2	4.9	3.8
G_y		3.7	0.9	2.4	0.5	3.4	2.5	2.0	1.7
G_Y	III	6.9	5.1	4.5	4.3	9.5	6.4	6.6	6.1
G_L		2.5	2.8	0.5	1.7	5.1	4.2	4.1	4.0
G_y		4.4	2.3	4.0	2.6	4.4	2.2	1.6	2.1
G_Y	IV	6.4	3.8	3.2	2.4	8.3	4.8	6.7	5.5
G_L		2.8	2.2	-1.4	-0.1	3.5	2.5	4.0	3.4
G_y		3.6	1.6	4.6	2.5	4.8	2.3	2.7	2.1
G_Y	V	4.7	3.8	0.6	1.5	5.2	3.7	4.7	4.1
G_L		0.7	1.7	-3.8	-2.1	1.0	0.3	2.5	3.1
G_y		4.0	2.1	4.4	3.6	4.2	3.4	2.2	1.0

Note: a. Because of the questionable statistical reliability of data for group I, subgrouping was not used.

Sources: See sources of Table 1.1.

---

alone for the aggregate (and in sectors, G_{yA}, G_{yI}, and G_{yS}) can be used as a simplified indicator of shifts in the production function assuming an unchanged capital-output ratio and an output elasticity of labor, β. We recognize the independent meaning of labor productivity change, but argue that as in growth accounting, there is a certain conceptual link with Chapter 1. The growth of the residual (G_R) or TFP is given as $\alpha(G_Y-G_K)+\beta(G_Y-G_L)$. In the case of $G_Y = G_K$, we have $G_R = \beta G_y$. The capital-output ratio and β actually changed over time during the two decades, both in terms of the aggregate and the sectors. However, in our average measurements for the period concerned, these are treated as being unchanged with other key terms.

the sectorwise performance of productivity growth, we see a regular pattern: in all three sectors for all country groups, the rate of productivity growth is greater for subgroup E than subgroup F. An exception is noted only in services of group III. This suggests that at least for the goods-producing sectors, agriculture and industry, common factors operate more strongly in subgroup E and more weakly in subgroup F. The common factors must be relevant to a nation's capability of realizing technological and/or organizational advance; but this is an important proposition of our original hypothesis.

Is the employment performance similar to all sectors? The answer is "no." For industry, the rate of employment increase, G_{LI}, is greater in subgroup E throughout all groups without exception. In agriculture, the reverse is true. The positive value of G_{LA} is greater in subgroup F than in subgroup E for countries with lower G_y, while its negative value is smaller in subgroup F for the countries with higher G_y. With respect to services, G_{LS} shows a pattern similar to that of industry; that is, G_{yS} is greater in subgroup E and smaller in subgroup F. However, the reverse case is distinctly clear for group V.

With regard to the relationship between productivity and employment, regularity is thus witnessed in the sectoral performance. In industry, both the rates of productivity and labor growth are greater for subgroup E than for F across all groups. This can be called a <u>concurrent</u> pattern. In agriculture the reverse pattern is witnessed, and we call this a <u>reversal</u> pattern. One may say that these patterns merely present what is already a well-known fact. However, what we are concerned with here is the quantitative relationship between productivity growth and employment increase, and the changes in this relationship as a country moves along the growth path of technological change. We believe that the difference between the two subgroups E and F, and the considerable variation in the quantitative relationship along the growth path is important. Unlike the goods-producing sector, the behavior of services is mixed. As is widely known, no distinct proposition has ever been made regarding the productivity-employment relationship in services. We contend that the effect of technological change on employment in the services sector is minimal, and employment is strongly determined by other factors depending upon the situation of total labor supply vs. demand for labor in the goods-producing sectors. Thus, the relationship between employment and productivity cannot be specified and can be referred to as being "mixed."

Table 4.4
Sectoral Contribution to Employment Growth [a] *(percentage)*

		Shares			Weighted Growth Rate		
Group	Subgroup	v_A	v_I	v_S	$v_A G_{LA}$	$v_I G_{LI}$	$v_S G_{LS}$
II	E	67.8	12.3	19.9	1.22	0.55	0.89
	F	73.6	9.3	17.1	1.40	0.30	0.65
III	E	53.8	19.0	27.2	0.27	0.97	1.39
	F	53.8	11.7	28.5	0.91	0.49	1.14
IV	E	34.9	25.6	39.5	-0.49	0.90	1.58
	F	32.6	24.2	43.2	-0.03	0.61	1.47
V	E	13.5	41.3	45.2	-0.51	0.41	1.13
	F	10.5	35.6	54.5	-0.22	0.11	1.69

Note: a. The terms v_A, v_I, and v_S refer to the employment share (proportion) of agriculture, industry, and services, respectively, to total employment. The percentages are an average for 1960, 1970, and 1980.

Sources: See sources of Table 1.1.

The last point leads to the issue of labor allocation between sectors. This is an area in which a great number of research results are available, particularly with respect to the historical records of currently developed nations. We add further to the existing knowledge by presenting simple figures of the combined results of different productivity-employment relationships which are assumed to operate in the three sectors (see Table 4.4). The order of magnitude of sectoral contribution of labor is supposed to change linearly with a rise in the Y/L level. However, this supposition is not necessarily endorsed by the data in the table. Although the weighted growth rate of the sectors does show that S>I>A for groups III, IV, and V, for group II, the order is A>S>I. After passing through the period of lower G_y, the contribution of the services sector is of a greater magnitude than industry for the major part of the industrialization process. This pattern is expected from our earlier discussion of group averages (Table 4.2).

The contribution of industry is greater for subgroup E than subgroup F without exception. In agriculture, the reverse is true. In an often-used two-sector model composed of industry and agriculture, one may be tempted to assert that the more labor-absorbing industry is, the more industry will contribute to enhancing productivity growth in agriculture by weakening underemployment pressure. Nobody would deny this assertion, and in fact, theoretical models of this type have emphasized the significance of this path of labor reallocation. We generally share this view to the extent that the type of technology is not specified. However, in order to accurately reflect the reality of developing economies, we must look at the performance of services in terms of the share of employment. A higher share of employment in services is witnessed for subgroup E in group II, and for groups III and IV, the employment share of subgroup F is higher than that of subgroup E. Despite these differences, the weighted growth rate ($v_S G_{LS}$) is greater for E than for F for all the developing nation groups, i.e., groups II–IV. Basically the same pattern is seen for industry ($v_I G_{LI}$). It is suggested that in enhancing labor employment, the activity in subgroup E is superior to that of subgroup F. This function is discussed in more detail.

"Underemployment" can be defined in a number of ways. In our framework, no rigorous definition can be used. For example, if the notion of a "lower marginal product of labor than prevailing wages" is adopted, then underemployment would be a general phenomenon that occurs in almost all of the developing economies. Rather than employing a rigorous definition, the term is used in a comparative sense. That is, compared to subgroup E, underemployment is supposed to increase in subgroup F at a faster pace in both services and agriculture. This implies backwardness in the long-term process of the sectoral reallocation of labor. This backwardness occurs in countries with lower rates with productivity growth.

II. Sectoral Characteristics of Technology

Further discussion is needed to clarify the characteristics of technological change and the relevant phenomena in the major sectors. The property of technology is different between the sectors and this is a basic factor in distinguishing the productivity-employment relationship. We are specifically concerned with the examination of the actual pattern of the sectoral performances for the postwar developing economies. Does agriculture lag industry in terms of technological advance as was asserted by the classical doctrine? To what extent can the pace of technological and/or organizational progress be considered to be associated or not associated with the Y/L level? Why does the rate of technological advance appear to have slowed down towards the 1970s? These questions illustrate the problems we face.

In this section, agriculture and then industry will be discussed. After the comparative observation of these two goods-producing sectors, services will then be taken up. Finally, a comparative observation will be attempted in integrating sectoral performance thus analyzed in terms of the two-sector approach, such as A vs. N (I+S), S vs. G (A+I), and I vs. A.

Agriculture

In the preceding section, productivity growth in agriculture was found to be higher than one would usually expect for most of the developing economies. It is our view that the significance of the "Green Revolution" should be emphasized even more than it is in the usual explanation. Our treatment of agriculture is focused on this point in relation to the "agricultural revolution" in the currently developed nations. Relevant data are arranged in Table 4.5.

The technological characteristic of agriculture is its dependence on land (or natural resources in the broader sense). Product per unit of land or simply "land productivity" is the important indicator to illustrate this point. The land constraint is crucial for many countries, and it is essential that the land constraint be taken into account in discussions of the type of technological change at the sectoral level and of farmers' behavior at the micro-level. Product per land, however, is not taken up explicitly in this analysis but is assumed implicitly[80] and is reflected indirectly by "labor productivity," a common term to all sectors of the economy.

For now, let us consider the rate of growth, G_{yA}. Several points draw our attention. First, G_{yA} exhibits a regular trend in the 1960s, increasing from 0.6 for group I to 5.3 for group V. The difference in G_{yA} between country groups is surprisingly wide and the sustained increase in G_{yA} with the increase in Y/L levels is impressive. In terms of subgroups, G_{yA} is greater for subgroup E in each group

---

80 $G_{yA} = G_{YA} - G_{LA} = (G_{YA} - G_B) + (G_B - G_{LA})$ where B refers to the land used in agriculture. G_{yA} is thus composed of two terms: the rate of land productivity increase and the rate of land increase for labor. As the variation of land is usually small, for simplicity we use $G_{yA} = G_{YA} - G_{LA}$, assuming $G_B = 0$. However, when and where $G_B < G_{LA}$, the pressure on G_{yA} is a serious problem in terms of underemployment.

Table 4.5
Growth Rates of Output, Labor, and Productivity: Agriculture, 1960–1980 (percentage)

Group &		1960-1970			1970-1980		
Subgroup		G_{YA}	G_{LA}	G_{yA}	G_{YA}	G_{LA}	G_{yA}
I[a]		2.0	1.4	0.6	1.4	1.3	0.1
II		3.3	1.6	1.7	3.0	2.0	1.0
	E	4.4	1.8	2.6	4.0	1.8	2.2
	F	2.2	1.4	0.8	2.0	2.2	-0.2
III		4.6	1.1	3.5	4.3	1.1	3.2
	E	4.6	0.6	4.0	4.4	0.3	4.1
	F	4.5	1.5	3.0	4.1	1.9	2.2
IV		3.2	-1.0	4.2	2.3	-0.4	2.7
	E	3.7	-1.5	5.2	2.6	-1.2	3.8
	F	2.7	-0.5	3.2	2.0	0.4	1.6
V		1.3	-4.0	5.3	0.8	-1.9	2.7
	E	1.0	-4.7	5.7	0.1	-2.9	3.0
	F	1.5	-3.3	4.8	1.5	-0.9	2.4

Note: a. For group I, no figures for subgroups are presented because of the insufficient number of countries for which both output and labor data are available.

Sources: World Bank, *World Development Report 1982*, World Development Indicators, Table 2.

without exception in the 1960s. During the 1970s, this pattern between the subgroups is maintained, but the tendency of increase of G_{yA} across groups changes. For example, both by group and by subgroup, the rate of productivity increase tends to slow down in group IV, after having attained a high rate for group III.

Second, G_{YA} follows the same pattern through the two decades: towards group III it tends to rise but from there turns to decrease towards group V, although the level itself shows a general decline from the 1960s to the 1970s. We believe this must be a strong phenomenon because the performances of subgroups E and F exhibit exactly the same pattern for both the 1960s and the 1970s without exception.[81]

Third, the different patterns found between productivity and output virtually stem from the performance of the rate of change in the labor force engaged in this sector, G_{LA}. For groups I to III, G_{LA} is positive through the two decades; a negative value appears in group IV, and G_{LA} becomes more negative towards group V. As is widely known, in the analysis of historical records of developed nations, this turning point from a positive to a negative value of G_{LA} calls for specific attention, because it is thought to be closely relevant to forming the "turning point" of development in terms of the basic change in the structure of the labor market mechanism. As for the postwar developing economies, the absolute number of the labor force engaged in agriculture has long followed an increasing trend, chiefly due to the high rate of increase in the labor force supply. This is represented by the change in G_{LA} of groups I–III for the two decades. After rising slightly from group I to group II, G_{LA} begins to decline as a result of the entire mechanism of labor

81 One may not be satisfied with the measures of agricultural output by added values. Physical measures, gross of intermediate goods supplied from nonagriculture in wheat unit are shown for selected nations of group V (see Appendix B and Hayami and Ruttan, *Agricultural Development*).

The rate of increase in the index (1960=100) is as follows:

Period	United States	Japan	Denmark	France	United Kingdom	Average
1890–1910	137	142	152	121	100	130
1910–1930	125	131	161	117	109	129
1930–1950	140	99	115	97	137	118
1950–1970	138	192	149	220	154	171
1960–1970	116	127	113	132	126	123
1970–1980	126	115	113	125	118	120

Two points are noted: (1) For the post-1960 period, a lower rate of output is witnessed but the rate is less distinct than for G_{YA} in the main text. Yet, excluding the United States, a simple average is 125 for 1960–1970 vs. 118 for 1970–1980; (2) an "unprecedented" magnitude of G_{YA}, mentioned earlier, is identified for 1950–1970 rather than 1960–1970 by this measure. Although with some exceptions (the United States and Denmark), the rate of index increase is generally unprecedented (see the simple averages). Measures for G_{YA} are not readily available for 1950–1960 for these countries. Yet, we contend that the agricultural revolution in the developed nations began to take place earlier than the 1960s and preceded the Green Revolution in the developing nations.

reallocation in the economy. Our concern here is the positive effects of this reallocation of labor on productivity enhancement.

These patterns cast a factual challenge for our interpretation. First of all, we wish to discuss the technological characteristics of agricultural production. A technological system of farming is composed of two elements: one is a biochemical (BC type) element and the other is a mechanical one (M type). The latter element is found in both agriculture and industry, but the biochemical element (BC type) is seen only in agriculture. The BC-type element is often called seed-fertilizer technology because a major component of this type of technology is the shift to improved high-yielding varieties of seeds to meet the complementary demand increase for fertilizers. Innovation of the BC-type is the core of the Green Revolution. Mechanization by M-type technology is supplementary. In another case, where the M-type of technology conforms to the core of the system of farming technology while the BC-type technology is supplementary, is illustrated in cases of small man-land ratios.

We share the view that the unprecedented high rates of productivity growth in postwar agriculture can be decomposed into the two kinds of innovations, the BC-type of innovation and the M-type of technology. In general, the BC-type innovation is relevant to developing economies and the M-type innovation is relevant to the developed economies. Some economists emphasize the importance of M-type innovations. However, in our view, the BC-type innovation is also of substantive significance. The occurrence of these two different types of innovation can be explained by economic factors: one factor is scale economies and the other factor is production factor combinations. Technology of the M type has benefits of scale economies while BC-type technology is scale-neutral. Rate of substitution of capital for labor through mechanization of farming is accelerated by increasing wages. Thus, when and where the land supply restraint is weaker, the M-type innovation takes place in a number of nations of group V. In the countries of groups III and IV, with a few exceptions, the land supply restraint is stronger and wage hikes do not strongly encourage factor substitution; given these conditions, technology of the BC type fits well. This view is not new but is, in fact, rather conventional.[82] We argue here that despite the different types of technology, the common yardstick for both cases is the speed of productivity growth,

---

82 For example, the differences in innovation types are comprehensively analyzed from the viewpoint of "induced innovations" (induced by different factor prices) in Hayami and Ruttan, *Agricultural Development*.

G_{yA}. Actually, the BC-type technology largely advanced in countries in groups III and IV while the M-type technology proceeded in a number of group V nations during the two decades under review. In other words, adoption of different types of technology makes acceleration of productivity growth possible, where the pace of productivity growth is almost parallel to the pace of adoption of technology. Recognition of this phenomenon is important.

However, with respect to the rates of diffusion of such innovations, differences occur in both the time-series and cross-sectional data. First, looking at the cross-sectional data, as a result of diffusion of high-yielding varieties, combined with augmented inputs of chemical fertilizers and pesticides, there is sizeable backwardness in groups I and II and subgroup F. Even under the sustained pressure of underemployment, this technology enables in principle a rise in the marginal productivity of labor on the farm, although the original productivity level in most of the countries was extremely low. Therefore, backwardness in those cases may not be essentially caused by a greater pressure on underemployment, but may be due to the organizational and/or institutional constraints, including physical infrastructure facilities. Second, a slowdown, which is generally recognized in the 1970s, may imply the operation of these restraints. Differentials in the changes over time are also wide. For example, subgroup E of group III sustains its 4 percent rate of G_{yA}, while subgroup F of group II shows a falling rate which even becomes negative. It goes without saying that the advance of technology of the BC type virtually stems from the local dissemination and diffusion of borrowed knowledge in the international context. Therefore, in addition to the restraints mentioned earlier, the difference in the human capability of diffusing technology is also an influential factor.

The performance of G_{yA} of group V in the 1960s endorses Kuznets' proposition regarding the "agricultural revolution" of the developed countries. It should be emphasized that G_{yA} is unprecedentedly high in developed nations in light of their historical records. This is essentially due to the rapid advance of technology of the M type, combined with the BC type, in addition to the effect of scale economies. The pace of the decrease in the labor force in agriculture accelerated in the 1960s. However, it slowed down during the 1970s. The rate of aggregate growth of the economy also slowed down in the 1970s and this may be relevant. The slowdown in the tempo of the decrease in the labor force in agriculture, particularly in subgroup F, is worthy of our attention

Perhaps further improvement of the effects of technological advancement of this type stagnated in these countries.

What suggestions can these examples give us for deepening our knowledge of historical changes or changes over time? The answer goes beyond the sectoral approach and pertains to the aggregate growth performance of the economy. However, let us first discuss a few specific points within this sectoral context. We have emphasized the significance of innovations through the advancement of the BC-type technology. This kind of technological improvements will remain valid for the nations in groups I and II in the future; this is important from the long-term perspective of productivity growth in agriculture. However, the role played by M-type technology may become relatively more important for the nations in groups III and IV, as there are limits on how far advances can be made with technology of the BC type. The successful adoption of this type of borrowed technological knowledge, of course, depends on its assimilation in the country to meet local requirements. This long-term perspective appears to be suggested in the cross-sectional data, particularly in comparisons between groups III vs. IV and V. Productivity growth in agriculture cannot be achieved without realizing a sustained decrease in the labor force engaged in this sector. It is quite difficult to say how and when the developing nations can meet this requirement. A long period of normal development will be required to answer these questions.

The beginning of negative values for G_{LA} shown for group IV does not mean that the agricultural labor force will tend to decrease in all countries in this group. That is, the negative acceleration towards group V does not necessarily imply the effect of a continuous change over time. For example, of the countries in group IV, the European countries such as Portugal, Greece and Yugoslavia, and some Latin American countries such as Uruguay and Argentina, show a negative value of G_{LA} in 1960–1970 and the term is even more negative in 1970–1980. For all of these countries, the rate of increase in the labor force supply had already slowed to levels close to those of developed nations. In all other countries in this group, the absolute number of the labor force engaged in agriculture still tends to increase. Our attention is drawn to a very slight decline in the 1970s for Brazil (group IV), Korea, Turkey, and the Syrian Republic (group III). These countries did not experience a very small rate of total labor force increase. Therefore, they illustrate the possibility of enhancing further the rate of productivity growth in agriculture. Nevertheless, a possible shift to the situation of realizing scale economies cannot be generalized on an equal footing. Even for these countries, arriving at the pivot point to an absolute decrease in the agricultural labor force is judged to occur at a much later phase of industrialization as compared to the historical records of currently developed nations. In this example, the basic difference between the postwar developing economies and the historical path of currently developed nations is evident.

Industry

It is widely recognized that differences in factor prices of advanced and backward nations (in terms of technological levels) discourage the less developed nations from assimilating industrial technologies from the more industrialized economies. This is a problem, especially when considering the "type" of technology, and involves factor substitution, which we treated in relation to the operation of human capability in Section I, Chapter 1. Simply put, the capability of substituting human work of improved capability for capital is, in our view, an important part of advanced technology. We recognize that the reluctance of less developed nations to borrow technology stems from differences in factor prices and this is incorporated into our capability approach. The rate of productivity growth in industry is sizeable in developing economies and the notion of a reluctance alone cannot explain what has occurred in reality. In the preceding section on agriculture, we emphasized the significance and usefulness of the "speed" aspect. Industrial technology can be categorized as being capital-using vs. labor-using or the like, if the technology is not being rigorously defined (e.g., neutral vs. biased). Our focus in this section is the speed of adaptation without discussing the differences in type directly, although the relationship between productivity and employment continues to be the important problem at hand.

The industrial sector is not homogeneous in its subsector components,[83] and yet we treat its characteristics by basically

83 Manufacturing is the major subsector but inclusion of construction, mining, and public utilities makes the behavior of the rates for industry mixed, although the heavy influence of mining is avoided by country selection. This defect may perhaps be weaker for country groups of higher Y/L level. For example, Japan's historical records show that in terms of the weighted growth rate, the proportion of nonmanufacturing to industry tends to increase slightly during prewar periods: the weighted growth rate was 40.8 in 1887–1904, 42.0 in 1904–1919, and 43.4 in 1919–1938, but decreases to 21.8 in 1955–1964. Transportation and communication are included in the nonmanufacturing sector and mining is included in manufacturing, though its share is very small. The share of output, however, shows no sizeable

Table 4.6
Growth Rates of Output, Labor, and Productivity [a]***:***
Industry, 1960–1980 (percentage)

Group &	1960-1970				1970-1980			
Subgroup	G_{YI}	G_{LI}	G_{yI}	η_I	G_{YI}	G_{LI}	G_{yI}	η_I
I	6.3	4.4	1.9	0.70	4.1	5.0	-0.9	1.22
II	8.4	4.5	3.9	0.54	5.0	3.1	1.9	0.62
E	9.1	5.2	3.9	0.57	6.5	3.7	2.8	0.57
F	7.6	3.9	3.7	0.51	3.5	2.5	1.0	0.71
III	8.1	4.3	3.8	0.53	7.8	5.1	2.7	0.65
E	10.2	4.8	5.4	0.47	8.7	5.3	3.4	0.61
F	5.9	3.7	2.2	0.63	6.8	4.6	2.2	0.68
IV	7.7	2.6	5.1	0.34	5.4	3.4	2.0	0.63
E	9.5	3.3	6.2	0.35	7.0	3.6	3.4	0.51
F	5.9	1.9	4.0	0.32	3.7	3.1	0.6	0.84
V	5.9	0.7	5.2	0.12	2.9	0.6	2.3	0.21
E	6.8	1.3	5.5	0.19	3.4	0.8	2.6	0.24
F	5.0	0.1	4.9	0.02	2.4	0.4	2.0	0.17

Notes: See notes of Table 4.5.
a. $\eta_I = (G_{LI}/G_{YI})$.

Sources: See sources of Table 4.5.

applying our original hypothesis of borrowed technology. The 1960s was a most prosperous period as the magnitude of technological knowledge and the speed with which the technology spread from leading industrial nations to latecomers was most remarkable. Towards the 1970s, the spread of technology decelerated somewhat.

The average growth rates of output, labor, and productivity for industry are presented in Table 4.6. From the table, the following points are worth mentioning.

First, the rate of productivity growth in industry (G_{yI}) shows a fairly regular trend, increasing at higher Y/L levels in the 1960s.

change over time: the prewar range is 67.4–68.3 and the postwar range is 66.3–66.5 for the manufacturing sector (Ohkawa et al., *LTES, National Income,* Vol. 1, Table 3.1:46 and Table 3.3:50).

But this is not the case for the 1970s. This over-time change is significant. During the 1960s, the rise in G_{yI} from 2 percent in group I to over 5 percent in group V is impressive. However, in the 1970s, the tendency to rise stops at group III and the rate of productivity growth for groups of higher Y/L levels, i.e., groups IV and V, is lower just slightly exceeding 2 percent. This trend can also be seen by subgroup observations, although the trend is less regular. The particularly slow rate of productivity growth of group IV in the 1970s is illustrated by the extremely low value of 0.6 percent for G_{yI} of subgroup F.

Second, the output growth rate of industry (G_{YI}) does not exhibit a pattern similar to that of G_{yI}. A declining trend is seen across the groups in the 1960s and in the 1970s; that is, G_{YI} exhibits an inverse U-shape pattern. In looking at the changes in G_{YI} over time, there appears to be a general slowdown from 1960–1970 to 1970–1980, though the degree of the slowdown differs between groups and subgroups. For example, the decline in G_{YI} from 1960–1970 to 1970–1980 is greater for group II whereas it is smaller for group III. For subgroup F of group III, G_{YI} even increased from 5.9 in 1960–1970 to 6.8 in 1970–1980. These dramatically different patterns found in the growth performances of productivity and output call for additional attention. The slowdown of G_{YI} between the two periods can basically be explained in the same way as was the trend in the aggregate residual (Chapter 2) using our hypothesis which focuses on the narrowing in the technology gap between developed nations.[84]

---

84 Data limitations prevent us from discussing this aspect specifically for the industrial sector. However, it seems possible to evaluate historically the unprecedentedly high rate of industrial growth (G_{YI}) in the 1960s and the significance of its slowdown in the 1970s, with respect to group V nations. Kuznets' data of G_Y combined with the product shares of the major sectors can be used to estimate G_{YI} (and G_{YS} for reference) in a historical perspective for the six Western nations. In terms of average annual rates of growth, these values are (dates used for the share estimates are in parenthesis when not the same):

		G_{YI}	G_{YS}	
France		2.3	2.6	1896–1963
Belgium		2.9	2.8	1900–1909 to 1963–1967 (1910 to 1963–1967)
Germany		4.3	4.4	1936 to 1963–1967 (1936–1938 to 1960)
Sweden		4.2	3.2	1861–1869 to 1963–1967
Italy	1)	1.3	0.8	1861–1869 to 1895–1899
	2)	3.8	3.0	1895–1899 to 1963–1967

Here, the problem at issue is the implication of the slowdown for G_{yI} of developing economies. A closer look at G_{yI} of the countries may roughly suggest what happened. The ratios of G_{yI} over the two periods for the groups (i.e., (G_{yI} of 1970–1980)/(G_{yI} of 1960–1970)) are as follows: Group I, 0.50; Group II, 0.50, Group III, 0.65; Group IV, 0.39, and Group V, 0.43.

In contrast to the slowdown in growth in the industrial economies, the slowdown in the developing economies tends to be smaller except for group IV. Group III is the least affected. Differences in the response of developing economies are more distinctly elucidated by analyzing by subgroups:

	Group		
Subgroup	II	III	IV
E	0.72	0.71	0.55
F	0.27	0.59	0.15

The higher ratio maintained by developing nations is virtually accounted for by the vigorous activity of subgroup E in each group. In direct contrast, the less active performance of subgroup F slowed down the group's productivity growth even below that of developed nations.

Our interpretation of this is as follows. In the 1960s, the unprecedented pace of technology diffusion for industrial economies contributed a great deal to enhancing its industrial productivity growth and this accompanied a high pace of diffusion of advanced technological knowledge to developing nations. In the 1970s, these effects were weakened, though not in a uniform fashion. Because of differences in the absorption capability of the developing countries, a higher rate of productivity growth was sustained by more vigorous nations on the one hand, while a noticeable slowdown in productivity growth occurred in less active nations on the other. We stand by our hypothesis that the possibility of absorbing advanced technological knowledge does not depend upon the gap in the

---

For group V, G_{YI} is 5.9 in the 1960s and 2.9 in the 1970s, and the average for the two decades is 4.4. An adequate comparison is not possible as figures for shorter intervals are not available from the historical records. Yet, we believe that G_{YI} in the 1960s is unprecedentedly high and that the slowdown of G_{YI} in the 1970s can be considered to be significant as it may be lower than the long-term average level for those nations.

technology level or changes in the technology level of the advanced nations, but essentially depends upon the nation's social capability. The subgroup comparison provides direct evidence of this. Time-series data may not contradict our hypothesis if we allow for the operation of certain disturbing factors, say, demand shrinkage, in the 1970s. In other words, the factors stemming from a nation's social capability operate as the major element in enhancing productivity growth of developing nations.

Turning to G_{LI}, let us look at $\eta_I(G_{LI}/G_{YI})$ in Table 4.6. This crude elasticity is often used, but in our view this is not appropriate for measuring the relationship between productivity and employment because it does not consider the speed of productivity growth. In the 1960s, η_I is generally higher for subgroup E than for subgroup F, with the exception of group III. During the 1970s, the pattern appears to be reversed. Except for group V, η_I is greater for subgroup F than for subgroup E. We do not share the view that η_I is stable in indicating the type of industrial technology. Rather, we would argue that the productivity-employment relationship in the growth path is affected by a shift in the labor demand function, if we want to follow the framework of the neoclassical approach. In an innovative growth process, the productivity-employment relationship is not primarily determined by the type of technologies adopted.

Let us first focus our attention on the subgroups to obtain evidence of this. For all groups, both G_{yI} and G_{LI} are higher for subgroup E without exception in both periods. The proposition that speedier productivity growth accompanies a greater rate of employment increase is affirmed by the data. The average values of these terms for the two periods are as follows:

	Group							
	II		III		IV		V	
Subgroup	G_{LI}	G_{yI}	G_{LI}	G_{yI}	G_{LI}	G_{yI}	G_{LI}	G_{yI}
Average	3.9	2.9	4.7	3.3	3.0	3.6	0.7	3.9
E	4.5	3.4	5.1	4.4	3.5	4.8	1.0	4.2
F	3.2	2.4	4.2	2.2	2.5	2.3	0.3	3.5
Difference (E–F)	1.3	1.0	0.9	2.2	1.0	2.5	0.7	0.7

The data imply important suggestions for characterizing industrial technology in terms of productivity-employment, and the "concurrent" growth relationship between productivity and employment generally appears to be confirmed insofar as the subgroups are concerned. What about the comparison by groups? The average G_{yI} for the two periods shows a straight line of increase from 2.9 for group II to 3.9 for group V. In contrast, G_{LI} shows two different trends. First, G_{LI} increases from group II to group III and then decreases from group III to group V. At the lower Y/L level groups, G_{yI} and G_{LI} move parallel to one another but at the higher Y/L levels, this is reversed. In Chapter 1, it was suggested that in the aggregate performance, capital intensity shows no sign of distinct increase from group II to group III, whereas from group III to group V, capital intensity increases dramatically. The two findings seem to be consistent despite the difference in scope, that is, the industrial sector vs. the aggregate. Thus, the subgroup comparison indicates a "concurrent" relationship, while the group comparison shows a partial "concurrent" and partial "reversal" growth relationship in productivity-employment. Is it possible to give a consistent interpretation of these findings? The answer is yes when it is recognized that the actual relationship is the combined result of two elements: one element operates to produce concurrent results while the other element operates to produce a reversed pattern. The first element relates to the "speed" of technological advance and the second element relates to the "type" of technological advance as it pertains to the substitution effects which stem from factor price changes. The relationship between subgroups E and F can be considered to operate under almost the same relative factor prices, so that substitution effects may not be the major factor. Rather, the speed of enhancing the innovative process is the crucial factor. To this end, we can use the difference in G_{yI} as a rough indicator. It is quite understandable that a faster G_{yI} accompanies a faster rate of shift in the labor demand curve. The effects of differences in terms of "type" are involved but are minor. On the other hand, we assume that differences in Y/L levels are associated with increase in wage rates relative to capital prices as we move from group IV to group V, perhaps partly including group III. The substitution effect thus operates more substantively in these country groups.

Quantitative evaluation of the operation of these two elements is impossible; nevertheless, we contend that "speed" of technological advance plays a more significant role in the earlier phases of development. For the lower Y/L country groups which showed an inverse U-shape path of development, the speed of technological advance remains a dominant factor in sustaining the concurrent relationship between productivity and employment.

Why was this possible? The answer is provided by the fast technological advances in the various light manufacturing industries which was largely made possible by assimilating borrowed foreign technological knowledge (although we cannot completely ignore the differences in capability between nations as illustrated by subgroup comparisons). The process of assimilation that took place differs between the individual countries, but we are convinced that the reluctance arising from factor price differences has largely been solved by local modifications in the production processes, when these are carried out successfully. The story appears to be different, however, for the higher Y/L country groups where capital intensification is required. This will be discussed further in Chapter 6.

Services

The characteristics of the services sector differ from the other two sectors in terms of both demand and supply. The services sector, which includes transportation and communication, is especially complex. Yet, from the supply-production aspect, a common characteristic can be found in the sector's "facilitating" activities when these activities are defined in a broad sense.[85] The activities of all services' subsectors facilitate the goods-production activities of the other two sectors. The speed and type of technological change is of less importance in services, while the demand-determined dimension is much more significant. The impact of technological advance on employment performance in services is to be analyzed from the viewpoint of these characteristics.

First, the operation of the facilitating function of the services sector depends heavily on variations in the demand for the sector's output. The growth relationship of productivity-employment

---

85 For example, public utilities (electricity, gas, and water), transportation, and communication are often classified together as "facilitating industries." This illustrates that public utilities, transportation, communication can be treated under a common category of facilitating activities on the one hand. Financing, commerce, trade, and services are treated as composing the sector of services in which transportation-communication is included (World Bank data) on the other. A broader scope of the facilitating function can thus be made possible by extending it so as to cover narrowly defined "services." Incidentally, in discussing physical infrastructure, the term "direct production activity" is often used. Facilitating activity in this sense may correspond to "indirect production activity." At any rate, such a classification is for the sake of convenience and depends on analytical purposes.

cannot be clearly identified. Second, viewed from the perspective of factor combinations, flexibility is a key feature of this sector. That is, the same facilitating function can be carried out by either a capital-using or a labor-using procedure. Such flexibility is not necessarily unique to this sector. In agriculture, and even in industry to a certain extent, flexible factor combinations can occur in the developing economies. However, flexibility is most significant in the services sector. Third, the productivity growth rates can be thought of as the combined result of demand variations on the one hand and labor employment performance on the other. Actually, because of the high rates of increase in labor supply, the flexible factor combination manifests itself by bringing forth a certain level of underemployment, the range of which depends upon the output demand situation. Fourth, the interrelationship between goods-producing and facilitating activities is not uniform, but varies between nations and between development phases. Therefore, what is mentioned above may actually appear mixed and less than clear-cut as compared to the goods-producing sector.

Let us look at the data in Table 4.7 which are arranged to correspond with the two preceding tables. The output growth rates of the goods-producing sector, G_{YG} and η_S (i.e., G_{LS}/G_S), are added.

First note that by group, G_{YS} appears to move in a fairly regular fashion: the magnitude of G_{YS} is relatively stable or varies within a small range relative to G_{YA} and G_{YI}, especially in the 1960s. Changes in G_{YS} over time from 1960–1970 to 1970–1980 do not show a distinct slowdown, unlike G_{YA} and G_{YI}. From group III to group IV, G_{YS} even increases. By subgroup, the same pattern that was observed in the goods-producing sector is witnessed: G_{YS} is higher for subgroup E than for subgroup F for all country groups except group V in the 1970s. However, the difference in G_{YS} between the two periods varies within a narrower range. Second, the rate of employment increase, G_{LS}, tends to rise from the 1960s to the 1970s; this is in sharp contrast to the case of output. The data in the table show this to be a strong tendency: by group there are no exceptions and by subgroup, only two minor exceptions occur. In the cross-sectional data, subgroup E maintains a higher magnitude of G_{LS} than subgroup F throughout both periods for the majority of developing economies. A reversed inequality is seen for group V. Third, as can be expected from the preceding discussion, G_{yS} decreased from the 1960s to the 1970s with the notable exception of group III. The lowest value of G_{yS} occurs for group III (subgroup F in

Table 4.7
Growth Rates of Output, Labor, and Productivity: Services, 1960–1980 [a] ***(percentage)***

(A):

Group &	1960-1970					1970-1980				
Subgroup	G_{YS}	G_{LS}	G_{yS}	G_{YG} [b]	η_S	G_{YS}	G_{LS}	G_{yS}	G_{YG} [b]	η_S
I	3.6	4.5	-0.9	3.0	-	5.2	5.3	-0.1	2.0	-
II	6.6	4.1	2.5	5.7	0.62	5.9	4.7	1.2	3.9	0.80
E	7.3	4.6	2.7	6.2	0.63	6.4	5.2	1.2	5.0	0.81
F	5.8	3.6	2.2	4.8	0.62	5.3	4.1	1.2	2.7	0.78
III	5.8	4.8	1.0	6.3	0.83	7.0	4.6	2.4	6.0	0.66
E	6.1	4.2	1.9	7.3	0.69	7.1	4.3	2.8	6.5	0.61
F	5.4	5.3	0.1	5.2	0.98	6.8	4.9	1.9	5.5	0.72
IV	6.0	3.2	2.8	6.3	0.52	6.2	4.2	2.0	4.4	0.67
E	6.5	3.3	3.2	7.5	0.51	6.9	4.6	2.3	5.5	0.67
F	5.5	3.0	2.5	5.0	0.55	5.5	3.7	1.8	3.2	0.67
V	5.3	2.8	2.5	5.3	0.53	3.5	2.8	0.7	2.6	0.80
E	6.0	2.6	3.4	6.7	0.43	3.4	2.4	1.0	3.4	0.71
F	4.6	3.1	1.5	4.5	0.67	3.5	3.1	0.4	2.3	0.89

(B): Difference in Over-time Changes

Group & Subgroup		G_{YS}	G_{LS}	G_{yS}
II	E	-0.9	+0.6	-1.5
	F	-0.5	+0.6	-1.0
Average		-0.7	+0.6	-1.3
III	E	+1.2	+0.1	+1.1
	F	+1.0	-0.4	+1.4
Average		+1.1	-0.2	+1.3
IV	E	+0.4	+1.4	-1.0
	F	0.0	+1.2	-1.2
Average		+0.2	+1.3	-1.1
V	E	-2.6	-0.2	-2.4
	F	-1.1	0.0	-1.1
Average		-1.8	-0.1	-1.7

Notes:
a. η_S is not computed because of doubtful data.
b. G_{YG} is the rate of output growth of the goods-producing sector, and is estimated by the weighted sum of G_{YA} and G_{YI} (the weights used are the average output).

Sources: See sources of Table 4.5.

1960–1970), but in the 1970s, group III has the highest values (the lowest values occur for group V). Thus while no regularity is found at the group level, a fairly regular pattern exists by subgroup: G_{yS} is greater for subgroup E than for subgroup F for both periods.

With these findings, we can now expand on the three mutually interrelated points: the role of demand in determining the output growth, the growth relationship between output and employment, and the insignificance of productivity growth.

Among the factors responsible for determining output growth in the services sector, variation in demand is the major factor, although other factors cannot be ignored. We try to give evidence of this here in a simple way. The output growth rate of the goods-producing sector (G_{YG}) is assumed to be an independent variable that causes variation in the demand for the output of services. Comparison of G_{YS} and G_{YG} suggest that these two terms move fairly closely with one another though with some difference between 1960–1970 and 1970–1980. For instance, consider the ratio G_{YS}/G_{YG} across subgroups. In the 1960s, this ratio varies around unity but in the 1970s, the ratio is higher and is considerably above unity. This suggests that factors other than G_{YG} operate on G_{YS}. These factors may not be so simple, with technological and/or organizational changes being involved. Given these qualifications, however, we think G_{YS} is fairly strongly determined by G_{YG}.[86] With respect to the majority of developing economies, we argue that there is a regular association between the growth of services and the growth of the goods-producing sector through demand.

The second topic, the growth relationship between output and employment, is also taken up from this point of view by looking at η_S (i.e., G_{LS}/G_{YS}) in the table. From 1960–1970 to 1970–1980, η_S increases considerably for all groups except group III. The conventional use of this crude elasticity for measuring labor demand may be meaningful in this case. Its usefulness, however, is limited to the extent that there exists underemployment. Let us make a comparison with industry. The cross-sectional data tell us that in both the services and industry sectors, employment increases to a considerable extent from groups of lower to higher Y/L levels.

---

86 The statistical measures of $G_{YS} = a+bG_{YG}$ give an almost equal value of b for both periods, if the deviations are excluded. For 1960–1970, $G_{YS} = 2.763+0.484G_{YG}$ ($R^2 = 0.842$) and for 1970–1980, $G_{YS} = 3.924+0.505G_{YG}$ ($R^2 = 0.942$).

Table 4.8
Comparison of Growth Rates: Employment vs. Productivity in Services and Industry, Average of 1960–1970 and 1970–1980 (percentage)

Group & Subgroup		G_{LI}	G_{LS}	G_{yI}	G_{yS}
I	Average	5.2	4.9	1.5	-0.5
II	Average	3.9	4.4	3.0	2.1
	E	4.5	4.9	3.4	2.4
	F	3.2	3.6	2.5	1.9
III		4.7	4.6	3.3	1.8
	E	5.1	5.1	4.4	1.5
	F	4.2	4.0	2.3	2.1
IV	Average	3.0	3.7	3.6	2.4
	E	3.6	4.0	4.8	3.7
	F	2.5	3.4	2.3	2.7
V	Average	0.7	2.8	4.0	1.6
	E	1.0	2.5	4.2	2.2
	F	0.3	3.1	3.7	1.0

Sources: See sources of Table 4.1.

However, the increase in employment takes place at the same time that the rates of productivity growth are distinctly different: productivity growth is higher for industry and lower for services. The productivity of services tends to decrease relative to that of industry. This suggests an important aspect of the employment function of this sector that is beyond the problem of underemployment. Table 4.8 may help us grasp the broad picture. The growth rates G_{LS} and G_{LI} move in close association with one another and are almost equal in magnitude except for group V for which new conditions emerge in the industry-services relationship. Some economists pay great attention to this concurrent pattern of employment, but we are more interested in the sharp contrast between this concurrent pattern and the distinctly different rates of productivity growth where $G_{YI}>G_{YS}$.[87] Despite such differences,

87 Additionally evidence can be drawn from data of both changes over time and subgroup comparisons with cross-sectional data. For the former, a greater η_S in the 1970s than in the 1960s may be enough evidence. With respect to the latter, although $G_{LI}>G_{LS}$ is maintained for subgroup E in the majority of developing economies, the differences between subgroups are

employment increases *pari pasu*. One may say that if the market mechanism operates in such a way that it tends to equalize the marginal product of labor, then G_{LS} would become smaller while G_{YS} would increase. Of course, the situation may vary among the services subsectors, but this would be the case for some of the subsectors in services. However, our attention is drawn to the fact that the sectoral level of Y/L for services is much higher than that for industry in group I. Although the difference between services and industry goes down with groups of higher Y/L levels, even for group IV, the sectoral level in a number of countries is still greater for services than for industry. This is directly relevant to the present discussion. The data suggest that at least at the sectoral aggregate level, the marginal productivity of labor in services is higher than that of industry. Thus, as an overview in comparison with the goods-producing sector, we emphasize these characteristics of the employment function of the services sector.

Does the above interpretation of productivity growth of the services sector imply that further discussion is unnecessary? The answer is no. The actual performance of G_{yS} deserves attention, particularly its relationship with the rate of employment change. And this is the final point. With respect to this relationship, what we have characterized as a "mixed" pattern implies that according to the given conditions, the pattern can be either one that is a reversal or a concurrent pattern, for which evidence is given in Table 4.7, Panel (B).

In terms of the group average, for groups II and IV, a distinct decline in G_{yS} is associated with a distinct rise in G_{LS}. For group III, G_{yS} increases with a minor decline in G_{LS}. Thus, these ratios exhibit a reversal pattern. However, we see a concurrent pattern for group V, though the decline in G_{LS} is minor. By subgroup, the subgroups in groups II and IV show a reversal pattern, but for group III, a concurrent pattern is recognized for subgroup E and a reversal pattern is exhibited for subgroup F. The reversal pattern for group III is a combined result of these patterns. As for group V, we contend that there is virtually no difference between subgroups as the change in G_{LS} is at any rate minor. Taken together, a reversal growth relationship is recognized as the dominant pattern in the majority of developing economies. However, we should hold some reservations for group III, insofar as the changes over time are concerned. With respect to the cross-sectional data, it is easy to see

---

negligible for groups III and IV despite remarkable differences in productivity growth rates between services and industry (Table 4.8).

that comparison by subgroups in both periods exhibits a concurrent growth relationship between G_{yS} and G_{LS} in each group. However, the notable exception of group III cannot be ignored. In this group, a reversal relationship is clear. This is significant relative to what we found earlier about the over-time performance of this particular group.

The challenge for us is how to interpret the different patterns, i.e., the reversal pattern in the over-time changes vs. the concurrent pattern in the cross-sectional data, preferably without ignoring the exceptional cases. All that can be said is that in the services sector, the rate of productivity growth is essentially determined as a combined result of G_{YS} and G_{LS}. The observed performance of these two growth rates from the combined cross-sectional and time-series data suggests the continuous operation of the two factors: one factor which produces the concurrent pattern and the other factor which produces the reversal pattern. The actual direction that is taken is determined by the various given conditions which cannot be specified.

Productivity Growth in Two-sector Comparisons

Before moving towards integration, let us make some two-sector comparisons in this section. Though sketchy, the exercise focuses on agriculture–nonagriculture (A–N) and agriculture-industry (A–I). These two-sector comparisons are, so to speak, an intermediary approach. The purpose of these comparisons is twofold: (1) to expand on the findings of the share approach in Chapter 3, and (2) to further clarify the intersectoral relationship. For the first comparison, the trend of convergence or divergence is the issue, whereas in the second comparison, we are particularly concerned with the significance of productivity growth in agriculture in a sectorwise context. We focus our attention on the cross-sectional pattern, which is shown by the data for the average of these two periods in Table 4.9, where the relevant terms are reproduced for convenience.

First, as is directly shown by the difference in productivity growth rates between agriculture and nonagriculture, the growth rate of agriculture is higher than that of nonagriculture—that is, $G_{yA}>G_{yN}$—for the majority of groups. This trend seems to be strong as it is also exhibited across subgroups. The technological progress in agriculture and the productivity performance of services both play an important role in forming the pattern $G_{yA}>G_{yN}$. In Chapter 3, the importance of the A–N approach through historical reasoning was emphasized. The trends found in the share approach

Table 4.9
Productivity Growth [a]***: Two-sector Comparisons of 1960–1980 Average (annual average rate, percentage)***

Group	Subgroup	G_{yA}	G_{yI}	G_{yS}	G_{yN}	G_{yA}-G_{yN}	G_{yA}-G_{yI}
I		0.4	0.5	-0.5	-0.3	0.7	-0.1
II		1.4	3.5	1.9	2.5	-1.1	-2.1
	E	2.4	3.4	2.0	2.8	-0.4	-1.0
	F	0.3	2.5	1.7	2.3	-2.0	-2.2
III		3.3	3.6	1.7	2.4	0.9	-0.3
	E	4.1	4.5	2.4	3.2	0.9	-0.4
	F	2.6	2.6	1.0	1.6	1.0	0.0
IV		3.5	3.6	2.5	3.0	0.0	-0.1
	E	4.6	4.8	2.8	3.7	0.9	-0.2
	F	2.4	2.4	2.2	2.2	1.7	0.0
V		4.0	3.8	1.6	2.5	1.5	0.2
	E	4.3	4.1	2.2	3.1	1.2	0.2
	F	3.6	3.5	1.0	1.9	1.7	0.1

Note: a. G_{yA} is from Table 4.5, G_{yI} is from Table 4.6, and G_{yS} is from Table 4.7. G_{yN} is calculated as the sum of $w_I G_{yI}$ and $w_S G_{yS}$, where w_I and w_S are output shares of industry and services, respectively.

Sources: See sources of Table 4.5.

are reconfirmed here: that is, there is divergence in the lower Y/L groups and convergence in the higher Y/L groups. This implies that the dominant factor is productivity and that the effects of relative price changes are limited to a certain extent.

Second, as is shown by comparing G_{yA} and G_{yI}, the rate of productivity growth in industry is sustained at a slightly higher rate than that of agriculture through groups II, III, and IV, followed by a reverse pattern in group V. As for the majority of developing economies, forces seem to operate such that G_{yA} and G_{yI} are nearly parallel with one another. At lower Y/L levels, the trend is weaker but at higher Y/L levels, i.e., groups III–IV, the trend is stronger. By subgroup, the rate of productivity growth of industry (G_{yI}) is definitely higher in subgroup E as compared to subgroup F. The rate of productivity growth of agriculture (G_{yA}) follows exactly the same pattern as that of industry in terms of comparison between the two subgroups. Except for group II, the difference between G_{yA} and G_{yI} generally appears to be small, suggesting that the two growth rates move parallel to one another.

This raises a problem that pertains to a two-sector approach of A-I in terms of both theory and measurement (this was touched upon in the preceding section). Since classical doctrine a lá David Ricardo,[88] the pace of technological progress has long been supposed to proceed behind that of industry in the long-term perspective. This implies a sustained trend of $G_{yI}>G_{yA}$. Actually, and as will be discussed later in the section that follows, the historical records of industrial nations generally follow this classical hypothesis. However, in our view, the postwar growth experiences demand a reexamination of this notion.

In this regard, the postwar agricultural revolution that was described earlier should be reconsidered in relation to postwar industrialization. The hypothesis of borrowed technology is often thought to apply to the industrial sector, but we emphasized its significance for agriculture as well. The international diffusion of advanced technological knowledge enabled the near equality of G_{yA} and G_{yI} despite the different types of technology in agriculture and industry. The classic notion of the relative backwardness of the speed of technological advance in agriculture thus is out of tune with reality. The problem can be seen in relation to industrial growth.

How can we interpret the almost parallel growth of industry and agriculture in groups III–V, and the pattern of $G_{yA}>G_{yI}$ in subgroup comparisons in the same groups? The answer is often given by analysis of the intersectoral relationship whereby industrialization affects agricultural development substantially, particularly by increasing supplies of modern intermediate goods for agricultural production. This seems to be an acceptable view. If we also accept this view, then in our subgroup framework, the different intensity of intersectoral flows of intermediate goods between subgroups E and F should provide the answer. The effects of the different intensity of intersectoral flows should primarily appear in differences in the rate of output growth of agriculture (G_{YA}) although no direct evidence is available. The data shown in Table

88 For the original notion, see David Ricardo, "An Essay on the Influence of a Low Price of Corn on the Profits of Stock," (1815), and "On Protection to Agriculture," in Piero Staffa, ed., *The Works and Correspondence of David Ricardo* (Cambridge: Cambridge University Press, 1966).

4.5 appear to roughly endorse this view. Another aspect relates to the effects of changes in the labor force engaged in agriculture. This aspect appears to be more critical. From the data of G_{LA} in Table 4.5, we can see that the difference is considerable: the pressure of overengagement (i.e., underemployment) of labor is noticeably weaker in subgroup E than in subgroup F due to the greater outflow of labor from agriculture to nonagriculture in subgroup E. Taking the average of the two periods under review, the difference between G_{LA} and G_{LN} is 1.0–1.3 percentage points. As a matter of fact, the pattern of $G_{yA}>G_{yI}$ for group IV and especially group V can be explained to a large extent by the higher rate of decrease in the labor force engaged in agriculture.

On the other hand, as explained in the preceding section, the rate of productivity growth of industry (G_{yI}) cannot be the factor responsible for forming the postwar pattern of $G_{yA} = G_{yI}$ because G_{yI} is higher in the postwar period as compared to the historical path of currently developed nations. However, the real problem pertains to the initial level of G_{yI}. Maintainance of the equality $G_{yA} = G_{yI}$ for developing economies cannot narrow down the wide sectoral inequality between agriculture and industry (that was identified in Chapter 3). In fact, even a moderate $G_{yA}>G_{yI}$ will not be effective in narrowing the initial gap. This symbolizes the "dilemma" of the postwar developing countries which emanates from their production structure characteristics. It is a dilemma because the policy of suppressing G_{yI} for the sake of achieving $G_{yA}>G_{yI}$ is not realistic. The magnitude of this gap may be greater for nations in the earlier phases of development and may be smaller for nations in the later phases of development (for an explanation of the phases, see the section that follows), although typological differences are an influential factor. This is not the result of a phase shift, but is due to differences in the initial inequality where the inequality is higher for the lower Y/L levels and lower for the higher Y/L levels. Nevertheless, taken as a whole, this dilemma cannot be ignored.

III. Integration Towards An Aggregate Pattern: Phasing

Detailed observation of the sectoral growth performances and elucidation of their characteristics are summarized in the integration of the aggregate performance of developing economies. It is our view that the long-term changes in the production structure that has been discussed can be integrated in terms of the major phases of development. This is done conceptually in terms of the shift in the production structure and empirically with the use of cross-sectional data that are supplemented by time-series data. Convergent vs. divergent performance, which was the subject of Chapter 3, can be incorporated into the analysis, separate from the changes in relative output prices which will be discussed in Chapter 5.

Demarcation of the Development Phases

The phase approach is a conceptual means of interpreting the changes in the pattern and the mechanism of economic development in a historical perspective. A certain phase is defined as a segment or interval in the long-term development path, characterized by a certain pattern and mechanism, that is distinguished from other segments. In general, with respect to the criteria and method for the demarcation and identification of the phases, no rigid requirement is given a priori. Depending upon the purpose of the analysis, different criteria may be chosen.[89] In this study, the productivity structure is the basic criterion. The aim is to confirm the nonlinear pattern that was discussed earlier in the analysis of the cross-sectional data which suggested possible phase demarcation of the development path in the aggregate dimension.

Compared to phasing in terms of changes in the trade structure which is treated in Chapter 6, phasing in terms of changes in the domestic economic structure has thus far been less developed. The discussion that follows is intended to be systematic in integrating the analytical results of Chapters 3 and 4. To do this, the most convenient indicator is obtained from the sectoral growth relationship between output and employment, evaluated and

---

89 For details, see Ohkawa and Kohama, *Lectures on Developing Economies,* Part I. Note, however, that our concept of phases differs from "stages." Kuznets criticized Rostow's "take-off" in terms of his five requirements in his theory of stages (W. W. Rostow, ed., "Notes on the Take-off," Chapter 2, *The Economics of Take-off into Sustained Growth,* (London: Macmillan, 1963). These requirements are often understood to apply to phases as well. However, we believe that these requirements are not necessarily indispensable for applying the notion of phases. Of course, requirements such as the distinction of phases in terms of empirically testable characteristics and identification of phases by testable terms in demarcating units or segments are essential for phasing, but we do not require a certain deterministic sequence for the pattern and mechanism of the growth path.

quantified from the viewpoint of aggregate productivity growth.[90]

This sectoral decomposition of aggregate productivity performance identifies each sector's contribution in forming aggregate performance. For example, an increased G_{yA} in group V contributes towards a direction of convergence; but the share of agriculture decreases to a very small magnitude in this group, both for output and labor, and its contribution to aggregate performance thus becomes minor.

Using shares as the weights for the growth rate of sectoral output and labor, we can measure sectoral productivity growth as components of aggregate performance. The formula for the three sectors is:

(1) $G_Y - G_L = (w_A G_{YA} - v_A G_{LA}) + (w_I Y_I - v_I G_{LI}) + (w_S Y_S - v_S G_{LS})$ [91]

Table 4.10 summarizes the relevant data which are reproduced from preceding tables. Attention is paid to the sectoral performances through the cross-sectional groups, taking into consideration changes over time. The characteristics of each sector are represented by two weighted terms: output growth and change in employment. Although most of the sectoral performance patterns are known from the preceding examinations, Panels (A) and (B) present a summary of the patterns in the two periods.

In Panel (C), the difference indicator between the output and employment terms is shown. The value of D_I for the average of the two periods exhibits a linear trend of productivity growth acceleration throughout all groups (from 0.5 to 1.5). If the growth path of developing economies is supposed to be dominated by industrialization, then D_I should show a linear acceleration. There may be no need to demarcate segments or intervals if the patterns

---

90 This does not necessarily reject phase demarcation using the pattern of output growth or employment performance. In fact, the point where the output shares of industry (or manufacturing) and agriculture are equal is used to set the target of the macro-development plan of the Thai government. Another example is the use of the output share of industry (say, 20 percent) by UNIDO. These criteria indicate that a notion of phase is required, though partially. Employment is touched upon later. Here the factors are integrated in a systematic framework.

91 This formula is related to the preceding formula for the ratio approach as follows, using the illustration of agriculture:

$$w_A G_{yA} = DY_A/Y_A \cdot Y_A/Y = DY_A/Y; \; v_A G_{LA} = DL_A/L_A \cdot L_A/L$$

while in the ratio approach we have:

$$Y_A/L_A \,/\, Y/L \text{ or } Y_A/Y \,/\, L_A/L$$

and mechanism are already distinguished from one another. However, such a supposition is far from reality. In fact, the pattern for services (D_S) appears to show a contrary tendency of deceleration (except for group I to group II), while the difference indicator of agriculture, D_A, almost follows a trend acceleration but with a decline in group IV. Furthermore, with respect to the two periods, the linear acceleration of industry, which is distinct for the 1960s, is not clear for the 1970s. This may illustrate the issue of changes in the role played by the industrial sector over time.

The degree of industrialization is often measured in terms of its output share in the aggregate (i.e., w_I in Panels A and B). Indeed, w_I shows a clear trend of increase in both periods. In addition, from 1960–1970 to 1970–1980, the share increases over time in all groups. Only one exception is found for group V; the reason for this was explained earlier. The labor share of the industrial sector, v_I, behaves likewise. Changes in this share over time have often been used to demarcate intervals in the development path. While the use of labor and output shares are simple and convenient, they cannot be used with a systematic explanation. We believe that industrialization is a driving force in economic development, but we also believe that its function should be interpreted from the standpoint of aggregate growth, in particular in terms of the relationship between productivity growth and labor employment reallocation (that is, the changes in the production structure which were defined earlier). For the following three intervals or segments of groups in the cross-sectional data, the difference (D) is observed as the indicator of structural change:

Primary phase—Groups I and II
Secondary phase—Groups III and IV
Developed phase—Group V

From the average magnitude, which is shown in Panel (C), the primary interval is characterized by a sharp divergence. The average indicator in a N–A (N = I+S) two-sector framework is -0.3 of A vs. 0.7 of N in group I, and -0.3 of A vs. 3.1 of N in group II. A negative value for agriculture vs. a positive value for nonagriculture characterizes these two groups, and is observed as a sustained pattern. The negative value of the indicator of agriculture ($v_A G_{LA} > w_A G_{YA}$) means that the effect of $v_A > w_A$ is greater than that of $G_{YA} > G_{LA}$. This has substantive implications beyond statistical measurement, as was noted earlier. From the cross-sectional data, we can suppose a path of development with acceptable reasoning. Any developing economy must pass through

Table 4.10
Sectoral Contribution to Aggregate Productivity Growth [a]
(average annual growth, percentage)

(A) 1960-1970

Group	Agriculture			Industry			Services			Sum
	G_{YA}	w_A	w_AG_{YA}	G_{YI}	w_I	w_IG_{YI}	G_{YS}	w_S	w_SG_{YS}	
I	2.0	52.5	1.1	6.3	13.0	0.8	3.6	34.9	1.3	3.2
II	3.3	33.5	1.1	8.4	21.0	1.8	6.6	45.4	3.0	5.9
III	4.6	29.4	1.4	8.1	23.4	1.9	5.8	47.2	2.7	6.0
IV	3.2	19.2	0.6	7.6	31.2	2.4	5.7	49.6	2.8	5.8
V	1.3	6.2	0.1	5.9	39.3	2.3	5.3	54.6	2.9	5.3
	G_{LA}	v_A	v_AG_{LA}	G_{LI}	v_I	v_IG_{LI}	G_{LS}	v_S	v_SG_{LS}	Sum
I	1.5	85.2	1.3	4.4	5.3	0.2	4.5	9.5	0.4	1.9
II	1.6	77.5	1.2	4.5	10.2	0.5	4.1	12.3	0.5	2.2
III	1.1	55.7	0.6	4.3	16.6	0.7	4.8	27.7	1.3	2.6
IV	-1.0	41.6	-0.4	2.7	23.8	0.6	3.3	34.6	1.1	1.3
V	-4.0	9.4	-0.4	0.7	40.9	0.3	2.8	49.7	1.4	1.3

(B) 1970-1980

Group	Agriculture			Industry			Services			Sum
	G_{YA}	w_A	w_AG_{YA}	G_{YI}	w_I	w_IG_{YI}	G_{YS}	w_S	w_SG_{YS}	
I	1.4	46.9	0.6	4.1	16.3	0.7	5.2	36.8	1.9	3.2
II	3.0	29.7	0.9	5.0	24.8	1.2	5.9	45.5	2.7	4.8
III	4.3	25.1	1.1	7.8	27.9	2.2	7.0	47.0	3.2	6.5
IV	2.2	14.7	0.3	5.3	35.5	2.0	6.1	49.8	3.1	5.4
V	0.8	3.7	0.03	2.9	37.4	1.1	3.5	58.9	2.1	3.2
	G_{LA}	v_A	v_AG_{LA}	G_{LI}	v_I	v_IG_{LI}	G_{LS}	v_S	v_SG_{LS}	Sum
I	1.3	81.4	1.0	5.0	6.6	0.3	5.3	12.0	0.6	1.9
II	2.0	68.1	1.4	3.1	10.7	0.3	4.7	21.2	1.0	2.7
III	1.1	47.3	0.5	5.2	21.2	1.1	4.6	31.5	1.6	3.2
IV	-0.4	33.2	-0.1	3.4	26.2	0.9	4.2	40.6	1.7	2.5
V	-1.9	3.3	-0.05	0.6	39.6	0.2	2.8	57.1	1.6	1.7

(C) Difference Indicators of the Groups and Subgroups

	1960-1970			1970-1980			Average		
	D_A	D_I	D_S	D_A	D_I	D_S	D_A	D_I	D_S
I Av.	-0.21	0.61	0.90	-0.38	0.35	1.30	-0.30	0.48	1.10

II	E	0.10	1.42	2.58	-0.05	1.25	1.78	0.03	1.35	2.18
	F	-0.30	1.06	2.17	-0.92	0.60	1.52	-0.61	0.83	1.85
	Av.	-0.09	1.25	2.42	-0.49	0.92	1.67	-0.29	1.09	2.05
III	E	1.06	1.34	1.70	1.06	1.22	1.68	1.06	1.28	1.69
	F	0.45	1.10	0.98	0.15	1.00	1.33	0.30	1.05	1.16
	Av.	0.74	1.22	1.34	0.62	1.11	1.52	0.68	1.16	1.43
IV	E	1.29	2.23	1.98	0.74	1.53	1.36	1.02	1.88	1.67
	F	0.62	1.45	1.44	0.16	0.54	1.52	0.39	1.00	1.47
	Av.	0.94	1.84	1.71	0.45	1.04	1.44	0.70	1.44	1.58
V	E	0.81	2.56	1.96	0.61	1.52	0.64	0.71	2.04	1.30
	F	0.43	1.68	1.08	0.42	0.28	0.26	0.43	0.98	0.67
	Av.	0.62	2.12	1.52	0.51	0.90	0.45	0.57	1.51	0.98

Note: a. The data in Panels (A) and (B) do not necessarily coincide with data in Panel (C) due to rounding.

the point where $v_A G_{LA} = w_A G_{YA}$. However, the trend of divergence comes to an end at this point. This is defined as the turning point from the primary segment to the secondary segment of the development path. In linking with the cross-sectional data, we can say that the turning point is located somewhere in the interval between groups II and III.[92]

The secondary segment is characterized by a trend of convergence in the A–N framework. First, the difference indicator of agriculture is definitely positive for groups III and IV. Using the average of the two periods, D_A is almost 0.7 for both groups. In the 1960s, D_A is 0.7 for group III and 0.9 for group IV, and in the 1970s, it

92 The equation $v_A G_{LA} = w_A G_{YA}$ implies a certain relationship in the productivity increase in agriculture to the aggregate Y/L level. In particular, $\Delta Y_A/Y_A \cdot Y_A/Y = \Delta L_A/L_A \cdot L_A/L$ is converted to $\Delta Y_A/\Delta L_A \cdot 1/Y = 1/L$ or $\Delta Y_A/\Delta L_A = Y/L$. At this point, an incremental increase in the product per worker in agriculture is equal to the incremental increase in product per worker in the aggregate.

For simplicity, observation by subgrouping is not tried here. Nevertheless, what follows may help us to understand the implication of linking cross-sectional with time-series phasing in this context. The indicator and components for subgroups E and F in groups II and III will be shown later. A turning point comes earlier for subgroup E and subgroup F lags behind. An average treatment here covers such a wide range of variance.

is 0.5–0.6 for the two groups. The stable sustained positive value of D_A over time is witnessed for these two groups. Second, in contrast, the indicator of the nonagricultural sector (I+S) in group II is smaller in the 1970s than that of the average for the two periods. Therefore, the value of the difference (i.e., the value of the N indicator less the value of the A indicator) is definitely smaller for groups III and IV as compared to group II which corresponds to the latter part of the primary interval. For example, the value of the indicator is 2.5 for group III and 2.3 for group IV vs. 3.4 for the average of the groups, though with some variations between the two periods.

The contrary roles played by industry and services within the nonagricultural sector are worthy of note: on average D_I increases while D_S decreases, though the trend in D_S appears somewhat blurred in both periods. Yet, it may be safe to say first that a trend of convergence in the A–N framework is brought forth by an increase in the A indicator and a decrease in the N indicator. Second, the decrease in the N indicator is a combined result of an increase in D_I and a decrease in D_S. Throughout these changes, however, we note that in both groups III and IV, the indicator pattern of S>I is generally sustained. Thus, the second pivot point towards the developed economies can be defined in this regard and it can be demarcated by an equality of the two indicators, that is, $D_I = D_S$.[93] After the economy passes through this point, a reversed inequality $D_I > D_S$ becomes the normal pattern. The cross-sectional data provide evidence for this view. For the average of the two periods, group V gives the relationship of the indicators between industry and services as 1.5 for I vs. 1.0 for S.

Taking these together, the cross-sectional data suggest that a developing economy has to pass through the second turning point, thus defined, somewhere between groups IV and V, if it is to succeed in shifting to a developed phase. The important implication of this new inequality is a greater rate of productivity growth in I than in S with a greater rate of employment increase in S than in I. Similarly, as in the case of agriculture, such an average pattern implies a range of variation, and the range is wider in the I–S relationship. Therefore, examination by subgrouping is important.

---

93 The equation $w_I G_{YI} - v_I G_{LI} = w_S G_{YS} - v_S G_{LS}$ can easily be converted to $(\Delta Y_I - \Delta Y_S)/(\Delta L_I/\Delta L_S) = Y/L$, as in the previous case of agriculture. In this case, however, at the incremental level, the ratio of the difference of product to that of labor is to be equal to the Y/L level.

The data are arranged in Panel (C) for the majority of developing nations. With this simple table, the need for a flexible application of the development intervals as was demarcated above is illustrated. As for the difference indicator, the inequality $D_S > D_I$ is clear for both subgroups E and F in groups II and III. In the case of subgroup E, however, D_I becomes closer to D_S in group III, and $D_I > D_S$ takes place in group IV. In the case of subgroup F, $D_S > D_I$ is still clear in group III and the difference is wider in group IV. These observations suggest on the one hand, that most vigorous developing economies may arrive at the second turning point at the Y/L level of group IV or even of group III. On the other hand, most backward developing economies cannot reach the turning point even after having arrived at the Y/L level of group IV.

The phases of development are thus demarcated by the two turning points between the primary and secondary phases, and the secondary and developed phases. This is our preliminary conclusion insofar as postwar development is concerned.

Characteristics of Phases and the Possibility of a Shift

In this section, the characteristics of each phase and the possibility of shifting from one phase to the other are discussed. As a preliminary note, the relationship between the inverse U-shape pattern that was identified earlier and the phase demarcation is explained. A consistent interpretation is possible in linking the two terms, output and labor, by sector. The sum of the two weighted terms in terms of growth rates are as follows (the data refer to the average for 1960–1980):

	I	II	III	IV	V
Output	5.2	5.4	6.3	5.6	4.3
Labor	1.9	2.5	2.8	2.3	1.5
Difference	3.3	2.9	3.5	3.3	2.8

The inverse U-shape is clearly identified for output, labor, and the difference, respectively. As is shown in Panel (A), Table 4.10, the output growth pattern in the 1960s is slightly blurred. Yet, we believe this is the basic pattern that is sustained throughout the two decades under review. Earlier discussions of this pattern have been partial and have identified the pattern from several specific points of view. Now, in a broad framework, the pattern seems to be consistently integrated with the phase demarcation in terms of the changes in the productivity structure. The pattern thus derived from the cross-sectional data confirms what was suggested earlier. The first point was that the shift from the primary to the secondary phase must come between the Y/L level of groups II and III. The second point was that the shift from the secondary phase to the phase of a developed economy must come between the Y/L level of groups IV and V. We are saying that the cross-sectional grouping at the macro-level cannot directly demarcate development phases, but it does provide useful knowledge for implying the pattern of residual growth found in Chapter 1.

Moving to the major topic, let us elucidate on the phase characteristics. The primary phase is characterized by the sustained negative performance of agricultural productivity growth in terms of its contribution to aggregate productivity growth. This situation continues throughout the primary phase. Of course, productivity does increase in agriculture, but its contribution becomes zero when the economy arrives at the turning point. As for the early interval of development, it is often asserted that "agricultural development is the basic requirement." Our analysis thus clarifies the basic reason for adopting this strategy.

The secondary phase is featured by an accelerated path of industrialization that is accompanied by a sustained positive contribution of agriculture to aggregate growth. The secondary phase is clearly distinguished from the primary phase, yet the contribution of the industrial sector to the aggregate productivity growth still remains smaller than that of services (which is also a property of the primary phase). Of course, the contribution of industry continues to increase relative to services during this phase. However, until the economy arrives at the second turning point, the inequality $D_S > D_I$ does not disappear. The term "semi-industrialization" is often used to denote this characteristic. Our analysis clarifies this characteristic in light of changes in the sectoral productivity structure of the postwar developing economies.

The entire growth path of developing economies is composed of these two major phases. The developed phase is characterized by a greater contribution of the industrial sector as compared to that of services. It is important to note that this phenomenon takes place despite the new activities which emerge in services.

These points are simple statements we can make regarding phase characterization. Discussion of the characteristics of each phase can be expanded by comparing the output and labor performances of industry and services for the average of the two periods using cross-sectional data. During the primary phase, the output component ($w_I G_{YI}$, $w_S G_{YS}$) increases rapidly from group I to group II both in industry (0.8 to 1.6) and services (1.6 to 2.9), but during the secondary phase, performance of output is almost stable between groups III and IV both in industry (2.1 to 2.2) and services (3.0 to 2.9). The labor component ($v_I G_{LI}$, $v_S G_{LS}$) during the primary phase shows a moderate increase from group I to group II for both industry (0.3 to 0.4) and services (0.5 to 0.8). In the secondary phase, the performance of labor decreases slightly in industry (0.9 to 0.6), while it is kept stable in services (1.4). Taken together, the cross-sectional data suggest that both components change rather moderately, or remain almost stable.

These trends are in sharp contrast to the performances of output and labor observed for the interval from group II to group III and for the interval from group IV to group V. Between these groups, distinct changes are often witnessed, though in some cases the pattern is stable. In the interval from group II to group III, the output component increases in industry (1.6 to 2.1) but remains stable in services (3.0). The labor component increases in both industry (0.4 to 0.9) and services (0.5 to 0.8). In the interval between groups IV and V, the output component decreases in both industry (2.2 to 1.7) and services (2.9 to 2.5). The labor component decreases in industry (0.6 to 0.3) and is relatively stable in services (1.4 to 1.5).

These observations may support the rationale for our phase demarcation and characterization. The economy in each phase tends to sustain a relatively stable performance, as characterized earlier, although the components of output and labor perform somewhat differently. Beyond that, with respect to the specific intervals, i.e., between groups II and III and groups IV and V, we see the location of the point of turn: the possibility of making shifts is suggested by these observations from the cross-sectional data.[94]

---

94 Particular attention is paid here to the phasing in terms of identifying the so-called turning point in shifting the mechanism of operation in the labor market. Without dealing with real wage behavior, it is impossible to deal with the labor market mechanism directly, but we believe that the economy may arrive at this turning point before it reaches the second point of turn to the third phase. This implies that the two turning points will not occur concurrently. The reason for this is as follows. As described earlier, industrialization is accelerated by an increase in the labor component to group III. Labor absorption increases as a proportion of G_L, the rate of increase in the total labor supply: from 16.0 in group II to 31.0 in group III. From group III to group IV, labor absorption also increases but not by much (the proportion is 41.6 in group IV). The labor engaged in agriculture decreases in group IV and this is brought about by a much higher and sustained increase in the proportion of labor engaged in

Now, let us discuss the crucial issue of the possibility of shifting from one phase to another. Without shifts from one phase to another, no development is possible in the long term. As stated in Chapter 1, our overall view is that the upgrading of a nation's level of social capability, which arises from the augmenting of its components, makes such a shift possible. We want to illustrate this proposal by providing certain empirical evidence using the results of our subgroup examination. Different time dimensions for making the phase shifts with respect to certain subgroups—that is, an "earlier" shift for subgroup E and a "later" shift for subgroup F—may be suggested from the data (such a possibility was suggested from earlier discussions as well).

With regard to the first turning point, the relevant data from Table 4.10 are arranged as follows (data are in percentage terms):

Group	Subgroup	$w_A G_{YA}$	$v_A G_{LA}$	D_A
II	E	1.3	1.2	0.1
	F	0.6	1.2	-0.6
III	E	1.4	0.3	1.1
	F	1.2	0.9	0.3

For subgroup E in group II, D_A becomes nearly zero. Recall that the first turning point was suggested to be somewhere between groups II and III. Therefore, subgroup E's arrival at this point is "earlier." In the case of subgroup F, the turning point is reached at the Y/L level of group III as indicated by the near-zero value of D_A. Thus, for subgroup F the turning point occurs "later." The average level of product per worker, Y/L, is $485 for group II and $874 for group III in 1970 (see Table 1.1). The difference in the Y/L level between the two groups is surprisingly large. Yet, the shift in the phases is not determined by the Y/L level; rather, differences in the nations' capability is the crucial factor. This difference in capability is

---

the services sector: the proportion of labor in services increases from 32.0 in group II to 73.7 in group IV. Demand for services is certainly more strongly sustained than for industry. We believe that industrialization of the labor-using type is an indispensable factor in bringing the economy to the turning point of changing the labor market mechanism. However, we believe the function of services cannot be ignored, although qualification should be made for underemployment in the data used above. As mentioned earlier, the time of arriving at the point of turn should be flexible. The same may be supposed for the labor-wage dimension.

however, is too long and a breakdown is not possible. However, the second turning point is believed to have been reached by these nations, including Italy, in subsequent years, as we see a reversed inequality, $D_I>D_S$. Thus $D_S>D_I$ was not sustained for a long period of time.

Table 4.12 summarizes the sectoral indicators for Japan for which the data are continuously available. As described earlier in Chapter 2, elimination of the effects of long-swings is indispensable and the intervals are demarcated to accommodate this, excluding the war years. With regard to D_A, no negative values are found in Japan, unlike the case of Italy, because of the very moderate increase in the labor force engaged in agriculture. The primary phase cannot be applied to nations whose positive value of G_{LA} is small. Similar cases are generally expected for Western nations, and the case of Italy may not be representative. Thus, between developing and developed economies, a basic difference is recognized in the initial interval.

In addition, in the initial interval of Japan (i.e., 1887–1897), $D_S>D_I$ is, like the case of Italy, very clear. However, this inequality is not sustained in the subsequent years. In 1897–1904, the inequality it is still sustained but in 1904–1919, D_S is very close to D_I. Occurrence of this equality in D_S and D_I appears surprisingly early. Since 1919, the $D_I > D_S$ (into 1919–1938 and 1955–1965) continues as Japan goes towards the end of its development path. Is Japan's case atypical? We do not think so. Japan does represent a strong case of the "earlier type" which we defined in the analysis of the postwar developing economies. If adjustment is made to include transportation and communication in the services sector, the point of equality, $D_S=D_I$, must be identified in the early part of the period 1919–1938. In a rough comparison with the postwar cross-sectional data, this point of equality would be found close to Y/L levels corresponding with group III. The adjusted output share of services (w_S) of Japan must have been definitely smaller than the adjusted output share of services of groups III and IV on average over the two decades. Moreover, w_S changes a little over time in Japan, whereas in the cross-sectional data, w_S tends to increase through the groups. The examination of Western historical records arranged by Kuznets suggests that a number of similar cases can be found in Europe in

Table 4.11
Sectoral Indicators of the Historical Changes in Selected Western Nations [a] (percentage)

(A) Italy (Initial years)

	G_{YA}	w_A	$w_A G_{YA}$	G_{YI}	w_I	$w_I G_{YI}$	G_{YS}	w_S	$w_S G_{YS}$
1861-1870	0.33	50.8	0.17	1.55	21.2	0.33	1.98*	28.0	0.55
	G_{LA}	v_A	$v_A G_{LA}$	G_{LI}	v_I	$v_I G_{LI}$	G_{LS}	v_S	$v_S G_{LS}$
1891-1900	0.60	57.0	0.34	0.75	27.0	0.20	0.42	16.0	0.07
Difference			-0.17			0.13			0.48

(B) Selected Nations [b] (Long term)

Italy	G_{YI}	w_I	$w_I G_{YI}$	G_{YS}	w_S	$w_S G_{YS}$
1895-1965	3.64	39.6	1.44	2.63	34.9	0.92
	G_{LI}	v_I	$v_I G_{LI}$	G_{LS}	v_S	$v_S G_{LS}$
	1.07	37.0	0.40	1.16	22.0	0.26
Difference			1.04			0.66
United Kingdom	G_{YI}	w_I	$w_I G_{YI}$	G_{YS}	w_S	$w_S G_{YS}$
1924-1963/69	2.16	55.7	1.20	2.47	40.3	1.00
	G_{LI}	v_I	$v_I G_{LI}$	G_{LS}	v_S	$v_S G_{LS}$
	0.48	56.9	0.27	1.28	36.7	0.47
Difference			0.93			0.53
France	G_{YI}	w_I	$w_I G_{YI}$	G_{YS}	w_S	$w_S G_{YS}$
1896-1963	2.44	48.6	1.19	2.54	37.7	0.96
	G_{LI}	v_I	$v_I G_{LI}$	G_{LS}	v_S	$v_S G_{LS}$
	0.47	37.5	0.18	0.63	30.0	0.19
Difference			1.01			0.77
United States	G_{YI}	w_I	$w_I G_{YI}$	G_{YS}	w_S	$w_S G_{YS}$
1919/29-1963/67	3.31	42.8	1.42	3.85	50.0	1.93
	G_{LI}	v_I	$v_I G_{LI}$	G_{LS}	v_S	$v_S G_{LS}$
	1.21	38.4	0.46	2.14	48.8	1.04
Difference			0.96			0.89

Notes: a. The selection of the intervals depended upon the availability of data for the shares of output and labor.
b. Panel (B) excludes agriculture for simplicity as this issue is not examined for these countries.

Sources: Kuznets, *Economic Growth of Nations,* Table 21 and Table 38; Maddison, *Phases of Capitalist Development,* Table A8 and Table C8.

terms of both percentages and trends.[96] Japan represents the case of a latecomer within the currently developed countries. In the postwar cross-sectional data of developing nations, a greater and increasing value of w_S characterizes their performance. This is one of the main phenomena recognized concerning the time of arrival at the second turning point.

By way of conclusion, we would like to add the following. First, as for methodology, our framework of phasing which is derived from the use of cross-sectional data can be applied in principle to the historical path of currently developed nations as well, if it is interpreted in a flexible way so as to consider the different historical conditions. In other words, a common framework of phasing can help to understand the similarities and dissimilarities of the performance of structural change between the two cases. Second, in order to do this, however, the phasing that was discussed earlier needs modification. More specifically, a three-phase framework is proposed for the developing economies: namely, the preprimary phase, the primary phase, and the secondary phase. The preprimary phase is characterized by a negative value for D_A, while the other two phases are the same as described earlier.

The preprimary phase is a common characteristic of the contemporary developing nations, whereas it is atypical for the historical path of currently developed nations. The primary phase is common to both cases, but tends to be sustained longer in the former than in the latter. The secondary phase generally starts earlier and is sustained longer in the latter group, whereas it comes much later for the developing nations. Here we see the discrepancy between "earlier" and "later" nations of the same Y/L level.

---

96 The initial percentage of w_S ranges between 31–39 for Denmark, Norway, Sweden, and Italy (transportation and communication are excluded and are directly comparable to Japan). The increasing trend over time is minor in these cases. These are latecomers in Europe (Kuznets, *Economic Growth of Nations,* Table 21:144–49).

Table 4.12
Sectoral Indicators of the Historical Changes in Japan[a] (percentage)

	Period				
	1887-1897	1897-1904	1904-1919	1919-1938	1955-1965
(A) Output					
G_{YA}	1.5	1.8	0.9	0.9	3.7
w_A	39.8	35.7	29.9	19.5	10.3
w_AG_{YA}	0.60	0.64	0.27	0.18	0.38
G_{YI}	3.9	5.3	7.1	7.0	18.1
w_I	13.0	17.7	25.1	44.3	42.8
w_IG_{YI}	0.51	0.94	1.78	3.10	7.75
G_{YS}	2.8	3.3	3.5	3.5	10.5
w_S	47.2	46.6	45.0	36.2	46.9
w_SG_{YS}	1.32	1.53	1.58	1.27	4.92
(B) Labor					
G_{LA}	0.1	-0.4	-0.1	0.20	-2.6
v_A	68.9	62.5	49.5	53.7	38.2
v_AG_{LA}	0.07	-0.25	-0.05	0.11	-0.99
G_{LI}	2.1	2.1	2.2	2.73	4.6
v_I	15.8	18.1	25.9	22.1	31.3
v_IG_{LI}	0.33	0.38	0.57	0.60	1.44
G_{LS}	2.1	2.1	2.0	1.81	3.1
v_S	15.3	19.4	20.0	24.2	30.5
v_SG_{LS}	0.32	0.41	0.40	0.44	0.95
(C) Difference Indicator					
Agriculture	0.59	0.89	0.32	0.07	1.37
Industry	0.18	0.56	1.21	2.50	6.31
Services	1.00	1.12	1.18	0.83	3.97

Note: a. The original estimates were revised for the present use. Data are based on a seven-year moving-average series (five-year moving-average series for postwar periods). In addition, the shares w and v in real terms are averaged for each interval.

Sources: Ohkawa et al., *LTES*, *National Income,* Vol. 1, Table 3.2:49, Table 3.3:50, and Table 25:227; Ohkawa and Kohama, *Lectures on Developing Economies,* Table 6.9.

Finally, we would like to offer the following remarks. On the one hand, a number of currently developing nations face difficulties in making a phase shift: first from the preprimary phase to the primary phase and from the primary phase to the secondary phase.

On the other hand, several developing nations did overcome these difficulties and made the successful shift from the preprimary phase to the primary phase and further to the secondary phase. Some of them even appear to be making the shift from the secondary phase toward the developed phase. The problem we face today is how to systematically recognize the wide range of differences in development phases which exists among the developing economies.

Epilogue: An Extension of the Phasing Concept

While this section duplicates parts of the main text, some readers may find it useful in gaining a clearer understanding of the phasing concept.

From Table 4.10, we see that on the one hand, the negative magnitude of D_A is greater and varies more widely in the 1960s as compared to the 1970s for groups I and II. In addition, the magnitude of $D_I<D_S$ is greater in the 1970s than in the 1960s for groups III and IV. On the other hand, with respect to group V, a shift to $D_I>D_S$ is clearly identified in both periods. Furthermore, as for the comparison of subgroups, the value of D_A is positive for subgroup E in group II in the 1960s but becomes negative in the 1970s. The magnitude of $D_I<D_S$ tends to be greater for subgroup F than for subgroup E for group III in both periods. The pattern of $D_I>D_S$ in subgroup E reverses to $D_I<D_S$ in group IV. As for group V, no difference is witnessed between subgroups with regard to the pattern shift.

Taking these together, we may say that the development path of shifting phases appears to be backwards from the 1960s to the 1970s insofar as the developing economies are concerned. With respect to the developed nations, as represented by group V, however, a normal shift pattern is identified, implying a forward structural change from full industrialization to the emergence of new activities in the services sector.

How can we interpret the appearance of a "backward" process in the developing economies? One may say that the 1960s is characterized by a most favorable international situation for development and hence the phase shift that is observed cannot be judged as a normal process and that the findings for the 1970s cannot therefore necessarily be labeled as being backward. Similarly, during the 1980s, perhaps further backward development can be expected. We recognize that this view is legitimate; as a matter of fact, the postwar experience is too short to permit a useful long-term phase demarcation in growth performance. In this sense, the term backward (or forward) should be used conditionally.

As was discussed in the latter part of this chapter, the validity of using the difference indicator for phase demarcation is endorsed by the historical experience of contemporary developed nations. Our assumption is that the pattern of phase shift should essentially be the same between developed and developing countries as they go through the long-term path of industrialization, although the case of contemporary developing nations has its own characteristic patterns. In this sense, our observations are also conditional, yet unavoidable, for the time being. Thus, our interpretation is as follows. The pattern shown by the 1960s indicates a forward process vs. a backward process of the 1970s. This is analogous to the patterns historically observed over the long-term path of industrialization in developed nations, in which "backward and forward" processes can also be identified. Here, the phase shift was not realized smoothly, though details cannot be discussed except for Japan. For convenience, the average of the two periods is thus shown in the table. This may generally indicate the forces that are operating to produce the phase shift that is seen over the two decades.

A second comment can be made of the characteristics of contemporary developing economies particularly for the agricultural and services sectors. First, a negative value of D_A is not exceptional for the developing economies, while it is found only in the initial phase of Italy in the case of developed nations. Because of the importance of this phase where $D_A<0$, we have proposed the demarcation of a preprimary phase. The validity of the preprimary phase is further endorsed by its larger negative magnitude and wider variation in the 1970s. This implies that the problem of realizing a shift to the primary phase in the developing nations should be understood as being quite a new challenge: the historical knowledge accumulated hitherto about developed nations cannot deal with this phase shift effectively.

Second, as repeatedly pointed out earlier, the significant role played by the services sector in contemporary developing economies draws special attention. In phasing, this important role of the services sector is represented by the difficulty countries face in making the turn from $D_I<D_S$ to $D_I>D_S$ through industrialization in competition with services. This is particularly witnessed in comparisons between groups III and IV and groups IV and V. From group III to group IV, the magnitude of the difference indicator declines sluggishly. The pattern of $D_I>D_S$ that is achieved by subgroup E of group IV in the 1960s disappears later, and in the group average, a remarkable gap still exists between group IV and V regarding the inequality of the difference indicators.

Third, the implications of these findings are suggestive, but can only be confirmed by applying the method to each individual nation as phasing is originally a historical analysis. Differences in the initial conditions, the typology, and country size would all help to bring forth dissimilarities in the time dimension of the phase shift. Because we are well aware of this, the average group observations were supplemented by analysis by subgroups; but even this may not be enough. Further analysis in terms of other criteria or groupings can be used. For example, using the abundance or lack of natural resources as the criterion may result in dissimilar patterns. Nevertheless, our primary intention here is to demonstrate that a general application of a common criterion to all the developing nations is possible and is useful in identifying major phases. In order to meet the objective, the indicator selected is simple so as to grasp the two inescapable points: $D_A = 0$ and $D_I = D_S$ which represent structural change in terms of product and labor combined.

Usually, a phase is demarcated in terms of changes in commodity trade and in Chapter 6, we will adopt this conventional approach in treating imports and exports. What about the relationship between the two approaches? It goes without saying that changes in the domestic production and changes in the pattern of trade are interrelated in all developing economies. However, what is needed is an analytical answer. Our proposal of phasing with respect to changes in the domestic production structure attempts to provide a systematic insight from a long-term perspective,[97] while the phasing of the trade structure pertains to a relatively short-term dimension often related to the option of relevant policies and measures. The two approaches are not exclusive but are rather complementary to one another. However, policy debates on import substitution vs. export promotion are apt to have been developed without paying due attention to the requirement of providing a long-term perspective.

We do not expect to see an exact matching of the dates of the phase shift from the two approaches, but a general correspondence is expected. According to the conventional demarcation approach using the trade pattern, five phases are typically derived: (1) the

97 A preliminary attempt was made in Ohkawa and Kohama, "Economic Development in Historical Perspective," Lecture 1, *Lectures on Developing Economies*. What was called the "industry approach" in that book is now revised and extended both in terms of concept and measurement.

primary (or traditional) product export phase, (2) the primary import-substitution phase, (3) the primary export substitution phase, (4) the secondary import-substitution phase, and (5) the secondary export substitution phase. For Japan, the date of phases (1) and (2) combined is 1870–1900, of phase (3) is 1900–1919, of phase (4) is 1919–1960, and of phase (5) is 1960–1975.[98] The difference indicators for changes in domestic productivity are reproduced from Table 4.12 as follows:

	1887–1897	1897–1904	1904–1919	1919–1934	1955–1965
D_A	0.59	0.89	0.32	0.07	1.37
D_I	0.18	0.56	1.21	2.50	6.31
D_S	1.00	1.12	1.18	0.83	3.93

Although data limitations do not permit an exact comparison of dates, we can illustrate the relationship between the two approaches. The first point where $D_A = 0$ is not identified (as stated earlier). The second point where $D_I = D_S$ is reached in 1904–1919 so that 1887–1897 and 1897–1904 combined correspond to phases (1) and (2) combined, forming the major part of the primary phase. The semi-industrial phase includes all subsequent years, including the first half of the 1970s and this corresponds to phases (3), (4), and (5) combined, each forming subphases of this long major phase.

In comparison with the cross-sectional observations for developing economies, particular attention is drawn to the possibility of shifting from one phase to another in the three subphases. Japan's process of increasing the magnitude of $D_I > D_S$ is sustained at a fast pace in the 1970s which is in contrast to the sluggish pace seen earlier in the cross-sectional data, particularly for group IV. Japan and the Asian NIEs are examples of countries with poor natural resource endowment, and the operation of stronger pressures for industrialization in such countries should be recognized. Nevertheless, we can identify the crucial problem: how to make a shift from subphase (4) to subphase (5) for the developing nations. It is our view that such a problem can be elucidated more clearly by a major phase demarcation in terms of domestic productivity performance.

98 John C. H. Fei, Kazushi Ohkawa, and Gustav Ranis, "Economic Development in Historical Perspective."

Finally, let us discuss the terminal point (a band of years) in the development process regarding the shift from semi- to full industrialization. We have pointed out that the cross-sectional data suggest that this shift occurs somewhere between groups IV and V, where $D_I > D_S$ appears to be distinctly realized. However, historical records of presently developed nations reveal that a turn to $D_I > D_S$ comes much earlier, suggesting that this point cannot be identified by this indicator. As a matter of fact, the significance of comparing the I–S indicator has been emphasized because of the specific characteristics of postwar developing economies. The indicator for the terminal turning point should be of a more general nature, rather than of a two-sector comparison. We are convinced that this is provided by the indicator of industrialization as expressed in terms of aggregate growth.

The weighted magnitude of product growth, $w_I G_{YI}$, and the weighted magnitude of employment growth, $v_I G_{LI}$, are two candidates for this indicator (the numerical values are given in Panels (A) and (B) in Table 4.10). Both terms decrease sharply from group III to group IV, but we assume that the normal pattern is characterized by a turn from a long-term trend of increase in both $w_I G_{YI}$ and $v_I G_{LI}$ that is sustained through the entire development path (this was illustrated earlier by Japan's case). This proposal is not utterly new but rather relevant to the difference indicator, D_I. As is shown in Panel (C) of Table 4.10, the magnitude of D_I on average suggests a normal decline from group IV to group V (a slight increase is due to the dip for group IV). This is nothing but the results of the decline in both $w_I G_{YI}$ and $v_I G_{LI}$.[99]

99 Is a technical qualification important in the selection of the product or employment indicator? We recommend that both indicators should be used, but if inconsistency between the two is at issue, the latter is safer to rely on because the problem of output prices is involved in the case of product, as is illustrated in Section II, Chapter 3.

Chapter 5

The Domestic and International Price Structures

Increasing international contact is an indispensable requirement for economic development. In the preceding chapters, this aspect was not explicitly dealt with despite its implicit involvement both in terms of growth and structural change in the economy. In Chapters 5 and 6, an explicit treatment of the international environment is attempted, but from a specific instead of a general view so as to be directly relevant to our analytical framework. The price system is the topic of this chapter, and merchandise trade is the topic of Chapter 6.

Economic development starts with the domestic price system of each nation through which economic activity works to affect both supply and demand. As development proceeds through increasing international contacts, the domestic price structure tends to change. This change is part of a long-term process, although both the speed and pattern of the path of "opening" the economy differ between nations. Changes in the relative prices of products over time are an important outcome of this entire process. In particular, the long-term shifts in the production structure cannot be clarified without elucidating its relationship with price changes.

Section I, which attempts to deal with this aspect, examines two dimensions: one is an extension of the discussions on the domestic production structure in the preceding two chapters, and the other is a preparatory discussion for dealing with international trade, which is the subject of the next chapter. The two dimensions are interrelated. With respect to the former topic, we are specifically concerned with the price effects in forming a trend of convergence and/or divergence, and this pertains to the relationship between change in output prices and productivity.

Efforts will be made to search for some regularity in this relationship through the growth path of developing economies. One may assert that the classical notion provides the answer: that is, prices of industrial goods will follow a decreasing trend, whereas prices of primary products will follow an increasing trend. We believe that this provides a basic insight insofar as the long-term path of productivity differential is concerned. Regularity is searched for in terms of the operation of the market mechanism. However, no *a priori* proposal will be made here in light of the characteristics of the postwar developing economies. The data by sector are not sufficiently available to do this. This lack of data explains why these two aspects have often been confused in theoretical arguments in terms of providing empirical evidence. In Section II, the relationship between changes in the domestic price and in export and import prices will be approximated. In this section, the pattern of change in the relative prices by major sector is examined. Note that this is not the same as the "commodity price" approach, but to the extent that selected commodities can be representative of sectoral outputs, the discrepancy between the two approaches can be assumed to be insignificant. With respect to the conventional terms-of-trade approach, brief comments will be made from the viewpoint of the nondichotomous structure of commodity trade.

We will discuss specifically the differences in domestic price systems in Section III using purchasing power parity data. First, in terms of the conventional dichotomy of tradeables vs. nontradeables, a general picture is given using our grouping by Y/L level. Then, by focusing on the sharp contrast in the price level of producer durables and services, the domestic price structure is analyzed in depth. Finally, with a heroic assumption, an assessment of the factor prices will be tried in view of the significance of the function of human capability in enhancing the absorption of borrowed technologies in the developing economies.

I. Changes in the Domestic Price Structure and Productivity Growth: The Competitive Mechanism

In examining the trend of convergence vs. divergence, the importance of the different sectoral performances in terms of current prices and constant prices was noted. In some cases, for example Japan, the difference was sizeable. Despite the importance of such an analysis, empirical knowledge of this subject is very limited. Since sectoral productivity growth was examined in terms of constant prices in Chapter 4, in this section we will focus our observation on the sectoral output prices from two aspects: changes in output prices over time and differences in the cross-sectional comparison. The procedures adopted in this study are explained in the text that follows.

First, in a simple analysis of long-term sectoral performance, output price and productivity are hypothesized to move opposite with each other to the extent that a competitive market operates on both output and production factors. The actual performance of output price and productivity in individual countries may appear different between the groups when classified by Y/L level and between the sectors. This raises an important factual challenge for our analysis. Output price data for the purpose of international comparison are not readily available, so that we need to make some approximation. Changes in sectoral output prices are found to be substantial. These results will be compared with changes in productivity which were found in Chapter 4. Second, despite limitations of the data, we will try to test our hypothesis. In so doing, a distinction is made between the goods (A and I) and the services (S) sectors. Between the phenomena that are found and the over-time performances found earlier, certain inconsistencies will be recognized. This is another challenge for our analysis.

Sectoral Output Prices: Cross-sectional and Historical Comparisons

The first task is to clarify the pattern of changes and differences in relative output prices of the major sectors using the cross-sectional data. Conceptually, the standard criterion for the development of sectoral output deflators should be the price of aggregate output, that is, GDP at factor cost or at market prices, which is estimated consistently with a deflator in national accounts for aggregate expenditure. In international comparisons, price indexes representing such sectoral deflators are not consistently available in terms of a uniform formula. Thus, in our analysis, the same World Bank output data that was used earlier to calculate $G_p = G_Y - G_{Y^*}$, where p, Y, and Y* refer to output prices, GDP at current prices, and GDP at constant prices, respectively, is used in this chapter. The same formula is applied to the three major sectors. The results are given in Table 5.1 (for simplicity, "G" is dropped in the table).

Panel (A) presents the average rates of change of sectoral output prices relative to changes in GDP price. The data show noticeable patterns. First, the relative output price of agriculture, P_A, tends to be moderately positive for groups I, II, and III (except for group I in period 1), but turns negative in groups IV and V for both the 1960s and the 1970s. Second, the relative output price of the

Table 5.1
Changes in Output Prices and Productivity Over Time by Major Sectors and by Group[a], 1960–1980 (percentage)

	1960-1970				1970-1980			
Group	P_A	P_I	P_M	P_S	P_A	P_I	P_M	P_S
(A): I	-0.1	0.7	2.4	-1.2	0.9	0.3	1.5	-2.0
II	0.1	2.0	-1.1	0.3	0.3	1.3	3.3	-1.2
III	0.1	-1.2	-0.5	0.3	0.1	1.0	-0.9	0.2
IV	-0.6	-0.3	-0.6	0.6	-0.5	0.3	-0.7	-0.1
V	-1.6	-1.3	-1.5	1.1	-1.2	-0.2	-0.5	0.3
	P_I-P_S	y_I-y_S	Sum	P_M-P_S	P_I-P_S	y_I-y_S	Sum	P_M-P_S
(B): I/S								
I	1.9	1.1	3.0	3.6	2.3	-1.8	0.5	3.5
II	1.7	2.4	3.1	-1.4	2.5	0.3	2.8	4.5
III	-1.5	3.2	1.7	-0.2	0.8	-0.5	0.3	-1.1
IV	-0.9	2.4	1.5	-1.2	0.4	1.5	1.9	-0.6
V	-2.4	2.7	0.3	-2.6	-0.5	1.8	1.3	-0.8
	P_A-P_S	y_A-y_S	Sum		P_A-P_S	y_A-y_S	Sum	
A/S								
I	1.1	0.1	1.2		2.9	0.0	2.9	
II	-0.2	-0.5	-0.7		1.5	1.0	2.5	
III	-0.2	2.2	2.0		-0.1	0.5	0.4	
IV	-1.7	2.0	2.5		-0.4	1.1	0.7	
V	-2.7	2.4	-0.3		-1.5	2.6	1.1	
	P_I-P_S	y_I-y_A	Sum	P_M-P_A	P_I-P_S	y_I-y_A	Sum	P_M-P_A
I/A								
I	0.8	1.0	1.8	2.5	-0.6	-1.4	-2.0	0.6
II	1.9	2.9	4.8	-1.2	1.0	-0.7	0.3	3.0
III	-1.3	1.0	-0.3	-0.6	0.9	-1.0	-0.1	-1.0
IV	0.3	0.2	0.5	0.0	0.8	0.4	1.2	-0.2
V	0.3	0.3	0.6	0.1	1.0	-0.8	0.2	0.7

Note: a. P_A, P_I, and P_S refer to the growth rates of prices in the agricultural, industry, and services sectors, respectively. Likewise, P_M refers to growth in the output price of manufacturing, which is a subsector of industry. All rates of change in output prices are presented, for simplicity, as the difference between the rate of change in aggregate prices estimated for GDP, as only changes in relative prices are needed. The rates of productivity growth, y_I, y_S, and y_A, are reproduced from Tables 4.3, 4.6, and 4.7. Note that G_y is measured by taking an "average annual" figure using a simple bridge, while G_Y is estimated by least squares. Some discrepancies between the two approaches are unavoidable, but such an approximation is inevitable given the limited availability of data.

Sources: World Bank, *World Development Report 1982,* Tables 2 and 3, and *World Tables,* 1983 issue, Economic Data Sheet 1.

industrial sector, P_I, is also positive in the lower Y/L groups and becomes negative in the higher Y/L groups. However, some difference is witnessed between the two periods: positive values occur for groups I and II in the 1960s, but in the 1970s almost all groups show a positive value of P_I, group V being barely an exception. Third, the price of services, P_S, is generally positive in the 1960s, except for group I, but changes somewhat in the 1970s. All of these data are useful but provide only a preliminary analysis. Let us now look at the performances of (P_I–P_S), (P_A–P_S), and (P_I–P_A), respectively.

P_I–P_S In the lower Y/L groups, the output price of sector I tend to increase at a faster pace as compared to that of sector S. However, the range of difference tends to narrow towards the higher Y/L groups, and in fact, the relationship is eventually reversed—that is, the output price of sector S increases faster than P_I. This pattern is distinct in the 1970s, while for the 1960s, the pattern is disturbed by group III. The reason for the differences in the pattern between the two periods cannot be specified and thus, the simple average for the two periods makes sense and is presented as follows:

	I	II	III	IV	V
P_I–P_S for 1960–1980	2.1	2.1	-0.3	-0.9	-0.2

In the lower Y/L intervals, $P_I > P_S$, but at the higher Y/L levels, $P_I < P_S$.

P_A–P_S The pattern of (P_A–P_S) appears to be similar to that of (P_I-P_S), but the pattern of (P_A–P_S) is different between the two periods. The simple average of (P_A–P_S) for the two periods are:

	I	II	III	IV	V
P_A-P_S for 1960–1980	2.0	0.7	-0.2	-1.1	-2.1

P_I–P_A The pattern of (P_I–P_A) over the two periods appears to be less regular and the simple averages for the two periods are:

	I	II	III	IV	V
P_I–P_A for 1960–1980	0.1	1.4	-0.2	0.5	0.7

The general pattern for prices of industrial and agricultural output appears to be $P_I > P_A$, but group III is a slight but noticeable exception. The heterogenous composition of output produced by the industrial sector may be one of the major causes for the lower degree of regularity. The two major subsectors—i.e., manufacturing and nonmanufacturing—produce different outputs, and the prices of these outputs may differ accordingly. This difference exists between each subsector and is particularly wider within nonmanufacturing which includes mining, construction, and public utilities. Rates of change in the output prices of manufactures, P_M, which are shown in Panel (A), are estimated from the same data. The performance of P_M shows that the difference between the two periods is in fact lower. What is witnessed for the 1960s and the 1970s is a mixed pattern of (P_M–P_A). Through all groups except group I, the degree of the inequality is greater in the 1970s than in the 1960s. As will be discussed later, the difference between the two decades deserves attention. The simple averages of (P_M–P_A) over the two periods are as follows:

	I	II	III	IV	V
P_M–P_A for 1960–1980	1.5	1.9	-0.8	-0.1	0.4

The output price of manufactures tends to go down relative to that of agriculture through groups II–IV, but at a slower pace; the sluggish decline is worthy of note.

A similar analysis using (P_M–P_S) is also made in Panel (B). The simple averages of this term over the two periods are:

	I	II	III	IV	V
P_M–P_S for 1960–1980	3.5	2.6	-1.0	-0.9	-1.7

The pattern of (P_M–P_S) shows little difference from (P_I–P_S) except for group V, perhaps due to stronger effects of the services sector.

At this stage no concluding remarks can be made because the data are indirect estimates and the range of variation is rather wide between the two periods. Nevertheless, what follows may be suggested as the broad pattern. First, relative price changes in output appear to produce differences between the lower Y/L intervals (i.e., groups I and II) and the higher Y/L intervals (i.e., groups III, IV, and V). This is particularly distinct in the comparison between the goods-producing sectors vs. services. Relative to the output prices of services, the output prices of industry (and manufacturing) and agriculture tend to increase in the lower Y/L intervals; but the output prices of the goods–producing sectors decrease relative to output prices in the services sector in the higher Y/L intervals. Second, within the goods-producing sector, a regular pattern cannot be distinctly identified, and yet some tendencies are suggested. In particular, a comparison between manufacturing and agriculture indicates that the output prices of manufacturing tend to decline for the majority of developing economies. This suggests that the issues need further scrutiny.

The second task is to compare the historical pattern of currently developed nations with the findings of the developing nations. Is there any consistency in the experiences of the two groups? Although observation over two decades is too short in examining the experiences of the developed nations and although the data available for the developing economies are also very limited, let us nevertheless try to answer this question.

Let us first begin with industry and agriculture, I/A. Since the time of classical doctrine, it has been thought that the output price of agriculture must show a long-term rising trend relative to that of industry; that is, $P_A > P_I$. We have used Kuznets' country data, which are available in both current and constant prices, and have applied the same procedure used earlier. The value of $(P_I - P_A)$ is as follows:[100]

United States	-0.7	1839–1899
	-1.8	1899–1929
Sweden	-0.2	1861–1870
Italy	-0.1	1861/70 – 1965/67

As was suggested by Kuznets himself, statistically these estimates may underestimate the actual changes in sectoral output prices. Nevertheless, we believe that $P_A > P_I$ is the normal pattern; this is further supported by Japan's case (Chapter 3). From the indexes of relative prices of output (I/A), which are shown in Table 3.4, the comparable value of the rates of average annual change (%) in $(P_A - P_I)$ is -1.0 for the entire period and -1.1 for the prewar period. It is our view that this illustrates the performance of successful latecomers in industrialization.

With respect to the domestic terms of trade between the two major goods-producing sectors, an important recognition is thus drawn. The output price of industry tended to decline relative to that of agriculture in the historical path of currently developed nations. As for the postwar developing countries, the same force seems to operate in the cross-sectional data. In the 1970s, $P_I < P_A$ and $P_M < P_A$ prevailed in almost all groups of developing economies. However, the path of its operation appears to be weak and at a sluggish pace, and is often reversed by the influence of stronger counterforces, as is shown in the 1960s.

In comparing industry with services (I/S), $(P_I - P_S)$ is estimated to be negative for all four Western cases during the same interval:

United States	-1.6	1839–1899
	-1.0	1899–1929
Sweden	-0.3	1861–1870
Italy	-0.9	1861/70 – 1965/67

For Japan, the value of $(P_I - P_S)$ which is calculated from Table 3.4 is -1.4 for the entire period and -0.6 for the prewar period.

100 Kuznets, *Economic Growth of Nations,* Table 21.

Taken together, another significant characteristic of postwar developing economies is recognized. The output price of services does not show a distinct tendency to rise relative to that of industry. The forces which raise the output price of services operate when the conditions are favorable, as illustrated by the performance of (P_I–P_S) through groups III–V in the 1960s. However, the forces are weak so that when the influence of counterforces becomes stronger, (P_I–P_S) is positive as is illustrated by groups I and II. A rise in P_S in group V is explained by the shift to a developed phase which is accompanied by new activities in the services sector.

Why are these characteristics found for developing countries? The answer lies in the combined results of two elements: the price structure that is indigenous to the developing country and the response to external influences through the enhancement of international contacts. Both elements have some relation to productivity changes in our hypothesis, which implies that the operation of the market mechanism differs, particularly for the lower Y/L groups.

Price Effects vs. Productivity Effects

The topic of this section is the relationship between price performance and productivity change by major sectors. It is hypothesized that the operation of the market mechanism brings forth a stronger negative association between the two, whereas weaker operation of the market mechanism tends to make for a smaller negative (or even positive) association. In this process there may appear "distortions" which we will define solely in terms of productivity growth.

Conventionally, we are accustomed to the simple argument that a productivity increase in producing a certain commodity means a rightward shift of its supply curve and this brings forth the possibility of a decline in its output price so long as changes in demand do not offset the effects of the increase in supply on the commodity price. An analogous argument is applied to the sectoral comparisons, say, agriculture vs. industry, in the long-term development process with respect to differences in the speed of productivity growth. As was discussed earlier in Chapter 4, the increasing trend in the relative price of agricultural products is explained by productivity growth: that is, a slower rate of growth in agriculture vs. a faster rate of growth in industry. We do not follow the classical notion in identifying the actual trends, but we believe the approach is valid insofar as the long-term perspective is concerned.

Let us revisit the problem of convergence and/or divergence in terms of product per worker. The convergence/divergence trend is composed of two elements: the price effect and the productivity effect which both operate in a mutually interrelated manner through the market mechanism. It is our view that a trend of convergence or divergence can be more fully grasped if these two effects are decomposed. The price-productivity relationship in this case is viewed in terms of the contribution to income produced per worker; thus, the dimension differs from the simple explanation given above. That is, the price-productivity relationship is the outcome of the simple explanation in terms of a sectoral comparison. In a two-sector comparison, the productivity and price effects may cancel each other out and neither convergence nor divergence will take place (i.e., the "neutral" case). Changes in the sum of these two effects, whether the sum becomes smaller or greater, measures convergence or divergence. In this sense, we want to explain the findings of Chapter 3 using the hypothesis of the sectoral pattern in the operation of the competitive mechanism.

Let us begin by comparing the sectoral performance of productivity growth (which was presented in Panel (B) of Table 5.1) and the behavior of sectoral prices. The procedure of approximating measurements is simple as illustrated below by an example of I/S in the 1960s. The sum of the price effects, i.e., $(P_I - P_S)$, and the productivity effects, i.e., $(y_I - y_S)$, is used to estimate the rates of change in product per worker in current prices (this procedure was also employed in Chapter 3 with θ_I/θ_S). The price effect is positive in groups I and II, but is negative in groups III–V; the productivity effect is positive through all groups. The sum of the two effects is positive for all groups as the productivity effects were larger than the price effects in the higher Y/L groups, and the magnitude of the sum is greater in the lower Y/L groups. In fact, a trend of divergence is identified during the 1960s across all groups. In Table 3.2, Panel (A), we noted a unanimous rise in θ_I/θ_S from 1960 to 1970 through all groups. The contribution shares of these two effects thus estimated here may be meaningful because it can suggest different ways in which the market mechanism operates. A clear example is provided by the different price effects witnessed earlier between the lower and higher Y/L groups. The latter's performance can be thought to have been brought forth by the operation of the market mechanism: a relative rise in the productivity of sector I is accompanied by a price decline. Why does this not take place in the lower Y/L groups despite a rise in productivity? These discrepancies are discussed in the following paragraphs.

With respect to I/S, during the 1970s, the productivity effect is negative in groups I and III as a result of the general decline of y_I relative to y_S. Nevertheless, a positive sum of the productivity and price effects is maintained for all groups since the productivity growth effect is greater than the price effect. This performance is sustained throughout the two decades under review. Response of the market mechanism may require longer intervals to operate effectively and a simple average observation through two decades though acceptable may not be enough to show the full response of the market. Nevertheless, we have tried using the three broader groups shown below[101]:

Combined Groups	P_I-P_S	y_I-y_S	Sum
I and II	2.1	1.2	3.3
III and IV	-0.6	1.7	1.1
V	-1.5	2.3	0.8

The expected negative association of the price effect operates most distinctly in group V, followed by a smaller negative magnitude for groups III and IV. With respect to groups I and II, however, the smallest rate of relative productivity growth in sector I is accompanied by a big rise in relative price. The contribution of the price effect to the total effect is around two-thirds. Certainly the trend of divergence is largely brought forth by the price effects in groups I and II. The trend of divergence becomes weaker in groups III, IV, and V, and the difference in price effects plays a dominant role as indicated by a comparison of the differences between groups (I and II) and groups (III and IV) vs. groups (III and IV) and group V regarding (P_I-P_S) and (y_I-y_S).

We are convinced that the findings are significant, but an interpretation of the results cannot be given without making some speculative supposition. There are four relevant elements: (1) essentially the industrial sector's output is tradeable while output of the services sector is nontradeable; (2) the profit-earning power of the services sector is very strong in the traditional economy; (3) the industrial sector, which is the newly introduced modern sector, cannot compete with the services sector in the domestic market without attaining considerable profit margins that

101 Because of limited data, phase demarcation is not possible, and this is a rough approximation.

are earned in the initial phase of industrialization; (4) hence, the industrial sector may not be able to be competitive internationally. If these assumptions are accepted, then the finding that $P_I>P_S$ despite $y_I>y_S$ for groups I and II may not necessarily imply that the industrial sector is noncompetitive internally if the strong effects of nontradeable activities are duly counted. As explained by the performance of difference indicators in Chapter 4, in the narrowing of the inequality $D_I<D_S$ in groups III and IV, the resistance of the services sector will gradually tend to weaken so that the industrial sector becomes more competitive internationally. Yet, until the economy arrives at the point represented by group V, the transitory situation may be sustained for a long period of time.[102]

A similar observation can be made with regard to A/S. In this case (except for groups I and II), the difference in the rates of productivity growth, (y_A-y_S) is generally significantly positive, though the value declines from the 1960s to the 1970s. The performance of output prices, (P_A-P_S), differs between the lower and higher Y/L groups: the value of (P_A-P_S) is positive in groups I and II, while it is negative in groups III, IV, and V. For the broader grouping, the average values over the two decades are as follows:

Combined Groups	P_A-P_S	y_A-y_S	Sum
I and II	1.3	0.3	1.6
III and IV	-0.6	1.9	1.3
V	-2.1	2.5	0.4

The pattern for A/S is virtually the same as that of I/S: a most active response in prices is found in group V, followed by a slightly negative response in groups III and IV, whereas in groups I and II the price effect is dominant despite almost no change in relative productivity. The direction towards divergence is sustained through groups (I and II) and groups (III and IV), but it appears to cease in group V. Output of the agricultural sector is essentially tradeable

102 This is not an exclusive interpretation. For example, one may assert that protective policies for early industrialization, as represented by import substitution measures, must be the major cause of the noncompetitive performance of the industrial sector. We have no objection to such an argument. What we wish to emphasize is the significance of nontradeable activity, and this will be discussed further in Section III.

and of the same nature as that of sector I, but it is different in two aspects: in the agricultural sector, the traditional structure prevails and land, which is the most important production factor in agriculture, is nontradeable. Nevertheless, the price-productivity relationship of the agricultural sector as compared with the services sector is found to be virtually similar to that of the industrial sector, and we are inclined to apply essentially the same interpretation.

The relationship between the two goods-producing sectors, I/A, is related to the problem of the domestic "terms of trade" in the conventional sense, although the literature on domestic terms of trade is not as rich as the literature on the international terms of trade. The operation of the market mechanism is expected to be more readily witnessed in I/A than in the preceding cases of I/S and A/S. Let us now look at the actual situation.

As was discussed earlier in Chapter 4, during the 1960s, the rate of productivity growth of industry was greater than that of agriculture, but in the 1970s, the rate of productivity growth in the two sectors was almost equal due to the slowdown in industrial growth. Despite these variations, the output price of industry increased relative to that of agriculture throughout the two decades. The averages for the broad groups are as follows:

Combined Groups	$p_I - p_A$	$y_I - y_A$	Sum
I and II	0.8	0.5	1.3
III and IV	0.2	-0.1	0.1
V	0.7	-0.3	0.4

Several points should be noted. First, the performances of groups (I and II) differ from that of groups (III and IV) and group V: both the price effect and the productivity effect are positive in groups (I and II), while in the other two groups, there is a positive price effect and a negative productivity effect. The different performances are also seen in the sum which is relatively large in groups (I and II), while it is minor in groups (III and IV) and group V. As compared to the preceding cases of I/S and A/S, the rates of change are much smaller, particularly for the intermediate groups (III and IV).

The assertion made earlier that the lower groups are characterized by peculiar features is thus endorsed by this evidence. In the nations belonging to these groups, a trend of divergence went on even within the goods-producing sectors over the two decades.

Viewed from the standpoint of agriculture, in which the majority of the labor force is engaged, relative product per worker tended to rise at a pace behind that of industry, combined with a tendency for the relative rate of the price increase to fall. This is exactly the pattern often asserted by critics of the primary import-substitution policy with regard to industrial goods. In our framework, the degree of output price increase, as compared to the services sector, is greater for the industrial sector than for the agricultural sector in the initial phase of industrialization. The market mechanism does not appear to operate between sectors I and A in that sense: in agriculture, despite its slower rate of productivity increase, its output price tends to decline relative to that of the industrial sector. Here, we must pay particular attention to the problem of underemployment in the agricultural sector which can arise as a side effect of government policy intervention. The possibility of bringing forth underemployment hampers the normal operation in the price-productivity mechanism. Keeping this in mind, let us look at the performance of other groups.

With regard to group V, the developed nations, the market mechanism seems to operate fully. The inequality, $y_A > y_I$, in the 1970s draws particular attention. A move towards a trend of divergence ceases because of the ending of the normal pattern of industrialization in developing economies. That is, a faster productivity increase in the agricultural sector occurs together with a slower rate of price increase. This is most impressive in $(P_M - P_A)$ relative to other groups, particularly in the 1970s.

As for the intermediate groups between the two extremes, a clear interpretation cannot be given for the average of groups III and IV combined, the performance of which appears to be somewhere between groups (I and II) and group V. The rate of productivity increase in the agricultural sector is slightly greater than that of the industrial sector and this accompanies $P_I > P_A$ to a slight degree. As seen in Panel (B) of Table 5.1, the price-productivity relationship actually appears to be rather mixed in both the groups and the periods. The "intermediate" performance means that the market mechanism does operate but with considerable effects of underemployment which is not witnessed for group V.

The factor responsible for the intermediate performance also affects the industrial sector, the output of which is essentially tradeable. However, as mentioned earlier, this sector includes nonmanufacturing, the output of which is nontradeable. The data in Panel (B) of Table 5.1 show that P_M tends to decrease relative to P_I: thus, $(P_M - P_I)$ is negative for all groups except group I. Therefore, between the agricultural and the manufacturing sectors, the competitive mechanism must be even more distinct, though productivity data are not available.

But even in this case, the significance of the existence of underemployment cannot be ignored. As was noted in the preceding chapters, the sustained difference in labor's marginal productivity across sectors, in our view, characterizes the structure of developing economies. Without considering the effects of underemployment, the conventional argument of the market mechanism process cannot be applicable. As was described in the case of agriculture, to the extent that underemployment exists, rightward shifts in the supply curve may not occur with increases in productivity. If the pace of the shifts in the supply curve of agriculture is slower than that of industry, we would ordinarily expect a relative rise in agricultural output. However, the pressure of underemployment would prevent this from occurring and a smaller or even no rise would be seen. Thus, the competitive mechanism does work, but with certain limitations. It can be generally suggested that even in the hypothetical case of a sustained path of $G_{yA} = G_{yI}$ for developing nations, we must also consider a fall in P_A relative to P_I. This idea is not considered in the classical notion.[103]

In concluding this section on the problem of sectoral inequality change, let us consider the two cases, I/A and S/A. These cases can be clarified further by decomposing the performance of (I+S)/A, which was discussed in Chapter 4 and was used as the most basic indicator of direction towards convergence or divergence. Table 5.2 summarizes the results of the measurements. Two topics are taken up: (1) industry vs. services, respectively, in relation to agriculture, and (2) price effects vs. productivity effects regarding agriculture vs. nonagriculture. For both topics, the data are approximated from the rates of changes estimated by the group average using a weighted sum as explained in the table notes. Two decades are dealt with on average so that differences between the 1960s and the 1970s are disregarded.

A move towards convergence prevails through all groups of developing economies except group I, as shown by the sum (i.e., the last column in the table). This is not a new finding but reconfirms

---

103 This point was argued in Kazushi Ohkawa, *Nogyo no Dotai Bunseki* [Dynamic analysis of agriculture] (Tokyo: Taimeido, 1954), particularly Chapter 2, "Long-term Changes on the Price of Agricultural Products."

Table 5.2
Decomposition of Changes in Product Per Worker in Two Comparisons [a] ***(percentage)***

	Industry/	Services/	Agriculture/Nonagriculture		
Group	Agriculture	Agriculture	Price Effect	Productivity Effect	Sum
I	-0.1	-2.1	-1.1	-0.7	-1.9
II	2.5	-0.4	0.0	0.5	0.5
III	-0.2	-1.1	0.0	0.7	0.7
IV	0.8	-0.7	0.9	-0.9	0.0
V	0.4	-0.4	1.8	-1.5	0.3

Note: a. Output shares were used to sum the price effects and labor shares were used to sum the productivity effects. For S/A, reversed rates of changes in A/S are used.

Sources: Tables 4.2 and 5.1.

what was identified earlier. Services contribute to sustaining the trend toward convergence, while the function of industry is mixed. In the agriculture/nonagriculture comparison, the productivity effects operate against the price effects for the higher Y/L groups. But for the lower Y/L groups (i.e., groups I and II), this relationship does not seem to work and it is reversed in the lowest level: that is, price effects appear to be stronger than productivity effects. These results are to be expected from our preceding findings, but we believe it is significant to reconfirm these findings in the basic framework of agriculture vs. nonagriculture. With respect to group V, the strong effects of both price and productivity operate to cancel the effects of each other out.

How can we characterize the performances of the postwar developing economies as compared to the historical records of currently developed nations? As has been recognized earlier, the paucity of available data prevents us from applying the decomposition procedure to the currently developed nations in terms of (I+S)/A. Nevertheless, it can be contended that for these developed nations, the price effects could not have been greater than the productivity effects (assuming a conservative supposition).[104] In

104 From the data referred to earlier, evidence can be given in terms of the average annual rates of changes. The productivity effect is greater

fact, similar to the performances of the postwar developing economies, the productivity effects were probably distinctly greater. If this is the case, what was described in Chapter 4 is essentially valid. The trend of convergence identified for the postwar developing economies, as opposed to the trend of divergence identified for the prewar historical path of developed economies, is a central issue. We believe two major factors need to be distinguished. The contribution of services in the primary phase and the contribution of agriculture in the secondary phase should be recognized as the major factors characterizing the trend of convergence for the postwar developing economies. These two forces had never operated in the historical path of currently developed nations. These are reconfirmed here by applying the decomposition procedure.[105]

II. International Dimensions

While we do not wish to discuss international output prices in general, we would like to deal with selected aspects to expand on

---

than the price effect, except for the United States in 1839–1899 for A/I.

		A/I		S/I	
		Productivity Effect	Price Effect	Productivity Effect	Price Effect
Sweden	(1959 price)				
	1861–1870	-0.36	0.23	-0.57	0.35
	1963–1967				
Italy	(1963 price)				
	1861–1870	-0.76	-0.13	-1.34	1.11
	1863–1967				
United States	(1859 price)				
	1839–1899	-0.31	-0.70	-1.21	0.82
	(1929 price)				
	1899–1929	0.91	0.04	-1.24	1.03

105 Though indirectly, the figures shown in Table 5.2 suggest the validity of these possibilities. In the higher Y/L groups (i.e., groups III and IV) a similar magnitude of negative effects suggests the positive common effect brought forth by agriculture. Of course, we should not ignore the relative slowdown of productivity growth in the 1970s in these groups. As for the lower Y/L groups, taken together with the data that compares agriculture and nonagriculture, the price effects of services can be supposed to be influential.

what was observed in the preceding section. This is what we refer to as the second dimension. Hence, the relationship between export prices and domestic prices is the focus of this section. The hypothesis is essentially kept: the association between productivity and price which works through the competitive mechanism will operate more fully in an international market. Note that import prices cannot be discussed due to paucity of data.

In looking at the postwar international relationship, the classic dichotomy in analyzing the productivity-price relationship for agriculture and industry cannot be readily applied as is widely supposed. This is because of the clear-cut classification of countries—that is, countries which export industrial goods and depend on imports of primary goods vs. countries that export primary goods and depend on imports of industrial goods—is no longer valid. This is, of course, due to the rapid industrialization of developing economies on the one hand, and the noticeable agricultural productivity growth in the industrial countries as well as in the developing nations on the other. In the dichotomic approach of the classic framework, consistent performance of the productivity-price relationship could be assumed between domestic and international markets. However, in the contemporary world, the relationship between the two requires particular investigation, even apart from the peculiar problem of fuel imports.

The "terms of trade" is still an important conventional term with regard to merchandise trade. This term usually refers to the net barter terms of trade and is calculated as the ratio of a country's index of export unit values (P_{ex}) to the country's index of import unit values (P_{im}), that is, P_{ex}/P_{im}. If the classic dichotomy closely reflected the real world, then for industrial economies and for primary goods-export economies P_M/P_A and P_A/P_M, respectively, can be used as a proxy for P_{ex}/P_{im}. But we cannot necessarily expect consistent movement between the terms of trade and commodity price indexes for manufactures vs. primary goods. Sometimes, the two may move even in a reverse direction. While it is true that if quantity were linked with the terms of trade, it would contribute to clarifying the effects of trade on income, this would not be directly relevant to our productivity-price approach by sector.[106]

---

106 In applying the competitive mechanism of the price-productivity relationship to the international scenario, our hypothesis should be modified to take the effects of labor-cost and wages into consideration. This is particularly important with respect to export competition between developing and developed nations. It is widely known that cheaper wages

The discussion that follows examines these problems by presenting empirical evidence in three areas: export prices of selected commodities, the domestic prices of the United States, and the relationship between the commodity approach and the terms of trade.

A Comparison of Manufactured and Agricultural Products

In order to understand the implication of our earlier findings about output prices of agriculture and industry (and manufacturing), we need to compare export prices. For this, we fully depend upon the research of Grilli and Yang (1988) which provide most reliable data that are comparable with our data. In Table 5.3, Panel (A), the data for the comparable postwar period are listed together with the long-run data in Panel (B). As noted in the table, our rearrangement procedure is simple and crude but is sufficient for our purposes. No index is directly available for prices of agricultural products; thus, two price indexes, a food and a nonfood price index, are used as proxies for the agricultural product price index. In addition, the index for primary nonfuel prices, which includes metals, is included in the table. Manufactured goods prices are compiled from the records of ten industrial countries and are used to represent international export prices.

For the two decades, 1960–1980, the prices of nonfuel primary goods and food increase by a considerable magnitude relative to prices of manufactures. Though relative prices of nonfood (which are mostly materials for manufacturing) go down, we believe that the relative price of agricultural products did in fact increase. For the same period, our earlier findings have shown that P_A increases in comparison with P_M for the majority of developing economies. Due to the crude estimates, it may not be possible to estimate the difference in the magnitude of the two; however, the performance seems consistent between domestic and international markets with regard to the relative changes in terms of direction. This appears to be a significant phenomenon and suggests the validity of the classical doctrine.

---

in the former strengthen their competitive power in international markets. We are dealing with the problem in terms of growth, so that the issue pertains to the rate of wage changes in relation to productivity increase: that is, the degree of the lag in wages relative to increases in productivity. This cannot be estimated but does in fact occur in the developing economies. To that extent, the price-productivity relationship will be distorted in the international market.

Table 5.3
Changes in the Export Prices of Selected Commodities [a]
(average annual rates, percentage)

(A): Postwar

	Manufactures	Nonfuel Primary	Food	Nonfood
1960-1890	6.28	6.80	7.24	5.70
1960-1985	4.43	3.64	3.70	3.14

	Primary Nonfuel/ Manufactures	Food/ Manufactures	Nonfood/ Manufactures
1960-1890	0.52	0.96	-0.58
1960-1985	-0.79	-0.73	-1.29

(B): Long-run

	1900-1918	1918-1932	1932-1949	1949-1973	1973-1985
Food/ Manufactures	1.82	-5.93	2.52	0.32	-6.08
Nonfood/ Manufactures	0.68	-8.90	4.24	-1.39	-4.52

Note: a. The rate of average annual changes is calculated by simply bridging the beginning and ending years of each period. The periods in panel (B) are roughly demarcated by the data contained in the sources.

Sources: Enzo R. Grilli and Alan Chang Yang, "Primary Commodity Prices, Manufactured Goods Prices, and the Terms of Trade of Developing Countries: What the Long Run Shows," *Economic Review* 2 (January 1988).

However, rather than providing a solution, this provides the starting point for analyzing the problems involved with (1) the difference in the performance of domestic and international prices and (2) the interpretation of this pattern from a longer-term point of view. With regard to the difference in the performance of domestic and international prices, let us recall the difference in $(P_M - P_A)$ between groups which is shown in Panel (B) of Table 5.1. If we reclassify $(P_M - P_A)$ into broader groups, then we find that $(P_M - P_A)$ is 1.3 for groups (I and II), 0.3 for groups (III and IV), and 0.2 for group V, though variations in the original groups and periods are not small. We can say that in general, for group V, the tendency is reversed: prices of agricultural goods decline relative to those of manufactures. Data for increases in productivity are not available, but (y_I-y_A) suggests that these can be associated with productivity performance. Under the seemingly consistent pattern of (P_M-P_A), such a differential is involved. Since the performance of price in the international market is greatly influenced by exports from group V nations, exports of developed nations have great influence on the exports from developing nations.

In the real world, through changes in the exchange rates and related measures, the influence of exports of developed nations operates in complex ways. Yet, we can hypothetically consider a case whereby the pattern $P_A<P_M$ emerges in international markets if the influence of developed country exports were to become much stronger. One may argue that although the pressure on prices of agricultural exports of developing nations would be sizeable, the pressure on prices of manufactured exports of developing nations would be sizeable as well.

The interpretation of this pattern from a longer-term perspective is not irrelevant to the hypothetical case mentioned above. The possibility of $P_A<P_M$ should be historically confirmed in light of the long-term records. In Panel (A) of Table 5.3, for the period 1960–1985, the direction of change reverses: all three comparisons show a decrease. This is caused by a sharp decrease that took place in 1980–1985. Both periods were listed in the table to illustrate the wide range of variation and to serve as a warning about the risk involved in judging productivity-price relationships. The data listed in Panel (B) may help in understanding this. The pattern of change in the relative prices of the two sectors varies a great deal by periodization. In short, the long-run performance of (P_A-P_M) consists of two parts: (1) a trend of increase of longer duration and (2) a tendency of a sharp decline of a relatively shorter duration. It can be expected that a secular trend, through upswings and downswings, may be identified as "increasing." The statistical procedure of identifying peaks and troughs should be refined, but the findings generally remain valid. Viewed from the productivity-price relationship, recognition of the secular trend of increase is significant. However, at the same time we are required to pay due attention to the effects of the forces which interrupt this trend, although the causes may vary from one period to another. A specific examination is needed to clarify the factors responsible for such swings. Yet, to provide a simple understanding we are inclined to offer the following explanation. Prices of agricultural products tend to rise relative to prices of industrial products during upswings, and

the prices of agricultural products tend to fall relative to prices of industrial products during downswings. The reason for this may be found in the different shifts in the demand curves within each sector rather than shifts of supply.

If this interpretation is generally accepted, it may be possible to characterize the postwar period 1960–1980 as showing an upswing pattern that is significantly influenced by an upsurge in food prices in the beginning of the 1970s. This interpretation does not ignore the characteristics of the faster pace of productivity growth of agriculture, which was emphasized in Chapter 4. Otherwise during the postwar upswing, the tendency of rise in agricultural product might have been at an even greater pace. During the downswing in the 1980s, the above-mentioned pressure would in fact cause a reverse pattern.

In order to expand our knowledge of the relationship between domestic and international prices, let us consider the case of the United States which is the lead country in terms of technological advance. Table 5.4 presents data of agricultural domestic prices. The data are specifically arranged to show the features of postwar performance as represented by the two decades, 1960–1980. In terms of the long-run pattern we see the same trend that was seen earlier. The procedure is simple and crude and yet the estimates suffice in providing evidence for our general observation.

The domestic sectoral price changes between manufacturing and agriculture show the same pattern as that of our earlier findings: that is, P_A tends to rise relative to P_M over the long-term path with the reverse occurring during shorter-interval downswings. With regard to the postwar period, 1960–1980, the magnitude of the sectoral difference appears rather moderate, especially in the case of all crops, A2. Although one would expect a reverse trend, that is, $P_M > P_A$, this is not the case. However, we note that compared to 1930–1950, the sectoral difference in 1960–1980 narrows drastically.

Data from Table 5.3 are rearranged in Panel (B) so as to be directly comparable with the data presented in Panel (A) of Table 5.4. The rate of change in the relative prices—that is, food vs. manufactures (F–M) and nonfood vs. manufactures (N–F)—exhibit somewhat different patterns from those described earlier. This is due to the method of periodization. Of concern here is the domestic prices in the United States relative to the export prices. The two prices do not appear to follow a similar pattern. While it is not easy to identify a consistently dissimilar performance between the two, we can say that the magnitude of the relative prices of manufactures and agriculture (or food and nonfood) is greater for the

Table 5.4
Prices of Manufactured and Agricultural Products (average annual rates of change; percentage)

(A): United States

	Manufactures	Agriculture [b]		Relative Prices	
Period [a]	Mu	A1	A2	A1-Mu	A2-Mu
1902-1920	4.32	4.80	5.08	0.48	0.76
1920-1940	-1.74	-2.21	-3.16	-0.47	-1.42
1930-1950	2.97	4.50	4.40	1.53	1.43
1960-1980	4.08	4.77	4.48	0.69	0.40

(B): International Exports

	Manufactures	Food	Nonfood	Relative Prices	
Period [a]	(M)	(F)	(N)	F-M	N-M
1902-1920	3.44	3.32	2.71	-0.12	-0.73
1920-1940	-2.00	-1.25	-2.42	0.75	-0.42
1930-1950	3.54	4.45	4.98	0.91	1.44
1960-1980	5.83	5.90	5.08	0.07	-0.75

Notes: a. Five-year averages that are centered on the registered year are used to make a simple calculation of annual average rates of change. The demarcation of period intervals is selected to be comparable to the period 1960–1980 in our analysis.
b. A1 is all commodities of agricultural production and A2 is all crops.

Sources: See sources of Table 5.9; Hayami and Ruttan, *Agricultural Development*, Table C.2:480.

United States than the international market in 1930–1950 and 1960–1980. For 1960–1980, the relative prices for United States are 0.69 and 0.40, while the relative export price is -0.34 (the simple average of F–M and N–M). For 1930–1950, the relative prices in the United States are 1.53 and 1.43, while the relative export price is 1.18 (again for the F–M and N–M average).

To a certain extent, national price policies must have affected these movements, but an analysis of this is beyond the scope of this study. Thus, we face some difficulty in consistent interpretation of domestic and export prices insofar as the performance of 1960–1980 is concerned. This can only be solved by further scrutiny of the individual nations' price performance. Nevertheless, we can suppose the following:

1. The pattern of $P_M>P_A$ for the group V nations consisted of two patterns, $P_M>P_A$ and $P_A>P_M$, where the former effect is stronger than the latter. As a matter of fact, it can be supposed that a number of industrial countries in group V have larger magnitudes of relative prices (P_A-P_M) than the United States; Japan, for example, may illustrate a large magnitude. On the other hand, a number of countries may have a reverse pattern of domestic prices: that is, prices of agricultural products increase more slowly than those of manufactures. Our earlier cross-sectional approach has shown that (P_A-P_M) is slightly negative on average.

2. The scope and intensity of export specialization vary among the nations, and because of this variance, there is some discrepancy between export and domestic prices. This is illustrated by A1 vs. A2 in the United States and F vs. N in international exports in Table 5.4. Here, we offer only a preliminary observation on these two points (further discussion on this is given later in Chapter 6 when dealing with export volumes). The divergence between domestic and export prices must be associated with productivity performance according to our analysis of sectorwise averages. However, the effect of specialization of resource allocation within a sector, which follows the comparative advantage rationale, must be noted.

What matters, however, is the recognition of the characteristics of the contemporary phenomena. Group V nations continue to play the major role in exporting manufactures, but their export activity of agricultural products is also very influential in the world market. The disappearance of the dichotomic specialization is, in our view, a factual challenge for the classical approach, which made no clear distinction between domestic and international prices. Our statistical exercise suggests the important question: how do we answer this challenge?

The answer to this question would require elaborate research and is beyond the objective and capability of this study. However, viewed from the aspect of developing economies, we can make several observations. Because of the discrepancies between domestic prices and international prices, a great number of developing nations are pressured to keep prices of agricultural products at a level demanded by the domestic productivity-price relationship.

However, this tends to aggravate the problem of underemployment. Thus far, the unprecedentedly rapid productivity increase in agriculture in developing nations could help to dilute this pressure; further enhancement in productivity via production specialization is required. Moreover, domestic prices of manufactures show a reluctance to fall in a great number of developing nations and another objective is for this to be realized more substantively. One may argue that enhancement of productivity growth by technological and/or organizational advance is needed concurrently both for agriculture and industry, though of course this should occur with due consideration for phases and typological differences. Nobody will disagree with this assertion. However, in pursuing this strategy in individual countries, the problem that is confronted is how to think about the relationship between the domestic and international market.

The notion of competitive efficiency is often applied without making a distinction between the two markets, and the policy towards "open" competition is apt to ignore the significance of the local market mechanism. Ample discussion on the operation of market mechanisms would cover a wide range of fields, both in theory and practice. Even within the specific scope of our study, a distinction must be made between the domestic and the international dimension. When one refers to market competition, the international market is usually implied. Protective policies and measures no doubt restrict international competition. However, the operation of market forces in the domestic economy has a different dimension. Even under the same protection policy, the degree of local regulation can differ. A competitive domestic market can prevail even if international competition is restricted. Despite the limits imposed by our analytical framework, we have also tried to emphasize the importance of distinguishing between domestic and international competition.

A Comment on the Terms of Trade

Finally, let us make a few brief remarks on the terms of trade. The ratio, P_{ex}/P_{im}, tends to decrease from 1960 to 1980 (Panel A in Table 5.5).[107] The rate of decrease is larger for countries of lower levels of per capita income. In Panel (B), the data are converted to our cross-sectional groups. The pace of decline appears to be associated with the level of Y/L; an exception occurs for group III, the reason for which is not clear. We are chiefly concerned with the

---

107 World Bank, *World Development Report 1982*, Annex, Table 8.

Table 5.5
Terms of Trade, 1960–1980

(A) (1975=100)

	Low-income Nations	Middle-income Nations	Industrial Market Economies
1960	111	100	98
1980	89	94	94

(B) [a] (%)

Group	1960	1980	Change (1960-1980)
I	118	87	70.3
II	109	83	76.2
III	97	94	95.0
IV	100	87	87.0
V	105	94	89.5

Note: a. Simple average for Panel B.

Source: World Bank, *World Development Report 1982,* Table 8.

reason for the general tendency. That is, why do we see a larger decline for the lower Y/L groups, which consists of countries exporting primary commodities? One may find some inconsistency between this and the previous finding which showed the tendency of $P_A > P_M$ for developing nations. Such a query is natural if we follow the notion of a dichotomy which assumes that the performance of import prices of manufactures can be approximated by that of export prices. Since no systematic data are available for import prices of manufactures, such a broad approximation may be unavoidable. Under this assumption, the simple answer would be shown by the effects of fuel price upsurge, which is peculiar to this period. Since the countries were selected to exclude oil-exporting countries, the impact of the upsurge in fuel prices undoubtably served to decrease the ratio P_{ex}/P_{im}. However, what is more important in finding the answer to our question is clarification of the differences in the shares of exports and imports through groups of different Y/L levels.

Table 5.6
Manufactures and Nonfuel Primary Commodities [a] *as a Share of Merchandise Exports and Imports: 1960–1979 Average (percentage)*

(A): Share of total

Group	X_M/X	X_p/X	$(X_M+X_p)/X$	M_M/M	M_p/M	$(M_M+M_p)/M$
I	8.2	91.2	99.4	67.2	15.5	82.7
II	14.3	67.8	82.1	62.5	21.3	83.8
II	16.5	61.2	77.7	60.9	23.5	84.4
IV	30.1	54.8	84.9	60.4	23.6	84.0
V	65.5	21.7	87.2	48.8	29.3	78.1

(B): Share of nonfuel total

Group	X_M/X'	X_p/X'	$(X_M-X_p)/X'$	M_M/M'	M_p/M	$(M_M-M_p)/M'$	$(M_p/M')-(X_M/X')$
I	8.3	91.7	83.4	81.3	18.3	63.0	10.0
II	17.4	82.6	65.2	94.6	25.4	69.2	8.0
III	21.2	78.8	57.6	72.2	27.8	44.4	6.6
IV	35.5	64.5	29.0	71.9	28.1	43.8	-7.4
V	75.1	24.9	-50.2	62.5	37.5	25.0	-37.6

Notes: X = Total merchandise exports.
M = Total merchandise imports.
X' and M' = Nonfuel exports and imports.
X_M and M_M = Exports and imports of manufactures.
X_P and M_P = Exports and imports of nonfuel primary commodities.

a. For nonfuel primary commodities, some inconsistency remains. For imports, we use the sum of food and other primary commodities other than fuel; for exports "other primary commodities" are taken, but this excludes minerals and metals together with fuels. Thus, in the case of exports, a certain amount of underestimation is involved in the data on nonfuel primary commodities.

Sources: World Bank, *World Development Report,* 1982 and 1983 issues, Tables 9 and 10.

The data in Table 5.6 are simple approximations of export and import shares based on averages. Changes in these shares over time are discussed in detail in Chapter 6. Attention is drawn to the following points. First, the values of $(X_M-X_P)/X'$ follow a regular declining trend as the Y/L level increases. The pace of the decline is sharp and for group V, the term is large and negative. Yet, the share of primary commodity exports is of a considerable magnitude for group V, accounting for one-fourth of total nonfuel merchandise exports. This case distinctly contrasts with the classical dichotomy.

On the other hand, the value of $(M_M - M_P)/M'$ follows a slower pace of decline as one moves towards groups of higher Y/L levels. The sustained high share of manufactures, M_M/M', may be a surprise. Despite the effects of import substitution, this share is reluctant to decrease. The sustained high capital goods import share contributes considerably to this pattern. Even group V has a high percentage of manufactured goods imports that is, in fact, close to that of group III. Thus we can say that P_{im} makes less of a difference through groups as compared to P_{ex}. In short, the effects of export and import prices are not symmetric but rather asymmetric as seen from the different phases of development.

The share difference shown in the last column of Panel B draws particular attention. For the developing economies of lower Y/L levels, i.e., groups I–III, the import share of primary products is greater than the export share of manufactures. This also indicates that the export share of primary products is greater than the import share of manufactures in this interval. The magnitude of this difference is not necessarily small, though needs some qualification.[108] As a result, the effects of changes in the terms of trade is reversed: in the case of increases in P_{ex}/P_{im}, countries in these groups are unfavorably affected, and in the case of declines in P_{ex}/P_{im}, the effects are favorable. Ironically, with regard to the countries in groups IV and V, this reverse in the terms of trade disappears.

Thus, the commodity price approach is desirable. Regretfully, however, for the period 1960–1980, it is not possible to obtain direct evidence. Thus, some speculative observations are employed. Earlier, it was found that during this period $P_A > P_M$. With respect to P_A, the annual rate of increase on average is 0.52 for nonfuel, 0.96 for food, but -0.58 for nonfood. If we assume that the export price of manufactures, P_M, is a proxy for its import price, we can suppose the possibility of $P_A > P_M$. Or to put it more conservatively, at least

---

108 The magnitude is subject to some underestimation. First, the term X_M/X' may be overestimated for reasons noted in the table. Second, because of the resource gap (M>X), this difference must be underestimated as our discussion simply assumes M = X in total.

$P_A < P_M$ might not be the case. We think this possibility can be included in interpreting the decline of the terms of trade discussed earlier which is strongly affected by the oil price upsurge.[109]

III. The Domestic Price Structure of Developing Economies

The influence of international elements on the local price structure tends to be increasingly broader and stronger. The influence of international elements cannot be grasped by a sweeping presumption as it varies among nations due to the initial conditions and government policies adopted in the subsequent periods as well as to differences in phases and typology. Observation at any given point in time would thus reveal a wide range of difference in the domestic price structures of developing nations. Yet, is there any regularity regarding the differences and patterns of changes between nations? We will try to answer this question using the limited cross-sectional data for the groups.

The discussion focuses on two dimensions which are relevant to our hypothesis of technology diffusion: (1) the role played by the low price level of human work and (2) the effects of the high price level of investment goods, in particular, producer durables (PDE). It is our presumption that the domestic price structures of developing economies are essentially composed of these two elements in varied combinations.

Ideally, price analysis should be approached from three dimensions: output, input, and expenditure. The conventional notion of tradeables vs. nontradeables pertains to output in the usual analysis, while our approach uses the input dimension. The majority of available data, however, are confined to expenditure. Thus, an empirical study of prices faces a number of difficulties, and without making heroic assumptions, we will not be able to proceed. First, by using data on purchasing power parity (ppp) we will give a general picture of the pattern of the major components. By doing this, the relationship between the expenditure data and output and input is clarified. Following this preliminary step, we will examine

---

109 The discussion on export and domestic prices in this chapter will provide indispensable knowledge for Chapter 6 on international trade and will not be restated. In particular, since for imports no systematic price data are available, the knowledge gained from the purchasing power parity data (Section III) combined with the changes in import shares will provide important information for import analysis.

the price structure in depth from the input dimension. The discussion will focus on investment goods, and in particular, producer durables. After dealing with the implications of technology, production capacity is discussed from the dimension of price structure by employing a heroic assumption of the "equivalence" assessment in evaluating human work.

The General Pattern of the Major Components

Dividing goods and services into two categories, tradeables vs. nontradeables, is a conventional but debatable issue, though no one disagrees with its necessity. We understand that no standard has been established to divide goods and services according to this dichotomy, both conceptually or empirically.[110] We have no intention of proposing any new idea in this regard, but would like to suggest a certain device which links the categories with the input approach used later. We begin with a preliminary discussion about the distinction between tradeable and nontradeable goods and "services" as presented by data on purchasing power parity.

The relevant data are presented in Table 5.7. Note that services are defined as including "categories in which expenditures are on goods that cannot be stored." All other categories of GDP are regarded as commodities. Qualification is needed in using these categories for the production dimension, otherwise discrepancies may be less for services than commodities. Despite a limited number of countries, fairly regular tendencies are found and the data generally exhibit a number of important features of the domestic price structure of developing economies.

First, as is well known, an exchange rate close to the purchasing power parity value of the currency is considerably stronger than that of market rates. In contributing to this phenomenon, services increasingly plays a more significant role than does commodities, as is indicated by the rising ratio, services/commodities. Second, through groups by Y/L level, we see that this tendency tends to be stronger at lower levels than at higher levels.[111] Though the data are limited, we believe that the pattern indicated by the data in

---

110 For example, see Rudiger Dornbusch and F. Leslie C. H. Helmers, eds., *The Open Economy* (EDI Series in Economic Development 1988).

111 No reliable data are available for examining this path because of the difficulty involved in identifying equivalence between commodities from the production aspect, though partial efforts have been made by some economists.

Table 5.7
Relative Purchasing Power [a]:
Services vs. Commodities by Group, 1975 (percentage)

Group [b]		Services [c]	Commodities	Services/Commodities
I	(2)	15.7	52.1	30.1
II	(5)	19.0	51.0	37.3
III	(4)	26.1	51.1	51.1
IV	(3)	44.1	75.5	58.4
V	(7)	84.8	103.7	81.8
United States		100.0	100.0	100.0

Notes: a. For the indexes of relative purchasing power, the United States = 100.
b. The number in parenthesis refers to the number of countries included from our earlier grouping. The total number of countries selected is 23, out of the 38 total that were selected for this project.
c. Services is chiefly composed of items pertaining to medical care, recreational, and educational activities in addition to the compensation by the government.

Source: United Nations, *World Product and Income,* International Comparison Project (ICP), Phase III (New York: UN, 1982), Table 6.12:196.

the table represent a general tendency of changes in the domestic price structure from one that is dominated by indigenous factors towards one that is increasingly internationalized.[112] If the cross-sectional data are assumed to broadly represent the long-term changes over time, the output price of services must tend to rise at a faster pace relative to that of commodities. This would mean a regular over-time change in the domestic price structure through development. This sketchy description may be enough to summarize the conventional knowledge from which we will start our analysis.

Let us first pose a simple question: is this observation of the relative performance of price of services consistent with the findings in the preceding section? P_S should increase continuously relative to

112 Technically, the assumption of equivalence for price indexes in estimating ppp for individual items may have some problems, but we believe this does not matter substantially for our purposes.

P_I and P_A in order to meet the requirement of consistency. Actually, this is largely met for the higher Y/L groups in the cross-sectional data. The Western historical pattern shows a relative rise in P_S and is also consistent. However, a reverse tendency is witnessed for the lower Y/L groups (i.e., groups I–II) of developing economies. Our interpretation is that this is due to the peculiar situation of the preprimary phase. As a matter of fact, the nontradeable output of the services sector was given an important role by this interpretation. However, the affirmative response thus given for the direction of relative price change cannot be a final answer but rather casts an important but difficult problem that we must deal with. Services in the ppp data are narrowly defined, whereas the coverage of output of the services sector in our data is much broader. The share of nontradeables involved in the economy and its changes are thus quite different. Without specifying services in estimates with a precise definition, our analysis cannot proceed. It is our understanding that the ppp performance of "services" also pertains to the input aspect which consists of an element of human work. This aspect is discussed later.

Turning to the second step of our general observations, the ppp data for the commodities are broken down to be fairly comparable to the time-series and cross-sectional data that have been analyzed in previous sections. Two categories are selected for this purpose: one is relevant to investment and the other is relevant to sectoral output. The first category uses three ratios: gross domestic investment/gross domestic product (I/G); producer durables/gross domestic product (P/G); and product durables/private consumption (P/C). The second category also uses three ratios: product durables/food (P/F); clothing and footwear/food (L/F); and clothing and footwear/product durables (L/P). These ratios are listed in Table 5.8 for 1980 and 1975 to link with the data in Table 5.7. The purchasing power parity exchange rate can be estimated using the ratio of the aggregate domestic price level in each economy relative to that of the United States. This exchange rate is almost twice as strong as the market rate in groups of lowest Y/L. We share the view that the broad pattern seems to be consistent through 1975 and 1980 in the original reports[113] and also for the selected countries in our framework. The pattern of the ratios across groups rather than

---

113 This can also be said for the relationship between services and commodities. Refer to the statement pertaining to this in II 7, p. 16, Part II of this report.

Table 5.8
Purchasing Power Parity: Ratios of Selected Items by Group Average, 1980 and 1975 [a] (percentage)

	I/G	P/G	P/C	P/F	L/F	L/P	GDP
<u>1980</u>							
I (4)	2.50	2.54	2.58	2.01	1.62	0.81	0.514
II (7)	2.17	3.32	2.25	2.38	1.01	0.42	0.531
III (8)	1.76	2.63	1.85	1.89	0.75	0.40	0.582
IV (7)	1.82	2.06	1.31	1.63	1.02	0.63	0.799
V (10)	1.10	1.24	1.09	1.03	1.12	1.09	1.197
Unites States	1.00	1.00	1.00	1.00	1.00	1.00	1.000
<u>1975</u>							
I (2)	1.62	3.92	3.93	3.06	1.48	0.352	
II (5)	1.66	3.47	3.49	2.07	0.89	0.409	
III (4)	1.25	2.56	2.64	2.09	0.81	0.385	
IV (5)	1.45	1.93	2.32	1.72	1.24	0.630	
V (7)	1.06	1.19	1.16	1.12	1.17	1.007	
United States	1.00	1.00	1.00	1.00	1.00	1.000	

Note: a. The original data are presented relative to the price level of the United States for various subaggregates of gross domestic product. Estimates for the breakdown into services and commodities are not available for 1980.

Sources: Sources of Table 5.3; United Nations, Commission of the European Communities, *World Comparisons of Purchasing Power and Real Product for 1980,* Part II (New York: UN, 1987), Table 9.

the changes in the absolute magnitude of the ratios are of concern here.

With respect to the ratios relevant to investment, we identify a straight tendency of decline in all three ratios, I/G, P/G, and P/C, from the lower to the higher Y/L groups. There are biases but these are very minor, particularly in 1980. The decline in the ratios of producer durables is especially remarkable. For the nations of lower Y/L levels, the price of investment goods, and in particular the price of producer durables, in the domestic price system tends to be higher. In the case of P/G, for example, the ppp of producer durables differs only slightly among the developing countries; the ppp of GDP is most responsible for making the difference in P/G. Of course, import dependency is greatest for producer durables and this is a crucial aspect of the process of internationalization of the indigenous price structure. All of this information can be used in our input approach, which will be developed later.

With regard to the ratios relevant to sectoral output, which aims at linking with the output approach, our approximation is less reliable because of the wider discrepancies between the expenditure and the output-production approach. Nevertheless, we can assume clothing and footwear (L) to be a rough indicator for the product of light manufactures, F to be a rough indicator for food produced in agriculture, and P to be a rough indicator for machinery produced in manufacturing. Both L/F and L/P show a U-shape pattern that is distinctly different from the ratios in the first category. In the lower Y/L groups, i.e., groups I–III, L/F and L/P tend to decrease but begin to increase in the higher Y/L groups, i.e., groups III–V. This pattern indicates a decline in output prices of light manufacturing relative to agriculture and a minor decline relative to machinery manufacturing. Incidentally, P/F exhibits a declining tendency, with a minor exception for group I to group II in 1980. This pattern is similar to that of the ratios in the first category.

In discussing the possible consistency of our findings based on the two sets of data, an explanation of the declining tendency in the relative price of manufactures is warranted. In general, there appears to be consistency between the cross-sectional approach which uses the ppp data and the historical pattern observed earlier. However, we must note the discrepancies between the individual developing countries. Table 5.9 presents the pattern of rise and fall in P_I and P_M in terms of the number of countries. As expected, the number of countries which exhibited a pattern of "rise" is greater for manufacturing than industry and is greater in the 1970s than in the 1960s. Although these figures seem less impressive than the average measures for the rates of change in relative prices, P_I and P_M, which were shown in Table 5.1, the data confirm that a normal path of industrialization had not been realized in a considerable number of developing nations. It goes without saying that import substitution and related policies are relevant in bringing about such a situation due to a lagged pace of technological and/or organizational advance. But, on the other hand, a number of nations, not only Asian NIEs, achieved a normal path of industrialization by realizing a declining trend of P_I and/or P_M.

Table 5.9
Performance of Output Prices of the Industrial and Manufacturing Sectors, 1960–1980 (number of countries)

	1960-1970				1970-1980			
	P_I		P_M		P_I		P_M	
Group	Fall	Rise	Fall	Rise	Fall	Rise	Fall	Rise
I	3	2	1	5	2	6	1	7
II	3	4	3	4	2	8	3	7
III	5	2	3	4	3	8	2	9
IV	7	4	3	8	6	5	4	7
V	11	1	11	1	4	6	5	5

Sources: See sources of Table 5.1.

The ppp data cannot provide a useful general interpretation concerning this problem, but the sustained high prices of producer durables is believed to be relevant to this phenomenon. Table 5.8 suggests that the normal path of decreasing P_M through industrialization, when witnessed, is almost always confined to products of light manufacturing.

Prices of Producer Durables and Investment Goods

Domestic investment is the crucial activity in testing our hypothesis concerning borrowing technology. Investment, as seen from the supply-production side, however, is not homogenous. Although investment goods are assumed to be the carriers of technological knowledge, this assumption virtually pertains to producer durable equipment and is less relevant to the other components of investment. Recognition of this point is particularly important in the analysis of developing economies. It is presumed that the price performance of producer durable equipment in developing countries must have a specific pattern that is strongly influenced by the international element. This presumption emanates from our original hypothesis, and examination of producer durable equipment prices can provide important evidence for testing its validity. As stated earlier, we believe that the price of services is representative of items that are most strongly influenced by the domestic market. Thus, these two, producer durable equipment and services, illustrate the two extremes. All other items are somewhere in-between these two and are more or less a mixture of the two elements. To the extent that the hypothesis is valid, what is the significance of the price structure in explaining the performance of developing economies? The search for the answer begins with an examination of the core role played by the producer durable equipment prices, and is then extended to related topics.

(1) Prices of Producer Durables: The Peculiar Pattern in the Domestic Price Structure

The most comprehensive data available are provided by the International Comparison Project, ICP, Phase IV.[114] Table 5.10 presents a summary of the average purchasing power parity, ppp (where the U.S. = 100), by country group. The group classification is the same as in Chapter 1, but other countries for which the ppp are available are added. In addition to producer durable equipment and investment, other relevant items are included in the table for comparison.

As is widely known, in the aggregate, the GDP price level of developing countries tends to be lower, or the ppp tends to be higher, as per capita income level becomes lower. Although our yardstick is product per worker (Y/L) rather than per capita income, this general trend is also seen for GDP in Table 5.10, although there is a slight reverse relationship between groups II and III. The same pattern is witnessed for consumption. What is conspicuous is that the ppp of producer durable equipment, shown in the first column, exhibits a peculiar pattern: the difference between groups appears to be the least for producer durables than all other relevant items in the table. In other words, its price level is not only high but differs only a little despite a remarkable difference in the Y/L level of countries. One may say that the relatively high level of ppp of producer durable equipment for lower-income countries is a well-known fact (this was touched upon earlier). In fact, this is one of the important findings of the ICP Report (Part Two, 16). However, we are more concerned with the significance of this peculiar performance. The components of producer durable equipment are machinery, electrical machinery, and transport equipment in the ICP estimates and include the "machinery" included in our following discussion on trade in Chapter 6. The majority of producer durable equipment in developing economies consists of imported goods; this is even true for the NIEs which recently began to export certain producer durable equipment. Price levels basically depend upon "border prices" which cannot vary much between countries, though

114 United Nations, *World Comparisons of Purchasing Power and Real Product for 1980,* Part II: Detailed Results for 60 Countries.

Table 5.10
Comparative Price Level of Producer Durable Equipment and Relevant Items in 1980 [a] (U.S.=100)

(A) Indexes

Group	Producer Durable Equipment	Gross Domestic Product	Investment	Consumption	Government
I (5)	137.9	47.9	126.2	47.4	18.6
II (12)	140.8	59.7	137.8	57.8	33.0
III (11)	155.7	58.7	100.0	56.8	36.3
IV (10)	152.1	75.5	100.5	74.8	59.7
V (15)	144.8	111.9	124.1	111.7	77.7

(B) Ratios

Group	Producer Durable Equipment/Gross Domestic Product	Producer Durable Equipment/ Investment	Producer Durable Equipment/ Consumption	Producer Durable Equipment/ Government
I	2.88	1.09	2.91	7.41
II	2.36	1.02	2.44	4.27
III	2.65	1.56	2.74	4.29
IV	2.01	1.51	2.03	2.55
V	1.29	1.17	1.30	1.86

Note: a. Figures in parenthesis are the number of countries selected. Some countries of the ICP are included in the table, so that the number of countries in this table is greater than that in Table 5.8. The figures are accordingly somewhat different. Panel (A) shows simple averages and Panel (B) are simple ratios of the data in panel (A). For some items, the average of the ratios for individual countries are tested, and only minor differences are found.

Source: United Nations, *World Product and Income*, Table 9.

they can be somewhat different as they are affected by local costs. This is the major reason for the unique performance of producer durable equipment prices, which reflects the international diffusion process of advanced technologies.

In the last column in Panel (A) of Table 5.10, the price of "government" is shown to illustrate the performance of the price of services.[115] The price exhibits another noticeable performance, which manifests itself in terms of very distinct differences in prices across groups. A regular increasing trend is witnessed throughout the groups, from group I to group V, without exception. This pattern is similar to what was described earlier, depending on the 1975 data for services. We believe that this is another important example of price performance in developing economies because its pattern of difference, not the absolute level itself, roughly indicates trends in the changes in prices of human work through the development path of increasing Y/L levels. This is discussed again later in the book when dealing with nondurables.

The price of "investment," which is shown in the third column in terms of gross domestic capital formation, behaves differently from the other terms; the price of investment tends to decrease through the groups of developing nations, and then turns to increase from group IV to group V. What is surprising is that this is not consistent but appears to be rather partially reversed in comparison to the performance of producer durable equipment. If these estimates are statistically acceptable, development analysis via the investment approach and via the producer durable equipment approach would arrive at quite different results. The major reason for this discrepancy is the price performance of construction which is very high for the lower Y/L groups. (This will be examined in more detail later in the book.) We must not forget that the price of investment also consists of two elements: the domestic and the international components. The domestic component is roughly represented by construction and the international component is roughly represented by producer durable equipment.

The ppp of GDP is an aggregate measure of the three major components: consumption, investment, and government. Due to the largest weight for consumption, the performance of the GDP price is very similar to that of consumption. Our discussion thus far, however, suggests a number of important problems in disaggregation. Panel (B) shows the ratios of these major categories in order to elucidate the position of producer durable equipment in the local price structure. While at first glance this data may appear to tell us nothing substantially new, speculative interpretation may have different implications.

---

115 The ppp of services are not available in Phase IV data in the report (Part Two: 16), but "government" can be roughly representative of services. We use government here as employee compensation which is estimated to be 75 percent of general government expenditure.

The factors specific to local circumstances contain the effects of national policy and its implementation. For example, import substitution policies increase the local price level of goods affected by the various protection measures. This is particularly true for producer durable equipment. In addition to the ordinary costs of transportation and distribution, the price of producer durable equipment must be higher than its U.S. level to the extent that these goods are produced in a protected market. The ICP data, shown in the first column of Panel (A), represent these effects: that is, the price of producer durable equipment is higher by some 40–50 percent as compared to the U.S. price level. If these additional costs were closely associated with the difference in the GDP price level, then the ratio PDE/GDP would be nearly equal across the groups. However, the actual figures of the ratio PDE/GDP differ to a considerable extent. A much smaller ratio for group V is understandable, but apart from that instance, the ratio does not show a regular tendency to increase towards lower Y/L level groups. For such deviations, the factors (which were discussed earlier) that are specific to local circumstances may be responsible.

The pattern of the ratio PDE/consumption is very similar to that of PDE/GDP, and further explanation will not be needed. The ratio PDE/investment shows a clear declining tendency for developing nations as one moves toward the lower groups. This pattern is expected from what was found earlier in discussing Panel (A). What we are particularly concerned with here is the ratio PDE/government because this is an illustrative indicator of the ratio of the two price elements: the international and the domestic markets. The price level of "government" cannot be used for a normal differential among groups because of certain possible distortions. Yet the ratio suggests a certain regularity to the extent that the difference of the two elements appears to be increasingly larger in the lower Y/L groups and to the extent that the magnitude of the ratio for group I is enormous. This is important since the problem of evaluating production capacity cannot be realistically solved without taking this fact into serious consideration. This important issue will be discussed later.

(2) The Expenditure of Producer Durable Equipment

Three topics are discussed below: producer durable equipment expenditure in GDP and its per labor input, the disaggregated performance of producer durable equipment, and producer durable equipment relative to total capital formation.

First, expenditure on producer durable equipment as a proportion to aggregate expenditure (as represented by GDP) must differ to a remarkable extent between the two measures, national currency vs. ppp. Per capita GDP in ppp terms is often used as the basic indicator for evaluating the real economic level of nations. However, we are also concerned with producer durable equipment per labor input, because it directly relates to production activities. All of these relevant data are presented in Table 5.11.

In the first column, it is clear that the ratio of expenditure on product durable equipment (PDE) to GDP in national currency differs very moderately across groups, although it appears to follow an inverse U-shape. In ppp terms (second column in the table), as expected, both the level and pattern differ markedly. The level of PDE/GDP in ppp terms generally decreases as we move towards group I, although with some deviation. In the third and fourth columns, the per capita expenditure of GDP and producer durable equipment in ppp are compared. A much sharper trend of decrease is witnessed for producer durable equipment as compared to GDP towards the lower Y/L groups, with no noticeable exception. Finally, in the fifth column the important term—that is, the PDE input per economically active population—is presented. Its pattern differs from the per capita measures because the labor force participation ratio differs considerably across groups (last column in table). Yet the sharp trend of decline towards lower groups holds firmly.

Viewed from our hypothesis, this pattern of producer durable equipment performance per unit of labor is of great significance. At the flow level, the ratio of producer durable equipment per unit of labor suggests substantive differences in what we call production capacity (PC) of the developing economies. However, expenditure on producer durable equipment cannot measure PDE intensity for these nations relative to that of the United States. Nevertheless, in reality, the life duration of the equipment may not differ much across countries. If this is accepted, then the flow measures can roughly approximate the PDE intensity per laborer in its relative order of difference among groups (though the actual life duration of the equipment may be shorter for groups of higher Y/L levels). Besides, the physical amount of producer durable equipment cannot be accurately determined, even by stock measurement, and one may assume that the ppp measures present the amount at least in a relative sense. If these presumptions are acceptable, the surprisingly low level of PDE intensity for developing economies in general and for nations of lower Y/L level in particular draws particular attention. We are requested to make the distinction between the two approaches to evaluating factor combinations: one approach is in terms of national market prices and the other

Table 5.11
Expenditure on Product Durable Equipment as a Share of GDP and Per Labor (percentage)

Group	Ratio of PDE Expenditure to GDP: In National Currency	Ratio of PDE Expenditure to GDP: In ppp	Per Capita Expenditure in ppp: GDP	Per Capita Expenditure in ppp: PDE	PDE Input per Labor in ppp	Ratio of Labor Force to Total Population
I	7.76	2.69	4.72	1.49	1.68	42.4
II	9.69	4.75	9.33	5.65	6.95	39.2
III	9.82	3.72	16.33	8.15	12.52	31.4
IV	9.09	4.57	32.84	19.44	24.17	38.8
V	8.93	6.92	76.54	63.33	70.09	43.6
United States	8.33	8.33	100.00	100.00	100.00	48.3

Sources: See sources of Table 5.10; and International Labour Organization, *Economically Active Population, 1950–2025.*

approach is in terms of ppp measures. Later we will return to this issue.

The second topic regarding disaggregation of producer durables is taken up to expand on the preceding discussion. The prices (in terms of ppp where U.S. = 100 and shares as a percentage of total producer durables in terms of ppp expenditures) of the three major components—machinery (M), electrical machinery (EM), and transport equipment (TE)—are presented in Table 5.12. Since PDE input per unit of labor in ppp terms is already included in Table 5.11 (column 5), just the share will suffice here. In view of our conventional treatment of industrial analysis, no doubt this is too simple a disaggregation as a great number of various items are included. Nevertheless, we believe that this treatment is enough for our purposes. We wish to examine three points. First, what will be the disaggregated performance of the product durable equipment prices? Second, what is the contribution of these components to the aggregate performance of product durable equipment per unit of labor? Third, how significant is this in terms of technological advance?

The PDE price level and its performance across groups are essentially not very different at the disaggregated level. A simple average of the price level for the developing nations (i.e., groups I–IV) is 145.7 for machinery, 143.6 for electrical machinery, and 172.8 for transport equipment. The price levels for machinery and

Table 5.12
Disaggregation of Producer Durables: Prices and Expenditure Shares

(PDE=100)

Group	Prices [a] (ppp)			Expenditure Shares [b] (%)		
	Machinery (M)	Electrical Machinery (EM)	Transport Equipment (TE)	Machinery (M)	Electrical Machinery (EM)	Transport Equipment (TE)
I	136.0	104.1	187.6	73.3	5.6	21.1
II	141.4	139.3	148.4	63.0	18.8	18.2
III	145.5	144.6	182.9	53.0	18.8	28.2
IV	159.9	186.4	172.3	58.4	18.4	23.2
V	141.6	144.4	156.1	55.3	18.6	26.1
United States	100.0	100.0	100.0	49.1	28.5	22.4

Notes: a. The price level of the aggregate of producer durable equipment is presented in Table 5.10.
b. The expenditure shares are the per capita expenditure of the three items in terms of ppp (U.S. = 100) for individual countries and its group averages are converted to the shares using U.S. shares.

Source: United Nations, *World Product and Income*, Tables 6, 7, and 9.

electrical machinery of the developing economies are close to the level of group V, though the price level for transport equipment is somewhat higher than group V's level. For this range, the price levels of both M and EM generally follow a similar trend of decline towards the lower Y/L groups, while TE appears to follow a zigzag pattern and exhibits no smooth tendency. However, a closer look reveals important suggestions about the path of technological advance. As pointed out earlier, a sequential process of successive progress is recognized with respect to borrowing advanced technological knowledge and its domestic assimilation. Viewed from this successive process, changes in the quality level of machinery would proceed from being inferior to superior corresponding to the development path from a less sophisticated to a more sophisticated level of technologies adopted. The trend of increase in the price of machinery indicates this process of quality upgrading: that is, the increase in the price level is more rapid for EM and is slower for M. In the lower Y/L groups, the lower price level in electrical machinery as compared to machinery may imply that their technology is relevant to less sophisticated electrical machinery and the sharp increase in the price level of EM towards higher groups suggests that the technology moves to a more sophisticated level. A decrease from group IV to group V may be due to a decrease in the import dependency ratio via the development of domestic production. The same suggestion is applied to machinery but with less intensity as its relevance is expected to occur earlier. We are convinced that there is no inconsistency between this explanation and the earlier one about the peculiar performance of product durable equipment prices, since the latter explanation is a comparative one of the price performance of other categories.

Let us now turn our attention to the expenditure shares. The share of machinery tends to increase towards the lower Y/L groups, while the expenditure share of electrical machinery shows almost no difference for the majority of developing economies (i.e., groups II–IV), though we see a sharp decline from group II to group I. The difference between the performance of expenditure shares for machinery and electrical machinery is not unexpected given the preceding discussion on relevance. It is widely known that the use of electrical machinery tends to increase later after the use of machinery (industrial or general) has been developed to a certain extent. With respect to transport equipment (TE), we can see no distinct tendency, the reason for which is not clear. However, we can say that the expenditure shares of transport equipment depend much more upon national requirements such as infrastructure, etc., and is thus less relevant to technological progress.

What is of importance is the direct ratio between electrical machinery and machinery. The ratio EM/M is 7.6 for group I, 29.8 for group II, 35.5 for group III, 31.5 for group IV, 33.6 for group V, and 58.0 for the United States. The rising trend of EM/M in the lower Y/L groups (i.e., groups I–III) of developing economies shows a pattern as usually expected, but in the higher Y/L groups (i.e., groups III–IV) together with group V of developed nations, there is no clear continuous trend. In examining the reason for this, the ratio averages are calculated for two subgroups in groups III, IV, and V, where subgroup A is the subgroup with the higher ratio and subgroup B is the subgroup with the lower ratio. The ratios of EM/M for the subgroups are as follows:

	Subgroup	
Group	A	B
III	32.4	8.6
IV	39.0	15.8
V	43.6	23.1

Note that the ratio of A in group III excluded the Republic of Korea. Because of the small number of countries in the analysis, we should be cautious in interpreting the data; nevertheless, we believe that there appears a continuous trend in the ratio for subgroups. If the ratio of subgroup A in group V is linked with the ratio of the United States (58.0), continuity is further confirmed.

Two points are suggested in a substantive sense. First, if expenditures are deflated by an internationally uniform price index, the deflated values of producer durable equipment can be used to indicate its physical quantities. The ratio M/EM in this case can be used as one of the convenient and useful indicators for ranking countries in terms of advancing industrial technology. Second, if this proposal is acceptable, then the technological level as reflected in the production capacity tends to move positively with the Y/L level. This is a general tendency, and what is important is to identify the fact that among the countries belonging to the same group, the ratio varies to a considerable extent. The level of industrial technology is not necessarily positively associated with the Y/L level. The combined effects of these two dimensions may be different between the lower and higher groups of developing economies.

The third topic is the relative position of producer durables in terms of gross domestic capital formation and its relationship to construction. Note that disaggregated discussion on construction is not our intent. While we do not ignore the importance of the respective functions of the components of construction (the ICP presents data for residential, nonresidential, and nonbuilding construction), this is due to our specific focus: that is, construction will be simply treated to the extent that it relates to producer durables, and discussion of components is only for illustrative purposes. From this viewpoint, the major results have already been derived from the earlier discussions insofar as producer durables are concerned. However, additional discussion is needed to reveal the comparative significance of producer durables. Data that are relevant to the discussion are presented in Table 5.13. The same

Table 5.13
Domestic Capital Formation and Construction: Prices and Expenditure* [a] *(percentage)

Group	Ratio of Domestic Capital Formation to GDP		Ratio of Construction to GDP		ppp of Construction	PDE	Construction Expenditures per Unit of Labor
	In National Currency	In ppp	In National Currency	In ppp			
I	19.66	7.48	11.90	4.14	1.377	99.9	4.33
II	24.78	12.39	15.09	7.29	1.424	98.9	7.43
III	21.76	12.80	11.94	8.65	1.808	51.9	26.75
IV	24.19	20.33	15.10	16.24	0.702	46.2	63.36
V	28.79	25.94	19.86	19.28	1.149	79.0	115.46
United States	18.19	18.19	9.84	9.84	1.000	100.0	100.00

Note: a. The ratio of domestic capital formation to GDP in terms of ppp is simply calculated by deflating the first column by the ratio of investment to GDP derived from Table M10.1. The ratio of construction to GDP in terms of national currency are from United Nations, *World Product and Income,* Table 1, Part Two. The same ratio in terms of ppp is simply calculated from the group average by deflating the ratio of construction to GDP in national currency terms by the ppp of construction, which is from the UN source, Table 9, Part Two. Construction expenditures per unit of labor are from the UN source, Table 7, Part Two. The per capita data are simply converted to per labor group averages by using the participation ratio in Table 5.11.

Sources: United Nations, *World Product and Income*; and International Labour Organization, *Economically Active Population, 1950–2025.*

qualifications regarding the use of flow data apply to this table. The investment proportion in terms of national currency (the first column) generally shows a similar pattern to the pattern observed in Chapters 1 and 2, though the pattern here is somewhat different.[116]

The investment proportion in terms of ppp (the second column in the table) shows a trend of increase across all groups from lower to

116 For example, in the lower intervals, the tendency of increase is not continuous and for the higher intervals, there is a distinct tendency of increase. Note, however, that the data is only from a single year, 1980, and the composition of groups is not the same.

higher Y/L levels without distortion. An analogous pattern is seen for construction: in national currency terms, the performance of construction is not necessarily continuous but in ppp terms, its trend is similar to that of investment. Since the share of construction is much greater than that of producer durable equipment, the aggregate pattern of investment is governed by the former. Changes in the expenditure ratio of construction to producer durables deserves particular attention.

	I	II	III	IV	V
In national currency	1.53	1.22	1.98	1.69	2.22
In ppp	1.54	1.53	2.33	3.55	2.79

The expenditure ratio of construction to producer durables in terms of ppp indicates no tendency of increase in the lower Y/L range, but it does for the higher Y/L range of developing economies; in contrast, the ratio in terms of national currency appears to show no regularity. We believe that this ratio in terms of ppp is a meaningful indicator of the relationship between producer durable equipment and construction. One may wonder, however, why such a remarkable difference is witnessed between the lower and higher Y/L ranges.

The price level of construction in terms of ppp is shown in the fifth and sixth columns of Table 5.13. The price level of construction in terms of ppp tends to increase in the lower Y/L range, while the ratio of the price level of construction to the price level of producer durables shows much higher levels for the lower groups (i.e., groups I and II) as compared to the higher groups (i.e, groups III and IV). One may be interested in comparing these ratios with the corresponding ratios listed in Panel (B) of Table 5.10. Looking at these ratios for the entire range of developing economies, the sharp difference in the ratio is not seen for any of the other items. Therefore, one may ask whether it is possible to explain the sharp difference in terms of the combined effects of national and international elements. While we are not qualified to provide a substantive answer to this question, what we can do is examine the ICP data from a regional perspective.

As is stated in the ICP Report,[117] the ppp of construction is extremely high in the African countries and we believe this to be

117 "The conclusion to be drawn is that capital goods are relatively high-priced in Africa, and when the amount of construction and producers

the major factor in answering the question. The Report explains this characteristic of African countries with regard to capital formation without mentioning the difference between construction and producer durables. In contrast, we emphasize the performance difference of the two in this region.[118]

Let us examine this point by looking at non-African averages. The value of ppp of construction (corresponding to the fifth column in the table) is 0.455 for group I, 0.540 for group II, and 0.698 for group III. The comparable figures for ppp of construction in Table 5.13 are much higher and the ratio of the two ppp data are 3.03, 2.64, and 2.59 for groups I, II, and III, respectively. Likewise, with respect to the ratio of the ppp of construction to producer durables of African countries to non-African countries is 34.5 for group I, 38.4 for group II, and 45.0 for group III. As the number of non-African countries is small, this comparison is just an illustration; yet, we can clearly see that the magnitude of the ratios is of a different order. If we assume the hypothetical case where the ppp of construction in Africa is equal to that of non-Africa, then the construction expenditure per unit of labor in terms of ppp would be much higher: a rough estimate would be 13.1 for group I, 19.6 for group II, and 31.0 for group III.

Taken together, we are inclined to say that the price performance of construction in the African countries seems to be atypical relative to the normal pattern that is internationally envisaged where the price of construction tends to increase relative to that of producer durables with economic development. If the ppp ratio, construction/producer durable equipment, which was tentatively calculated for the non-African countries is linked with the ratios for groups IV and V, then a normal pattern is suggested.[119]

---

durables are valued at world prices, the real quantity is much less. The difference between investment effort and likely capacity creation as a result of that effort is quite significant" (United Nations, *World Product and Income*, 2).

118 The "core countries" that are used to link regional prices with world prices are Kenya and Senegal. The ppp (U.S. = 100) of construction is 2.579 and 1.696, while that of producer durable equipment is 1.447 and 1.600 for Kenya and Senegal, respectively. The price level of producer durables does not differ much from the group average for group II (Table 5.10).

119 Actually, the evidence for this is not sufficient. As is widely known, output prices of construction are difficult to estimate because of the unavoidable use of the cost approach. Just for illustration, the historical trends of the United States and Japan are shown For a comparable prewar interval (from 1879–1888 to 1929–1938 in the United States and from 1882–85

Substantive explanation of this peculiar pattern is beyond the scope of this book. However, a general speculative observation would be that, in forming the price level of construction, the domestic element operates more strongly than the international element. Construction has often been treated as a nontradeable. Yet "construction materials" are imported for developed nations and it is believed that this is particularly important for nonresidential and nonbuilding structures (the domestic element is believed to operate more strongly in residential construction). This supposition does not appear to apply to Africa.[120] As for the import dependency of construction materials, no information is readily available for African countries, but we suppose it may be of considerable magnitude.[121] Otherwise, such high prices cannot be explained. To that extent, prices of construction become analogous in nature to those of producer durables. On the other hand, the view which stresses the importance of the domestic economy is partly endorsed by much lower prices of nonbuilding (structures), which may depend much more upon local inputs including cheap labor.

Significance of the Factor Price Difference: An Illustration

In winding up the preceding investigation, we would like to focus the discussion on the structural differences in factor prices.

---

to 1932–1935 in Japan), the average annual rate of price rise is 1.74 for producer durable equipment and 2.10 for construction in the United States and 2.01 for producer durable equipment and 4.17 for construction in Japan. (For details, see Ohkawa et al., *LTES,* Prices, Vol. 8 especially Chapter I, Section III.)

120 The core countries' ppp is as follows:

	Residential	Nonresidential	Nonbuilding
Kenya	284.5	214.6	136.0
Senegal	257.9	213.5	126.8

The relatively low value for construction that was shown in the preceding footnote is due to nonbuilding.

121 The import ratio of construction materials to domestic production is estimated to be 19.4–27.8 percent for the five-year averages for 1905–1929 in Japan. Since then the ratio has become much smaller (Ohkawa and Kohama, *Lectures on Developing Economies*). Much higher ratios are implied for African countries in the description in the main text.

This is because of our recognition that the domestic price structure and its changes in the developing economies cannot be explained without elucidating on the structural differences in factor prices.[122] However, no data are readily available for systematically dealing with factor prices for these economies in our input approach. Therefore, both in concept and measurement, we are obliged to rely on rough proxies and on heroic assumptions in order to assess reality.

Let us begin by recalling the issue that was left unsolved earlier. When the producer durable equipment expenditure per labor was estimated in ppp terms, we found that its magnitude was extremely small for the lower Y/L groups (Table 5.11). Together with the case of construction (Table 5.13), the domestic investment expenditure per labor can be easily estimated (I/L, in our terms). The ppp of investment that is used (cf. Table 5.10) is not precisely the price of capital and the labor input is not evaluated by labor prices because no wage data are available in the ppp approach. To break through this bottleneck, we use investment goods prices as the proxy for capital prices. Analogous to commodities, the application of the equivalence assumption is tried for human work in order to assess factor price differences by group. From the preceding investigation, we know that a characteristic of factor prices in developing economies is the higher capital price vs. the lower labor price. Rather than proceeding along the line of ppp measures dealt with above, we want to estimate and explain its movement through different Y/L levels in the context of the general pattern of growth described in Chapter 1.

Our average treatment of the period 1960–1985 depends upon the grouping of nations by Y/L levels estimated at 1970 in U.S. dollars (cf. Table 1.1) and the investment proportion to GDP, I/Y, 1960–1980 (cf. Table 2.8 on average for the entire period). In order to illustrate the basic pattern, the domestic investment expenditure per labor, I/L, is estimated by I/Y·Y/L, using these figures. The term I/L cannot represent the factor combination in the conventional

---

122 One may say that an alternative is the use of expenditure data as proxies to the output approach depending upon the dichotomy: tradeables vs. nontradeables. The ICP, Phase III, adopts this procedure, assuming that the NTr is the sum of services and construction. According to this approach, numerical values are shown in Table 6.11. The price level relative to that of GDP is 66.6 for NTr and 134.5 for Tr for the lowest group (corresponding to groups I and II in our grouping), taking the highest group as 100 (corresponding to group V in our grouping). The differentials between NTr and Tr are much narrower as compared to the case of our input approach.

sense. We used the incremental term, $\Delta K/w\Delta L$ or K/Lw, in our analysis in Chapter 1. Keeping this in mind, I/L is taken as a first approximation to be priced. With regard to labor, the expenditure of wage payment by entrepreneurs can be approximated by the income share of labor, β (cf. Table 1.1, i, and Technical Notes). Expenditure for wages can thus be obtained by $Y/L \cdot \beta$ measured in U.S. dollars.

Now, the issue is how to evaluate the wage level that corresponds to the ppp measures. As is suggested in the preceding discussion on the government ppp measures (cf. Table 5, column 5), we are convinced that the answer can be obtained by assuming an equivalency in human work between developing and developed economies (taking the U.S. as the standard). "Services" in the private sector may provide more acceptable data for our purposes because of the lower number of distortions. Price of services as reported in the ICP Report, Phase III, Table 5.7 were adopted since data for 1980 are not available. Our presumption of equivalency in human work allows for a wider range of measurements, on the one hand. However, on the other hand, one may say that it is impossible to think about an equivalent level of activity of human work because of the wide differences in technological levels between nations in which human work is incorporated. Skilled labor is one example. An acceptable view is that skilled labor is composed of these two elements: on the one hand, skilled labor can be treated on the assumption of equivalence and on the other hand, it cannot. With regard to the latter case, it might be safer to omit any further evaluation. Yet, for our purposes, some sort of rough approximation is needed in order to obtain an aggregate evaluation of wages. Our bold assumption is that the labor price of the second category can be evaluated in association with the level of labor productivity which is evaluated by Y/L in terms of ppp.

Finally, the composition of the two categories of labor are assumed to be approximately given by the shares of producer durable equipment and nonproducer durable equipment in domestic investment. This stems from the notion that labor of the first category works with no producer durable equipment, while labor of the secondary category does work with producer durable equipment. With this presumption, our task of evaluation can be made operative. We emphasize that this procedure aims to provide meaningful illustrations of our proposition, rather than to give evidence of it. The preliminary figures are summarized in Table 5.14.

Table 5.14
Investment Per Labor Compensation, 1960–1980

(V=100)

Group	Y/L	I/Y	I/L	w	I/Lw	I/Lw [a] (in terms of ppp)	Price Ratio
I	2.6	70.3	1.8	2.3	78.3	10.6	13.5
II	7.0	78.5	5.5	6.0	91.7	15.2	16.6
III	12.6	100.0	12.6	10.3	122.3	42.9	35.2
IV	29.1	96.8	28.2	24.8	113.7	70.0	61.5
V	100.0	100.0	100.0	100.0	100.0	100.0	100.0

Note: a. I/Lw (in terms of ppp) is calculated using the rate of the price level of wages in investment goods in terms of ppp which was estimated from Table 5.15.

Sources: See sources of Tables 1.1 and 2.8.

The figures are derived from Table 1.1 and are expressed as indexes, where the level of group V=100. The level of investment per labor compensation (I/Lw) shows no large differences between groups, although we see a U-shape pattern, as expected. In contrast, the ppp measure shows a sharp increasing trend from lower to higher Y/L groups. If I/L were calculated by evaluating investment in terms of ppp, the values of I/L would be 1.8 for group I, 5.0 for group II, 15.6 for group III, 34.8 for group IV, and 100.0 for group V. This is very similar to the pattern of producer durable equipment per labor. It is clear that such an extremely low level can disappear by using a wage evaluation that appropriately corresponds to ppp measurements. What is appropriate is the problem at issue. The relevant figures used in the procedures are listed in Table 5.15.

The table is meant to explain the background data used to derive the ratio of price levels that were applied in Table 5.14. The estimates derived in the table may be called "semi-hypothetical." The price level of labor inputs is a hypothetical estimate, but the ppp for investment (1980) and services (1975) are actual data and the shares of investment and wages to GDP are average data that are centered on 1970, instead of 1980, to avoid fluctuations due to a single year. (For more details, see the notes of the table).

Table 5.15
Semi-hypothetical Estimates of Factor Price Difference

(V=100)

	Weighted Sum				Weighted by Shares [e]			Residue	
Group	w_1 [a]	w_2 [b]	w [c]	I [d]	w	I	Sum	Share [f]	Price Level [g]
I	18.5	5.4	13.7	101.7	7.3	17.4	24.7	29.8	61.6
II	23.4	12.1	18.4	111.0	9.4	21.4	30.8	29.7	45.6
III	30.8	22.6	28.4	80.6	13.9	19.8	33.7	26.4	61.8
IV	51.9	42.6	49.8	81.0	25.4	19.3	49.7	23.2	81.1
V	100.0	100.0	100.0	100.0	60.0	24.6	84.6	15.4	63.9

Notes:
a. w_1 (wages of the first category) are from "services" in Table 5.7.
b. w_2 (wages of the secondary category) are from Y/L in terms of ppp from Table 1.1 (Y/L) and Table 5.10 (ppp of GDP).
c. The weight used for w is the proportion of producer durable equipment and nonproducer durable equipment in gross domestic investment from Tables 5.11 and 5.13.
d. The ppp of I is from Table 5.10.
e. The weight is β for w (from Table 1.1) and I/Y in terms of national currency which is from Table 5.13.
f. The share of the residue to GDP is approximated by deducting the sum of shares of w and I from 100.
g. The price level of the residue is estimated by ppp of (GDP, column (7)) /column (8).

Without dwelling upon the technical details of the procedure, let us note the broad patterns that are indicated. First, the wage level which was obtained by assuming equivalence (w_1), shows a much smaller differential as compared to w_2, which was obtained using the productivity assumption between groups. We hope that these estimates illustrate the pattern and are not far from the reality. Second, no data are directly available to check the plausibility of this procedure. Rather, we try to examine it indirectly by calculating the price levels of the residue, the portion not dealt with by expenditures of investment and wages. This chiefly consists of consumption expenditures from nonwage incomes, although if a resource gap exists (i.e., I–S), its effects cannot be ignored. The pattern appears rather stable, showing neither an increasing nor decreasing tendency across groups, although the variation is not very small. The residue is the unexplained portion, and this pattern appears to generally support the plausibility of our bold approach: at least our treatment of investment goods and wages seems to be able to explain the major part of the ppp differentials of GDP. Third, if this is acceptable, we would like to argue that the different price structures of developing economies can essentially be explained by the combined effects of the two price levels: that is, the price level of investment goods and of human work.

How can we interpret the derived supposition, particularly the statement about the wage performance? Let us try to answer the question from two aspects: one aspect pertains to NTr, nontradeables, and the other aspect pertains to the human capability in relation to SC and PC.

First, viewed from the output-expenditure approach, the price level of the so-called wage goods is an issue. It is impossible to obtain this price level from the conventional ppp approach. In the output approach, wages have often been used together with commodity prices. One example was the estimation of real exchange rates, but here the real meaning of the procedure was not specified. The possibility of linking the price level of labor can be found in its relationship to the price level of wage goods. We are convinced that the lower price level of labor in lower Y/L groups corresponds to the lower price level of wage goods. The major reason for the lower price of wage goods can, in turn, be explained by the low price level of human work because wage goods are chiefly composed of NTr. As the Y/L level goes up, the proportion of Tr of higher prices tends to increase. This must be the major reason for its tendency to increase toward higher Y/L levels. This is relevant to the performance of output prices of the services sector that was discussed in Chapter 3.

Second, our hypothesis has some relevance to the concept of human capability (HC) which was defined in Chapter 1. Enterprises employ workers by paying prevailing wages which are essentially determined by the operation of the market mechanism. It is assumed that workers can socially contribute to enhancing productivity beyond the marginal product of labor, the degree of which depends upon the relationship between production capacity (PC) and social capability (SC). The assumption of equivalence applied to human work pertains to the human capability in this conceptual framework. Actually, HC is estimated with regard to wages on average for the groups, without paying attention to possible differences, which could be shown in subgroup observations if the data so permitted. In view of the findings from the subgroup observations of subgroups A and B in Chapter 1, it is speculated that the same wage level can imply higher activity of human work in subgroup A as compared to subgroup B.

In general, given the extremely high price level of investment goods, particularly the price of producer durable equipment in developing economies, the potential capability of human work is our key concern. The assumption of equivalence originally stems from this concern, and we attempt to illustrate the degree of discrepancy between wages and possible capabilities. It is hoped that the discrepancy is significantly recognized as the key element in achieving technological advance in the developing economies.

Chapter 6

Growth and International Trade

This chapter discusses the problem of international trade. However, the purpose of this chapter is not to analyze trade independently, but rather to examine trade from the aspect of growth under the hypothesis of technological progress (this approach underlies all preceding chapters). This naturally implies that the discussion is focused on merchandise trade, imports, and exports that are related to the performance of the domestic economy, in particular, to changes in the domestic production structure and productivity growth. Therefore, the approach cannot be the same as the conventional trade analysis. This is not due to our peculiar preference, but rather the result of our efforts to respond to the challenges presented by the reality of contemporary developing economies.

In Sections I and II, imports and exports are dealt with, respectively. However, we believe it is important to pay particular attention to what we call the "sequential interrelation" between imports and exports. In Section III this sequential interrelation is discussed in terms of phasing. Efforts are made to clarify the relationship between development phases and exports and imports. Development phases were discussed with respect to domestic structural changes in the previous chapters. In this chapter, development phases are discussed with respect to trade pattern changes.

I. Imports: Requirements and Substitution

Imports are discussed from the perspective of domestic requirements and substitution. From our hypothesis of technological progress, two dimensions naturally arise. First, borrowing and assimilating advanced technological knowledge requires imports of investment and intermediate goods. Growth stimulates further increases of this requirement. Second, technological progress contributes to enhancing productivity growth and domestic production, thereby shifting supply curves to the right. This in turn leads to a reduction in the import requirements as the domestic economy successfully achieves import substitution.

The ex post records of import performance reflect the combined results of these activities and understanding the process is not a simple matter. A number of research work on imports of developing economies have, however, shed some light on this mechanism. We hope to add to this knowledge through additional scrutiny, with a special focus on the link between imports and the performance of the domestic economy. Because the effects of the various activities cannot be singled out effectively for analytical purposes, an indirect approach is used to obtain an overview. The topics of discussion are, first, an aggregate conceptual framework; second, a disaggregated treatment; and third, the relationship between imports and domestic industrial growth.

A Specific Framework

We do not attempt to provide a comprehensive treatment, but rather a partial and selective treatment of imports and growth. In dealing with imports (M), the focus is on merchandise trade, and the relationship with gross domestic capital formation (I) must be the first topic. This follows the basic line of "growth and investment" in Chapters 1 and 2, and differs from the conventional approach where the relationship of imports with income (Y) is examined through the import function, M = f(Y), where price effects are incorporated. We recognize the general usefulness of the conventional approach and believe that such a specification of imports is not necessarily contradictory. It goes without saying that merchandise imports are composed of a variety of goods for consumption, capital formation, and intermediate use. In the conventional import function approach, these disaggregate treatments have been developed beyond the macro-setting. In our framework, it is necessary to clarify the direct relationship between imports of capital goods and gross domestic investment at the macro-level. This is particularly so in light of the changes in this relationship through the various levels of developing economies. Thus the analysis proceeds with a disaggregation of imports of manufactured goods which are categorized into capital goods and noncapital goods.

In the discussion at the macro-level, the term $M/\Delta Y$ is used. This is called the "incremental import–output ratio" or IMOR. The IMOR describes the quantitative relationship between an incremental increase in the output and the required imports in a simple form. This term corresponds to ICOR, $I/\Delta Y$, a familiar term used in Chapters 1 and 2. The term IMOR is a product of $I/\Delta Y$ and M/I, the ratio of imports to domestic investment,

(1) $M/\Delta Y = I/\Delta Y \cdot M/I.$

Regarding two economies of equal ICOR, for example, the ratio M/I and, accordingly, IMOR may differ. A larger (smaller) value of IMOR is associated with a greater (smaller) amount of required imports for a specific level of output. Thus, the effects of domestic capital formation, despite an equal ICOR, may differ on the resource balance and hence the position of foreign payments. This may illustrate different effects on the import dependency of domestic investment. On the other hand, for two economies of equal M/I, IMOR may differ due to different magnitudes of the ICOR. A larger (smaller) value of IMOR again suggests that a greater (smaller) amount of imports is required for a specific level of output. This may illustrate the different effects of investment efficiency in the domestic economy and its relationship to import dependency.

Formula (1) is an identity describing the macro-relationship, and the ratio M/I is not a behavioristic term. But we believe the formula can be made more realistic if it is specified for the relationship between domestic investment and investment goods imports (M_I) as follows:

(2) $M_I/\Delta Y = I/\Delta Y \cdot M_I/I.$

Our concept of requirement is represented by M_I/I (i.e., the import coefficient of investment). This coefficient is assumed to be given technologically: so with a given magnitude of ICOR, the term IMOR or $M_I/\Delta Y$ is determined. The option of the type of technology used by the domestic industry and its investment efficiency are thus at issue under given prices of output and factors.

In the conventional approach, we have a well-known problem of the ex ante inequality, between $I–S = M–X$ (where S = savings and X = exports). Here the converted inequality between $I–M$ and $S–X$ is the point at issue. (Exports will be discussed in the section that follows.) In treating the balance $I–M$, the above-mentioned formula (1) (and later formula 2) are actually introduced. In the discussion that follows, ex post data, i.e., $I–M = S–X$, are used. But no assumption will be made a priori regarding either constancy or change in a certain direction of the ratio $M/\Delta Y$. Instead, we are concerned with the problem of whether the term has a certain regularity of change.

Table 6.1
Relationship Between Imports and Domestic Investment [a]
(percentage)

		IMOR (M/ΔY) [b]			ICOR (I/ΔY)			M/I		
		1960–1970	1970–1980	Ave.	1960–1970	1970–1980	Ave.	1960–1970	1970–1980	Ave.
(A): By Group										
I	(9)	6.8	6.6	6.7	5.2	5.2	5.2	1.30	1.26	1.28
II	(11)	6.5	7.5	7.0	4.6	5.2	4.9	1.41	1.44	1.43
III	(11)	4.3	4.6	4.5	3.1	3.4	3.3	1.39	1.35	1.37
IV	(11)	4.0	4.9	4.3	4.5	4.5	4.5	0.89	1.08	0.96
V	(12)	5.1	10.8	7.7	5.1	7.9	6.5	1.00	1.37	1.19
(B): By Subgroup [c]										
I	E	3.6	4.8	4.2	4.5	4.8	4.7	0.79	1.01	0.89
	F	7.8	5.0	6.4	6.2	5.8	6.0	1.25	0.86	1.06
II	E	3.2	4.1	3.6	2.8	3.6	3.2	1.13	1.14	1.14
	F	10.9	10.7	10.8	6.8	7.0	6.9	1.60	1.53	1.57
III	E	3.5	3.3	3.8	3.0	3.0	3.0	1.17	1.11	1.27
	F	4.2	5.5	4.9	3.2	3.9	3.6	1.31	1.41	1.36
IV	E	2.9	3.7	3.4	3.8	4.5	4.2	0.77	0.82	0.80
	F	4.9	4.4	4.7	5.6	4.4	5.0	0.87	0.99	0.93
V	E	6.1	10.1	8.1	4.8	8.1	6.4	1.27	1.25	1.26
	F	6.1	11.0	8.3	5.4	7.7	6.5	1.13	1.43	1.28

Notes:
a. Imports refer to imports of goods and N.F.S. The number of selected countries is in parenthesis in Panel (A). A slight difference from previous tables is due to data availability.
b. IMOR (M/ΔY) is estimated by I/ΔY·M/I. The term M/I is calculated by M/Y / I/Y using simple averages; 1960 and 1970 figures were used for 1960–1970, and 1970 and 1980 figures were used for 1970–1980.
c. Subgroups E and F are the same as that used in Chapter 4. Figures for subgroups do not necessarily add up to those for the group due to exclusion of some countries because of data unavailability and rounding.

Sources: World Bank, *World Development Report,* various issues; *World Tables,* various issues.

Table 6.1 is a summary of the relevant aggregate data by Y/L groups with subgroups E and F. The data provide preliminary factual knowledge. Let us first look at the data by group (see Panel A). The performance of IMOR, M/ΔY, which is our major concern, shows a fairly regular trend of decline throughout all groups of developing economies, though it increases sharply from group IV to group V, the group of developed economies. Let us look at the two components. The term M/I is greater than unity (M>I) for all groups with the slight exception of group IV. For the majority of developing economies, from group II to group IV, a trend of decline of M/I seems to be the normal pattern. This is witnessed both for the 1960s and the 1970s without exception. In the early interval, from group I to group II, and in the last interval, from group IV to group V, however, M/I increases. These reversed performances require additional attention. The ICOR changes regularly, forming a U-shape pattern: from a high value, the ICOR declines and then turns to increase toward group V. The ICOR appears to have increased from the 1960s to the 1970s, but the shape is kept unchanged in the cross-sectional data.[123] The range of variation among the individual countries is not necessarily small, particularly for the countries in the lower Y/L groups. Nevertheless, we believe this pattern is regular and of substantive significance with respect to import behavior.

Following formula (1), we can say that the pattern of M/I is composed of two components, IMOR and ICOR. The difference in magnitude between IMOR and ICOR plays a significant role. It is very wide in the early interval, but tends to narrow towards group IV. At the lower level, developing economies face a dual handicap: on the one hand, they must cope with a lower efficiency of domestic capital formation and on the other hand, a greater dependency on imports in terms of incremental output. Such disadvantages must be reduced if development is to proceed normally. In fact, our average group observations, if not for the individual countries, suggest that a successful shift is made in realizing a normal path, both in terms of increasing investment efficiency and reducing import dependency. This path is identified through groups II and III. Suppose only one of the two requirements are fulfilled, either for investment or for imports; then a shift of development phase would be impossible. The data are cross-sectional, but we are convinced that such a broad suggestion can be safely made. As was indicated earlier in Chapter 2, ICOR may be unchanged or even increased in the earlier interval of development in a number of countries. Moreover, the increase in ICOR to group IV is another issue to be noted. As is illustrated by an increase in IMOR from group I to II in the 1970s, a smooth path is not easy. Yet, we believe these suggestions are substantive, even though

---

123 Note that the inverse U-shape pattern of the reciprocal ($\Delta Y/I$) is identified for 1960–1985 in Table 1.1.

at this stage we do not actually know the behavioristic nature of $M/\Delta Y$ and M/I.

Second, moving to the subgroups (Panel B), let us recall our classification in terms of the rate of aggregate productivity growth (greater for subgroup E vs. smaller for subgroup F). This is a device for observing growth and structural change at a dimension that is not associated with the change in the Y/L level. At this dimension, we want to examine how the import performance differs relative to differences in productivity growth.

We can see a regular pattern in the different magnitude of the two terms M/I and $I/\Delta Y$ and, accordingly, IMOR ($M/\Delta Y$). The magnitude of the terms are greater for subgroup F than for subgroup E without exception for both periods. Some deviation occurs in each period, but these are minor. These regularities can provide us with preliminary information. Because of the property of formula (1) which describes only the results of past performance, it is not possible to explain these regularities without making an assumption of one kind or another. For example, let us assume that the reciprocal of ICOR, $\Delta Y/I$, can be used as an indicator of productivity growth of the sector output which pertains to imports. In fact, we do have some knowledge for such an assumption. Regarding the industrial sector, the rate of its productivity growth, G_{yI}, shows a fairly close association with $\Delta Y/I$ in the aggregate data (Table 4.5). The regularity presented above by subgroup is interpreted to suggest that the effect of replacing imports by enhancing domestic production may be realized more intensively and speedily in subgroup E than in subgroup F. Assuming the same import requirements, this is mainly because the rates of productivity growth, and hence the pace of the rightward shift in the supply curve, are greater in subgroup E than in subgroup F. If the requirement for imports is greater for subgroup F than for subgroup E, for example, by a "big push" policy in the former, the difference would be intensified. This is just a problematic suggestion which is to be examined further in the approach that follows.

Disaggregation: Imports of Investment Goods

So much for a macro-approach as a preliminary discussion. Disaggregation is required both for imports and investment. For our analytical purposes, the focal point is capital goods. The first step is to decompose imports of manufactured goods into capital goods and noncapital goods. Second, in terms of domestic capital formation, fixed investment needs to be specified and broken down into producer durables (PDE) and construction. Imported capital goods virtually correspond to producer durable equipment, though construction materials are included. The item on which we focus is producer durable equipment because this is the carrier of modern technological knowledge which is borrowed from economically advanced countries, as was discussed in depth in Chapter 5.

These conceptual points can be only approximately implemented due to the limitation of readily available data. For example, with respect to merchandise imports, machinery (Ma) vs. nonmachinery (NMa) is used, assuming that investment goods are minor in nonmanufactured imports. The readily available data include transport equipment in the machinery category without specifying it as capital goods and noncapital goods (e.g., passenger cars). Precisely classified data can be estimated using other sources, but they cannot cover the entire time span under review. As for gross domestic fixed investment, even the simplest breakdown into producer durable equipment and construction cannot be obtained consistently for the entire period for all selected countries under consideration. To obtain an overview by group observation, we assume changes or differences in the proportion of these two components are not so large as to seriously disturb our examination.[124] With these reservations in mind, we use Ma as a proxy of M_I.

Table 6.2 summarizes the estimates of Ma, NMa, and Ma + NMa (i.e., the sum). The following points should be noted. First, panel (A) shows that total imported manufactures as a share of total merchandise imports decreases regularly from lower to higher groups of developing economies over the entire period covering a quarter of a century. This is the expected pattern brought forth by industrialization. The reversal in this trend from group IV to group V in 1985 is due to the shift to services discussed earlier.

Second, in contrast, the percentage share of machinery imports, Ma, tends to increase while that of nonmachinery imports, NMa, goes down in association with the decline of the share of total manufactures imports. These patterns are regularly witnessed in the

124 According to United Nations, *World Product and Income*, the proportion of producer durable equipment to the sum of PDE and construction estimated at domestic prices in 1975 is as follows, on average, for the groups: group I (2), 53.5; group II (5), 50.2; group III (3), 45.6; group IV (4), 45.7 (the number of countries for which data are available is shown in parenthesis). The data are limited but differences between groups do not appear too large. The proportion of construction material in total investment goods imports is much less. In addition, the property of this item as a carrier of advanced technological knowledge is much less significant than producer durable equipment. For further reference, see Chapter 5, Section III.

Table 6.2
Machinery (Ma) and Nonmachinery (NMa)
as a Share of Manufactures and Manufactured Imports [a]:
1960, 1965, 1980, and 1985 (percentage)

	1960			1965			1980			1985		
Group	Ma	NMa	Sum	Ma	NMa	Sum	Ma	NMa	Sum	Ma	NMa	Sum
(A): In Merchandise Imports												
I	20.7	47.0	67.7	25.0	46.9	71.9	32.3	41.3	73.6	28.1	39.4	67.5
II	22.4	41.4	63.8	27.2	40.9	68.1	29.9	34.5	64.4	27.2	32.4	59.6
III	24.0	39.9	63.9	27.2	39.7	66.9	28.3	35.3	63.6	27.2	33.3	60.5
IV	33.7	27.4	61.1	30.4	35.7	66.1	29.3	31.7	61.0	25.6	31.1	56.7
V	20.9	26.7	47.6	23.7	31.3	55.0	25.4	33.4	58.8	29.7	34.3	64.0
(B): In Manufactures Imports												
I	30.6	69.4	100.0	34.8	65.2	100.0	43.9	56.1	100.0	41.6	58.4	100.0
II	32.0	68.0	100.0	39.9	60.1	100.0	46.4	53.6	100.0	45.6	54.4	100.0
III	37.6	62.4	100.0	40.7	59.3	100.0	44.5	53.6	100.0	45.6	54.4	100.0
IV	55.2	44.8	100.0	46.0	54.0	100.0	48.8	51.2	100.0	45.1	54.9	100.0
V	43.7	56.3	100.0	43.1	56.9	100.0	43.2	56.8	100.0	47.9	52.1	100.0

Note: a. All shares are in current prices.

Sources: World Bank, *World Development Report,* 1983 and 1987 issues.

1960s except for group V in which Ma decreases. In the 1980s, however, the pattern appears to change and the percentage shares of Ma tend to decrease slightly.

Third, panel (B) presents the relative changes in the proportion of Ma and NMa. At first glance, there seems to be no regularity. A closer look, however, reveals that two different tendencies are involved: in the lower Y/L groups (i.e., groups I, II, and III) there is one trend and in the higher Y/L groups (i.e., groups IV and V) there is another trend. A typical case of the former example is identified for group III. The percentage share of Ma for group III increases from 37.6 in 1960 to 45.0 in 1985 (that of NMa decreases from 62.4 to 55.0 in the same years). This is exactly the pattern of the increasing trend recognized earlier in the cross-sectional data in Panel (A). This pattern is also generally seen for groups I and II, although with a slight decrease from 1980 to 1985. The other trend is found for group IV. In 1960, the percentage share of Ma was distinctly greater than that of NMa (55.2 vs. 44.8) in contrast to all other groups. It tended to decrease towards 45.1 in 1985, which was smaller than that of NMa, 54.9, although between 1965 and 1980 a reverse movement is seen. This pattern is not witnessed for any other group of developing economies. Group V of developed economies shows virtually unchanged percentage shares of Ma and NMa throughout 1960, 1965, and 1980. The increase in the share of Ma (a smaller share of NMa) to 1985 deserves attention because of the shift to services in the domestic production structure.

Fourth, as a result of these varied performances over time, the percentage share of Ma and NMa in 1985 becomes closer (Ma=45 percent, NMa=55 percent) in the majority of developing nations (i.e., groups II, III, and IV). As compared to the shares in 1960, Ma increases by 23.6 points in group II, by 7.4 points in group III, and in sharp contrast, decreases by 10.1 points in group IV. (Incidentally, group I increases by 19.0 points which is similar to that of the former groups.) As a ratio to total merchandise imports, the percentage point of increase in Ma is 7.4 points for group I, 4.8 points for group II, and 3.2 points for group III, while the decrease in group IV is 8.1 percentage points.

These statistical findings have important implications for our discussions of import requirement and substitution. First, changes in the shares of imports of machinery and nonmachinery over time cannot provide direct evidence of import substitution of manufactured goods, but do indicate the replacement performance of importing these goods. A decrease of the Ma share, for example, suggests secondary import substitution by promoting domestic production of machinery. Such a performance took place particularly for those countries with a high Ma import share in 1960 as is illustrated by group IV. Because of the initially high degree of import dependence on machinery, import substitution in these countries had to be carried out in subsequent years. The policy of secondary import substitution cannot be understood separately from the historical developmentpath followed by each nation. The originally high share of machinery imports was brought about by the big push strategy. This pertains to the aspect of "requirement." That is, the domestic investment push was realized by imports of machinery which in turn depend upon the given coefficient of imports dependency that was almost fixed by the technology involved. Our average group observations indicate this to be what in fact occurred. The investment proportion I/Y in group IV is very high in the 1960s (22 percent for group IV vs. 17 percent for group III) with a high value of ICOR (4.5 in group IV vs. 3.1 in group III). The M/I is low for group IV, so that its high Ma import dependency was not realized by increasing merchandise imports in general, but probably by replacing NMa imports.

An illustrative comparison of import requirement and substitution implies an important issue. If import requirements are large, but the effects appear reluctantly due to a retarded follow-up that comes from a slow place of productivity enhancement, we have the phenomenon of the so-called prolonged secondary import substitution. What we must be concerned with is not only the productivity effects but also the historical causes brought forth by a too large requirement. This will be discussed in detail later in the chapter.

Second, a few words on the normal pattern of import allocation need to be said. A long-run path of replacing NMa imports by raising the share of Ma imports can be judged as being normal. The performance of groups II and III are broadly representative of this pattern, although with certain accompanying resource imbalances. Though the evidence is indirect, this suggests a path of primary import substitution through the promotion of the domestic light manufacturing industry. To that extent, foreign payments resources could be increasingly allocated to Ma imports. As for group I, this path is not necessarily recognized. An increase in the Ma share between 1960 and 1980 appears to have been brought about by an increase in the ratio of manufactures imports to total imports which is associated with greater pressure on resource imbalances. Not only in group I, but in group II and even group III, a normal path is not necessarily realized. As suggested earlier by the subgroup observations, the performance of imports in relation to domestic production differs considerably between subgroups E and F. In achieving a normal path, subgroup F may face greater difficulties and hence may fall behind subgroup E.[125]

---

125 One may have the impression that the discussion on import substitution is biased towards manufactures, in particular, investment goods. This may not be a legitimate criticism as efforts have also been made for agricultural products, particularly food. As shown earlier in the import shares, agricultural products are of considerable magnitude for nonfuel primary products through the groups, including the lower Y/L levels. During 1960–1980, the share of agricultural products decreases and this indicates the results of import substitution of agricultural product by successful promotion of domestic production.

In Table 5.12, Panel (B), the ratio of nonfuel primary goods to nonfuel merchandise imports, M_p/M, decreased from 1960 to 1980 through all groups :

Finally, an important suggestion seems to be drawn from a comparison of groups IV and V. Apart from the different situation which emerged in the 1980s, as pointed out earlier, the shares of both Ma and NMa are required to decrease as the economy moves towards completing the path of economic development. Actually, this possibility is indicated by a decline in the ratio of manufactures in total merchandise imports. This must be implemented by fulfilling the minimum condition of completing secondary imports substitution. However, the required condition is not to make a drastic fall in the Ma import share. As is indicated by the performance of group V, the Ma import share is kept considerably high in developed economies with mutual interdependence in the machinery trade.

With these preliminary estimates and observations, we can use formula (2) as a framework to analyze the related movements of the three major terms. It is hypothesized that the import dependency of domestic capital formation, M_I/I (Ma has been used as a proxy in previous discussions), is another independent variable, together with $I/\Delta Y$, in determining the IMOR (i.e., $M_I/\Delta Y$). The reason for this hypothesis is as follows. In Chapter 1, the decision-making process involved in undertaking domestic investment was assumed to be the basic element in forming the production capacity (PC) of the economy, assuming a certain level of technology had been incorporated. Depending upon the domestic supply, the required level of imports of investment goods is determined by this process. Of course, the requirement can be met subject to the condition of possible foreign payments and this may raise some doubt as to whether M_I/I can really be treated as an independent variable. For the time being, we would like to simply assess the core of the problem and will reserve this conditional aspect for later clarification.

Table 6.3 and Figure 6.1 show important patterns. For the majority of developing countries by group, M_I/I shows a tendency of decline, with an increase from group I to group II. A U-shaped

	I	II	III	IV	V
1960	20.2	26.1	28.3	31.8	44.8
1980	15.8	23.8	25.5	24.1	25.6

This is another aspect of nondichotomous features of postwar performance that requires further scrutiny.

Table 6.3
Relationship Between Investment Goods Imports (M_I) and Domestic Investment (I), 1960–1980

Group	Subgroup	M/I(%)	M_I/M(%)	M_I/I(%)	I/ΔY	M_I/ΔY
I	E	0.89	27.8	24.7	4.7	1.2
	F	1.06	21.1	22.4	6.0	1.3
	Average	0.98	24.5	24.0	5.4	1.3
II	E	1.14	29.1	33.2	3.2	1.1
	F	1.57	22.0	34.5	6.9	2.5
	Average	1.36	27.4	37.3	5.1	1.9
III	E	1.28	25.7	32.6	3.0	1.0
	F	1.36	25.3	34.4	3.6	1.5
	Average	1.32	25.5	33.7	3.3	1.1
IV	E	0.80	28.8	23.0	4.2	1.0
	F	0.93	32.5	30.2	5.0	1.5
	Average	0.89	30.7	27.3	4.6	1.3
V	E	1.26	17.2	21.7	6.4	1.4
	F	1.28	25.3	32.4	6.5	2.1
	Average	1.27	21.3	27.1	6.5	1.8

Sources: See sources of Tables 6.1 and 6.2.

pattern, however, is witnessed both for $M_I/\Delta Y$ and $I/\Delta Y$ with a trough at group III and with a deviation for $M_I/\Delta Y$. The incremental import-output ratio ($M_I/\Delta Y$) changes through the majority of groups in close association with the performance of ICOR, rather than with M_I/I. The term M_I/I even moves in a contrary direction in the interval of higher Y/L groups. However, for the lower interval groups (i.e., groups I–III), $M_I/\Delta Y$ moves in association with M_I/I.

Our broad interpretation of these trends in terms of the phases demarcated in Chapter 4 is as follows. During the preprimary phase in which agriculture is dominant, the pattern may differ: that is, $M_I/\Delta Y$ is strongly determined by the increasing tendency of M_I/I. In the primary phase during which industrialization proceeds, the performance of ICOR tends to be increasingly influential in determining $M_I/\Delta Y$, first in terms of a declining tendency and then in terms of an increasing trend. The latter pattern continues as further industrialization takes place and as the

Figure 6.1
Comparison of M_I/I, $I/\Delta Y$, and $M_I/\Delta Y$ by Group: Averages for 1960–1970 and 1970–1980

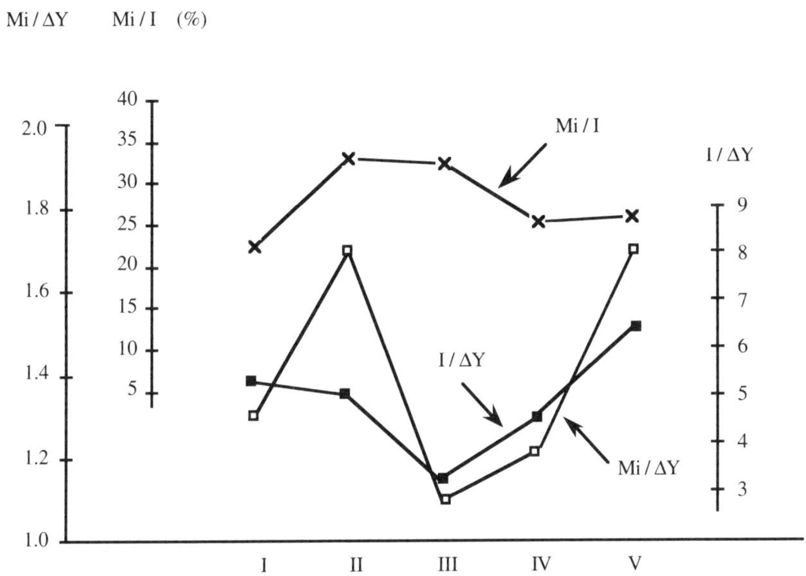

Sources: See sources of Table 6.3.

economy proceeds through the secondary phase. Because of the use of rough indicators, reservations and qualifications are necessary. Nevertheless, we believe these suggestions are of substantive significance.

By subgroup, we examine further the possibility of interpreting formula (2). In comparing the two subgroups, E and F, what is most noticeable is that $M_I/\Delta Y$ is smaller in subgroup E than in subgroup F throughout the entire range without exception. In addition, the effects of the two components, M_I/I and $I/\Delta Y$, in forming this significant pattern differ between the former range, i.e., groups I–III, and the latter range, i.e., groups IV–V. In the former range, almost no difference is witnessed for the magnitude of M_I/I between subgroups E and F, whereas in the latter range, the magnitude of M_I/I is distinctly smaller in subgroup E. We know the pattern of $I/\Delta Y$ which is smaller for subgroup E and greater for subgroup F, so that $M_I/\Delta Y$ is dominantly influenced by $I/\Delta Y$ during the former range, while during the latter range $M_I/\Delta Y$ is affected by both M_I/I and $I/\Delta Y$.

Taking the group and subgroup observations together, the following can be deduced: (1) during the primary phase (including the preprimary phase), a sharp increase in the M_I/I requirement is unavoidable to pursue initial industrialization for all nations, but even at this phase, the investment efficiency of domestic production as indicated by $I/\Delta Y$ is a problem. This point is illustrated by the comparison of $I/\Delta Y$ and $M_I/\Delta Y$ between subgroups E and F. As the economy proceeds through industrialization and undergoes shifts, the subgroup difference is sustained as indicated by the magnitude of $M_I/\Delta Y$. In addition to the sustained influence of $I/\Delta Y$, the difference in M_I/I becomes more influential. As shown in Table 6.3, the magnitude of the difference of M_I/I between the two subgroups tends to be greater towards group V.

What we wish to point out is that the pattern of the actual operation of formula (2) can be grasped in terms of phase demarcation, rather than applying it straightforwardly to the entire path of development. In order to expand on this argument, we need to know more about the factors behind the differences and variations in M_I/I.

Relationship between Imports and Domestic Production of Investment Goods

Through domestic capital formation, M_I is absorbed into domestic production. This aspect has been discussed above in terms of M_I/I which determines $M_I/\Delta Y$. Naturally, we are concerned with its relationship to the domestic production of investment goods (D_I). This is relevant to the so-called secondary import substitution (IS). One may say that import substitution of investment goods implies simply a trend of M_I decrease vs. a trend of D_I increase. In light of the M_I/I performance hitherto identified by group and subgroup observation, we suspect that such a supposition of simple regularity is not realistic. In order to explain $M_I/\Delta Y$, the term M_I/I has been treated as an independent variable with certain conditions. Now we try to explain the way in which M_I/I is in fact determined. Among the relevant factors, technological advance is singled out in relation to the performance of investment goods imports.

To clarify these issues, Table 6.4 presents the growth rates of M_I (G_{MI}) and D_I (G_{DI}). By group average, $G_{MI}<G_{DI}$ seems to be the pattern for the majority of developing economies; by subgroup, the

Table 6.4
Growth of Import and Domestic Production of Investment Goods [a, b]
(percentage)

		Group				
		I	II	III	IV	V
G_{MI}						
	E	2.5	9.3	9.0	3.5	7.8
	F	7.9	1.9	5.8	3.3	7.6
	Average	5.3	5.6	7.4	3.4	7.7
G_{DI}						
	E	2.5	11.4	15.6	12.0	5.1
	F	3.9	10.8	8.8	9.3	2.4
	Average	3.2	11.1	12.2	10.7	3.8

Notes: a. The growth rates are simply calculated by bridging the shares at the beginning and the ending years. G_{MI} is for 1960–1980 but G_{DI} is for 1970–1980 due to lack of data for 1960.

b. Machinery and transportation equipment (Ma) is used as a proxy for M_I.

Sources: See sources of Tables 6.1 and 6.2.

same pattern is seen. These appear to be acceptable as it indicates a long-term tendency in the import substitution process for investment goods. The subgroup comparison, however, may not be readily acceptable, because both G_{DI} and G_{MI} are greater in subgroup E (and smaller in subgroup F) for the majority of developing economies. One may pose the question: why is it that a higher growth of D_I in subgroup E cannot bring forth a smaller growth of M_I as a result of import substitution? We believe this is the important question that must be answered. Is it possible to explain it?

We believe that an answer is possible. In the path of absorbing technological knowledge borrowed, the performance of M_I and D_I are interrelated. This is not to deny the usual relationship of substitution in terms of the supply of goods. But it is our view that from the perspective of technological advance, additional discussion is needed. On the one hand, an efficient transfer of technological knowledge is usually carried out through the investment goods imports specific to the industry concerned. On the other hand, the process of assimilation of the borrowed technology can be effectively pursued domestically through the promotion of domestic production of the specific investment goods concerned. Otherwise, the technology specific to the industry cannot be fully absorbed. This usually takes place with a certain time lag. An important point in the longer-term perspective is that such an interrelationship between imports and domestic production occurs in sequence and is stimulated by the process of upgrading the level of required technological knowledge. In proceeding from a lower to a higher Y/L level, developing economies require further imports of investment goods and a corresponding domestic process of assimilation will also be needed.

Such a sequential path differs in magnitude and speed among nations. In our framework, we would say that investment goods imports can augment the production capacity (PC) of all developing nations concerned, but the varied level and component of social capability (SC) make the difference in realizing the sequential path. A comparison of subgroups E and F illustrates this difference in capability. The data suggest that more vigorous activity in subgroup E results in greater rates of both D_I and M_I, while a less active performance of subgroup F shows smaller rates of growth of both terms. These associations are shown in the cross-sectional data and cannot provide direct evidence; yet the respective difference between subgroup E and subgroup F at each Y/L level is informative.

In order to supplement our interpretation, let us consider two aspects.

1. Level Comparison of M_I and D_I

An exact comparison of M_I and D_I is only possible through individual country research. The use of the World Bank data listed in its *World Development Report* is a rough approximation for the level comparison in 1980 (Table 6.5). The most important item is the ratio, D_I/M_I. Conventionally, import substitution performance can be dealt with in various terms, but here this ratio is used as supplementary information to confirm the preceding interpretation. The ratio D_I/M_I rises from the lower to the higher groups in both subgroups E and F. This is the result (as seen in 1980) of the over-time performance found in the earlier treatment of G_{MI} and G_{DI}. As for the developing economies as a whole, import substitution went on but its pace is moderate on average, even including group IV, for which D_I cannot surpass M_I. This is not a surprise because at the starting point, the level of M_I and D_I was low so that G_{MI} and G_{DI}

Table 6.5
Comparisons of Import and Domestic Production of Investment Goods in 1980 [a] *(percentage)*

		I	II	III	IV	V
D_I/M_I						
	E	—	13.8	16.5	21.7	227.9
	F	—	—	12.4	18.8	110.0
	Average	—	(6.9)	14.4	20.3	194.0
D_I/Y_M						
	E	—	5.9	6.5	12.0	23.5
	F	—	—	3.6	3.3	28.0
	Average	—	(2.9)	5.1	7.6	25.7

Notes: — = Near zero.
a. D_I, M_I, and Y_M are in 1980 dollars.

Sources: See sources of Tables 6.1 and 6.2.

in the earlier discussion appeared considerably high. In view of our interpretation stated above, we are specifically concerned with the following points: (1) by subgroup comparison, a higher (lower) ratio is witnessed for subgroup E (subgroup F) regularly over the entire range. As is shown by the ratio, D_I/Y_M, the share of investment goods production in manufacturing products is distinctly higher (lower) for subgroup E (subgroup F). These may endorse our argument regarding the difference in capability. That is, the larger (smaller) amount of investment goods imports in subgroup E (subgroup F) is associated with a greater (smaller) allocation of resources to investment production in subgroup E (subgroup F) in manufacturing. Certainly the process of assimilation of borrowed technology thus went on more (less) vigorously in subgroup E (subgroup F); (2) we cannot discuss each group individually, but a few words on groups IV and V are worthy. The problem of import substitution is often discussed with respect to group IV developing economies. The data suggest that differences between the nations are wide. For example, D_I/Y_M differs to a considerable extent between subgroups E and F in group IV. The latter's level is almost close to that of the group III average, reflecting similarities in domestic activity of production and technology diffusion. With respect to the performance of group V, its remarkably high ratio of D_I/M_I is, of course, due to high exports of D_I product. The reason is the same for the much higher ratio of D_I/Y_M in this group. We see almost no difference in D_I/Y_M between subgroups E and F in group V which is definitely greater than D_I/Y_M of group IV.

2. The Initial Situation of Investment Goods Imports: The Effects of Import Policies and Subsequent Changes

In order to examine the effects of import policies, a simple indicator is used: the ratio of investment goods imports, M_I, to other (noninvestment) manufactured goods imports, M_N, that is, M_I/M_N. An overview was given earlier in terms of Ma/NMa in Table 6.2. The same data are used as proxies of M_I/M_N. Determining this allocation is, of course, not a simple matter, but we believe this ratio is useful for measuring the initial situation and to track subsequent changes. Its magnitude, shown earlier in terms of Ma/M in Table 6.3, shows a certain regularity. The ratio ranges rather narrowly between 20–30 percent for the majority of developing economies, with greater values for subgroup E in the lower Y/L groups, i.e., groups I–II, and with greater values for subgroup F in the higher Y/L groups, i.e, groups IV–V. The distribution of imports of manufactures varies a great deal between individual developing nations (see Table 6.6). The initial situation is approximated by the year 1960, though this should be earlier for some countries. The ratio M_I/M_N varied a great deal in 1960 among nations in the entire range of developing economies. Let us first look at those countries for which the ratio is extremely high. Ten countries listed in the table were selected using the criteria, M_I/M_N>50 percent (column 1). For reference, the average value of the ratio of the subgroups to which these countries belong is also shown for 1960 (column 3). Over-time changes are observed between 1960 and 1980. The ratios in 1980 are listed for both the selected countries and the subgroups (columns 2 and 4).

The distribution of these designated countries is biased towards group IV, although we have one or two countries from the other groups as well. In all of these countries, the ratio tended to decrease from 1960 to 1980 except for Peru, but the tempo of the decline differs. Comparison of the ratio between 1960 and 1980 reveals two patterns. One pattern shows that the ratio becomes lower than that of the corresponding subgroup in 1980 and in the other pattern, the ratio is sustained at a higher level in the same comparison. Four out of the five Latin American countries listed in the table show the

Table 6.6
The Ratio of Investment to Noninvestment Goods Imports (M_I/M_N) of Selected Countries (percentage)

Country		Selected Countries 1960 (1)	Selected Countries 1980 (2)	Subgroup Averages 1960 (3)	Subgroup Averages 1980 (4)
India	(I, E)	66.7	33.3	38.3	43.2
Philippines	(II, E)	51.4	35.8	41.5	44.5
Turkey	(III, E)	61.8	41.9	37.3	46.6
Colombia	(III, F)	58.1	52.8	40.1	46.6
Brazil	(IV, E)	66.7	44.2	55.9	51.7
Portugal	(IV, E)	55.3	49.0	55.9	51.7
Yugoslavia	(IV, E)	62.7	50.0	55.9	51.7
Peru	(IV, F)	50.0	54.8	51.4	49.0
Mexico	(IV, F)	61.9	60.2	51.4	49.0
Argentina	(IV, F)	60.3	52.0	51.4	49.0

Sources: See sources of Tables 6.1 and 6.2.

latter pattern, while the former pattern is witnessed for the nations which belong to subgroup E.

Admittedly, our data treatment is crude, but these findings may suggest an important aspect concerning the effects of import substitution. The extremely high ratio of investment to noninvestment imports may be the result of import policies that are strongly biased towards investment goods which were required for carrying out the so-called big push plan and policies. Such an initial move did greatly influence the subsequent pattern, but the influence is not uniform. The simple ratio indicates that a number of countries did in fact achieve the import substitution effects. Although its relationship with the performance of domestic production remains to be confirmed by individual country research, the fact that all of the countries exhibiting the former pattern belong to subgroup E merits particular attention. Insofar as the subgroupings are concerned, subgroup E shows a better performance of productivity growth. This is illustrated distinctly in group IV, where a sharp contrast is seen: countries in subgroup E are of the former pattern—that is, the ratio declines over time—whereas countries in subgroup F are of the latter pattern—that is, the ratio is sustained at a high level. For these F countries, together with Colombia, we see no evidence of import substitution, as the high ratio of 1960 is kept high in subsequent years. The problem of secondary import substitution policy has often been a subject of debate for these nations. In a comparative sense in our framework, the import substitution effects of investment goods appear to have fallen behind.

We believe that these findings enrich our empirical knowledge of the performance of investment and imports. The current magnitude of M_I/M_N is the combined effect of two factors: initial policy decisions (i.e., the historical factor) and the pursuit of an import substitution strategy to enhance productivity in subsequent years.

II. The Causes and Effects of Export Growth

Exports are the topic of this section. We share the view that merchandise exports can basically be promoted by realizing a rightward shift of the supply curve for the individual commodities concerned. This view stems from our hypothesis on technological advance. However, one may argue that demand expansion through export promotion should be the driving force of development. "Outward-looking" rather than "inward-looking" policies have often been recommended accordingly. These terms are not necessarily defined precisely and are not used in the same sense. Yet, nobody can deny the fact that expectations of export expansion provide strong incentives for enhancing domestic investment and growth. No doubt the demand-expenditure aspect is important. Therefore, with the use of ex post data one may naturally be able to derive an associated growth performance between exports and productivity, but such a statistical finding may not provide evidence that export growth is always the cause of productivity growth.[126] Thus, we must explore the causes and effects of exports with great care.

Unlike the case of imports, external factors play an important role in determining the export performance of developing economies. The conventional approach which measures parameters of income

---

126 For example, a close association between the rate of residual growth (as measured in Chapter 1) and the rate of exports growth (which will be presented in Table 6.9) may be identified. In this regard, consider World Bank, *World Development Report,* which says: "It shows that total productivity increased much faster in the strongly outward-oriented economies than the strongly inward-oriented economies" (p. 92), and a similar argument in Bela Ballasa, "Essay in Development Strategy," *Occasional Papers,* No. 5, International Center for Economic Growth (San Francisco: ICEG, 1988), 27–35.

effects and price effects in measuring the export function for individual economies is beyond the scope of this analysis, but it does have some relevance to our treatment. "Income effects" represent changes in the world on demand. Different values of elasticities of individual countries provide useful information for response behavior. However, the responses may vary due to differences in the composite factors among the countries and the countries' development phases such as the degree of competition with internal demand (size and country involved), the structure of domestic production, and the institutions and/or organizations built for export activities, etc. Thus, we focus our concern on the internal factors. In this sense the differences in policy on trade between the nations must be an important factor responsible for determining export performance. Our stress on the significance of the productivity-supply aspect would not deny its positive effects.

As a matter of fact, we examine the growth relationship between productivity and exports, but viewed from our original hypothesis on borrowed technology, the examination includes information on output price behavior. We will try to give an overview on this subject of price competition illustrated by the sector approach with a focus on a comparative analysis of manufacturing and agriculture. The method is to compare the rates of productivity growth and export growth using cross-sectional data and time-series data, relying on the findings in Chapter 5. Finally, the balance of payments regarding sectoral trade will be discussed with the focus on manufacturing for the long-term changes in imports and exports.

All of these factors pertain to the supply-production aspect. As for the demand aspect of exports, no systematic treatment is attempted. Just a specific problem—that is, the relationship between exports and domestic investment—that has often provoked debatable issues in characterizing the pattern of developing economies is considered. This aspect is briefly discussed later in the chapter.

Exports vs. Domestic Investment: The Demand Aspect

Let us compare export growth (G_X) and GDP growth (G_Y) by Y/L levels. If a regular association exists between the two terms, G_X would show an inverse U-shape instead of a linear pattern because of the G_Y pattern found in Chapter 1. In fact, G_X does not necessarily show a close association with movements in G_Y. Through the 1960s and 1970s and for 1980–1985, the group average of G_X is as follows (%, average annual rates):

	I	II	III	IV	V
1960–1980	3.2	3.0	6.3	5.3	7.7
1980–1985	3.5	3.2	3.9	2.3	4.4

A dip in group II in 1960–1980 deviates from the inverse U-shape pattern that is generally found for the developing economies, and in particular, a remarkable increase is seen from group IV to group V in both periods. In developed economies, despite a slower rate of output increase, the rate of exports growth is distinctly greater than that of the developing economies. A similar trend is observed for the ratio of exports (including nonfactor services) to GDP, X/Y. The group average of X/Y for the two periods is as follows (%):

	I	II	III	IV	V
1960–1980	15.5	22.4	21.6	18.1	26.5
1980–1985	13.1	23.9	25.7	26.0	35.8

An inverse U-shape pattern cannot be identified, and we also see a large value of the ratio for group V, although the contribution of nonfactor services is not specified. In the 1980s, when G_Y declined, X/Y tends to be increasingly higher towards the groups of higher Y/L levels. This is the reverse of the tendency found in 1960–1980 for a wider range, insofar as the majority of developing economies is concerned.

By subgrouping, G_Y is greater for subgroup E than for subgroup F. The values of G_X are as follows (%):

		I	II	III	IV	V
1960–1980	E	2.6	5.3	7.9	6.3	7.5
	F	4.1	0.7	4.7	4.3	7.2
1980–1985	E	-0.9	2.2	8.6	3.4	4.5
	F	-4.0	4.3	0.8	1.3	2.2

Though with some deviations, that is, group I in 1960–1980 and group II in 1980–1985, the inequality $G_X^E > G_X^F$ is generally found for both periods including for group V.

These patterns suggest that the relationship between export expansion and macro-growth cannot be simply grasped and needs

Table 6.7
Comparison of Growth Rates of Exports [a] and Domestic Investment (percentage)

Group & Subgroup		1960–1970 G_X	1960–1970 G_I	1970–1980 G_X	1970–1980 G_I	1980–1985 G_X	1980–1985 G_I
I		3.3	6.0	2.9	4.9	-2.5	-4.0
	E	2.3	7.4	1.8	5.0	-0.9	-5.1
	F	4.3	4.6	3.9	4.8	-4.0	-2.9
II		7.0	7.1	0.4	5.0	3.2	-8.1
	E	8.4	10.0	3.8	6.3	2.2	-7.8
	F	5.6	4.1	-3.0	3.7	4.3	-8.3
III		7.9	8.4	5.1	9.3	3.9	1.1
	E	9.6	9.4	7.0	10.1	8.6	3.9
	F	6.1	7.5	3.2	8.4	0.8	-2.0
IV		5.0	6.2	5.7	6.9	2.3	-6.6
	E	5.9	8.2	6.6	5.9	3.4	-3.4
	F	6.1	4.6	4.7	7.8	3.2	-9.7
V		9.0	6.4	5.8	1.3	4.4	1.7
	E	10.1	7.4	5.9	1.2	4.5	0.9
	F	7.9	5.5	5.6	1.4	4.4	2.5

Note: a. G_X is merchandise exports. Nonfactor services are not included due to the limits of data availability.

Sources: World Bank, *World Development Report,* various issues, and *World Tables,* various issues.

further investigation both from the supply and demand aspects. The discussion below focuses on the supply aspect. However, a selected aspect of demand is also touched upon—that is, the relationship between exports and domestic investment—because investment plays a central role in our framework. The term X/Y is one of the major items of aggregate demand, together with I/Y and C/Y (where C=consumption) where both can be disaggregated into private and government components. However, we do not deal with all of these demand components, but insofar as the demand effects are concerned, G_I or $\Delta I/I$ is a legitimate term to be compared with G_X or $\Delta X/X$. Table 6.7 compares G_X and G_I in three periods.

First, the group average shows a distinct pattern of $G_I > G_X$ for all groups of developing economies in the two periods 1960–1970 and

1970–1980. However, in 1980–1985, the pattern reverses and $G_X > G_I$. As for group V, the developed nations, the latter pattern $G_X > G_I$ is witnessed throughout all three periods. The difference in the performance of group V as distinguished from that of the developing economies deserves attention. This may be relevant to the earlier observation of the $G_X = G_Y$ relationship. Views on interpreting these patterns may differ among observers. If the notion of an export-led economy is argued, which essentially pertains to the demand aspect of exports, we should see a sustained pattern of $G_X > G_I$. On the other hand, if the role of the growth engine is played by domestic investment, we should see the reverse, that is, $G_I > G_X$. What is indicated in the table suggests that neither $G_X > G_I$ nor $G_I > G_X$ is always sustained. Instead, in general both patterns are possible depending upon the varied situations of economic growth. At this juncture, however, it is possible to say that $G_I > G_X$ is a general pattern for the developing economies, while $G_X > G_I$ is the pattern for the downswing periods. With respect to the historical records of economic development of industrial nations, a pattern of $G_X > G_I$ is witnessed in the downswing, particularly for the depressed interwar period. This is the reason for the argument of the "contemporary role" played by domestic investment and export expansion in enhancing macro-demand, because during upswings the pattern of $G_I > G_X$ prevails. The complementary role can conceptually be applied to the contemporary developing economies, although a much longer duration of $G_I > G_X$ is actually witnessed due to the long-sustained upswing of postwar growth.

A somewhat detailed observation reveals the following. The rate of increase in domestic investment (G_I) in 1960–1970 and 1970–1980 exhibits an inverse U-shape pattern that is associated with the pattern identified earlier for output growth and other relevant terms. In the less favorable growth situation of the 1970s, the decline in G_I from group III to groups IV and V was drastic and the vitality of growth deteriorated. The decrease in the rate of export expansion from the 1960s to the 1970s in group V is considerable, but the decline is much less than that of G_I. Thus, the inequality $G_X > G_I$ was witnessed through all periods. This was mainly possible because of the mutual dependence of export demand among the industrial market economies (some 70 percent of world merchandise exports). However, this export demand among the industrial nations decreased from 1960–1970 to 1970–1980 with the decline of G_Y in this group in the latter period. A more limited level of less than 30 percent for the exports of developing economies was actually sustained in the two decades under consideration. Therefore, the different patterns identified between developing and developed nations are quite understandable.

Second, by subgroup we are concerned with the endorsement of what was revealed by the group averages and with suggesting possible qualifications. The normal performance of developing economies in terms of $G_I>G_X$ has some exceptions by subgroup. In the cases such as group II, subgroup F; group III, subgroup E; group IV, subgroup F in 1960–1970; and group IV, subgroup E in 1970–1980, $G_X>G_I$ is mostly due to an extremely high rate of export expansion in one or two countries. For example, in the case of group III, subgroup E, $G_X>G_I$ is caused by the value of G_X being 34.1 percent vs. 23.6 percent for G_I in Korea in the 1960s. This is an "atypical" case. If Korea was excluded from the group, then G_X of group III, subgroup E would be 7.4 percent, and would thus conform to the normal pattern. Due attention must be paid to such cases. The Asian NIEs are small countries and are largely of a similar pattern in achieving success in trade competition. The size of country is an important factor. It is our view that such cases cannot present a normal pattern in general for all developing economies. Thus, an important suggestion is to clarify the specific factors responsible for forming such a pattern.

In referring to a "normal" pattern, we imply a performance of development that can be generally pursued through the realization of the internal growth mechanism discussed in Chapter 1. But what about the subgroup behavior? Between 1960–1970 and 1970–1980, certain differences are witnessed between subgroups E and F. A sustained performance is generally indicated in their period averages as shown in the following:

		I	II	III	IV	V
G_X						
	E	2.1	6.1	8.3	6.3	8.0
	F	4.1	1.3	4.7	5.4	6.8
G_I						
	E	6.2	8.2	9.8	7.1	4.3
	F	4.7	3.9	8.0	6.4	3.5

With only the exception of group I, both G_I and G_X are greater in subgroup E than in subgroup F. A greater rate of augmenting domestic capital formation is associated with a greater rate of export growth at the same level of Y/L. We think this suggests a general pattern of the development mechanism in a long-term perspective. Subgroup E represents the stronger group. In the lower Y/L levels of the inverse U-shape path, vigorous investment activities are augmented in subgroup E, associated with a realization of an accelerated rate of export growth. Such a mechanism manifests itself as being weaker in subgroup F, although the mechanism cannot be fully evaluated by the discussion confined to demand alone. It is often asserted that exports and investment operate in a mutually enforcing way. A favorable perspective for exports in the future provides better incentives for domestic investment, while augmented capital formation provides a higher level of production activities for the sake of export promotion. Nobody can deny this interactive relationship. The point is the strength of such forces in the actual operation. In our view, the strength or the weakness depends not on any one specific activity, but depends on the nation's level of capability as explained earlier by the overall difference between subgroups E and F.

Finally, a few words are required to qualify our observation which depends solely upon the simple data for G_X and G_I. Technically, the analysis should be extended by taking two factors into consideration: one factor is the relative share of exports and investment to aggregate demand, and the other factor is the import leakage in aggregate demand. The difference between groups and the changes over time in the relative shares affect the magnitudes of the so-called incremental contribution of exports and investments. With respect to domestic investment, we emphasized the significance of the import dependency of investment goods. The magnitude of import leakage should be duly evaluated. In this simple overview, however, we must say that even with these qualifications, what is found above would essentially remain intact, although the relative magnitude of G_X vs. G_I would differ to a certain extent. Table 6.8 presents evidence in terms of $\Delta X/Y$ and $\Delta I/Y$, where adjustment for import leakage is included in the latter term. During the 1960s, $\Delta X/Y > \Delta I/Y$ for the lower Y/L groups, i.e., groups I–III, and $\Delta I/Y > \Delta X/Y$ for the higher Y/L groups. In the 1970s, $\Delta I/Y > \Delta X/Y$ is witnessed for all groups except group V. Simple averages of the two periods indicate a regular pattern of $\Delta I/Y > \Delta X/Y$ for all groups of developing economies, whereas a reverse pattern, $\Delta X/Y > \Delta I/Y$, is identified for group V.

By way of concluding remarks, we would like to say that a simple comparison of single factors such as exports and investment

Table 6.8
Contribution of Exports and Investment to Aggregate Demand Growth [a] *(percentage)*

	1960–1970		1970–1980		Average	
Group	ΔI/Y	ΔX/Y	ΔI/Y	ΔX/Y	ΔI/Y	ΔX/Y
I	0.71	0.94	0.61	0.26	0.68	0.50
II	0.91	1.01	0.78	0.01	0.85	0.54
III	0.81	1.13	1.49	1.11	1.18	1.10
IV	0.97	0.87	1.19	1.05	1.08	0.96
V	1.27	1.15	0.20	1.01	0.74	1.42

Note: a. In calculating the export share of period averages, nonfactor services are not excluded. For calculating leakage resulting from the import of investment goods, a proxy of Ma/I (the ratio of machinery import to investment) is used (Table 6.3).

Sources: See sources of Table 6.7.

and the argument derived thereof, such as export-led growth, cannot be acceptable and that even confined to the demand aspect, the positive role played by domestic investment should legitimately be identified. In search for a possibility of sustaining the normal operation of the interaction mechanism of development, we wish to examine below the supply-production aspect related to export activities.

Growth Relationship of Exports and Domestic Productivity

Our discussion on the supply aspect of export activity is, to begin with, concerned with its relationship to the productivity performance of the domestic economy at the macro- and major-sector level. The hypothesis is that productivity enhancement provides the basic condition for promoting the export growth of individual countries. To precisely test the validity of this hypothesis, we would need an elaborate analysis using adequate data. In this section, just an overview is attempted.

Table 6.9 summarizes the relevant data. For the sake of convenience of direct comparison, some data are reproduced from previous tables. For 1980–1985, estimates are impossible because of the nonavailability of labor employment data. Let us begin with the possible relationship between G_X and G_y, the rate of increase in

Table 6.9
Growth Rates of Merchandise Exports and Domestic Productivity, 1960-1980 (percentage)

Group & Subgroup	1960–1970 G_X	G_y	G_{yA}	G_{yI}	1970–1980 G_X	G_y	G_{yA}	G_{yI}	Average G_X	G_y	G_{yA}	G_{yI}
I	3.3	1.0	0.6	1.9	2.9	1.4	0.1	-0.9	3.1	1.2	0.4	0.5
E	2.3	1.1	—	—	1.8	2.9	—	—	2.1	2.0	—	—
F	4.3	0.9	—	—	3.9	-0.1	—	—	4.1	0.4	—	—
II	5.7	2.6	1.7	3.9	0.4	2.6	1.0	1.9	3.1	2.6	1.4	2.9
E	8.4	4.0	2.6	3.9	3.8	3.3	2.2	2.8	6.1	3.7	2.4	3.4
F	2.9	1.2	0.8	3.9	-3.0	-0.2	-0.2	1.0	0.0	0.5	0.3	2.5
III	7.7	3.3	3.5	3.8	6.1	3.5	3.2	2.8	6.9	3.4	3.4	3.3
E	9.6	3.7	4.0	5.4	9.0	5.1	4.1	3.4	9.3	4.4	4.1	4.4
F	6.1	2.8	3.0	2.2	3.2	1.8	2.2	2.2	4.7	2.3	2.6	2.2
IV	5.0	3.5	4.2	5.1	5.7	2.7	2.7	2.0	5.4	3.1	3.5	3.6
E	5.9	4.7	5.2	6.2	6.6	2.5	3.8	3.4	6.3	3.6	4.5	4.8
F	4.1	2.9	3.2	4.0	4.7	2.6	1.6	0.6	4.4	2.7	2.4	2.3
V	9.0	4.0	5.3	5.3	5.8	2.1	2.7	2.3	7.4	3.1	4.0	3.8
E	10.1	5.1	5.7	5.6	5.9	2.9	3.0	2.5	8.0	4.0	4.4	4.0
F	7.9	2.8	4.8	4.9	5.6	1.3	2.4	2.0	6.8	2.1	3.6	3.5

Note: — = data are not available.

Sources: Sources of Tables 4.1, 4.5, and 4.6; World Bank, *World Development Report*, 1982 issue, Table 8.

product per worker, Y/L (qualification of which was discussed in detail in Chapter 3).

First, there appears to be a certain positive association between the two terms G_X and G_y, but the association is not so strong as far as the group average is concerned. For both G_y and G_X, the basic pattern was identified earlier, though features of the 1960s and the 1970s are added here. For the performance of G_y, a shift is seen from a straight trend acceleration in the 1960s to a pattern of inverse U-shape in the 1970s. The performance of G_X does not show a similar change especially due to the different features between groups IV and V. These features are also seen for the average of the two periods .

Second, a much closer and intensive association between G_X and G_y manifests itself in the subgroup comparisons. For all groups, both G_X and G_y are distinctly greater for subgroup E as compared to subgroup F. (The only exception is group I, where the supply factor does not appear to operate explicitly.) The difference between the subgroups is much more sizeable than the preceding group comparison. The averages of groups II, III, and IV, are as follows:

	1960–1970		1970–1980		Average	
Subgroup	G_X	G_y	G_X	G_y	G_X	G_y
E	7.0	4.1	6.5	3.6	6.8	3.9
F	4.4	2.3	1.6	1.6	2.5	1.9

We believe the data have significant implications. From our subgrouping, the inequality of G_y between subgroups E and F is seen at the outset. The point is that nations with higher rates of productivity increase realize higher rates of expanding merchandise exports insofar as the majority of developing economies are concerned. To that extent, the difference in Y/L levels between countries does not appear to be the responsible factor in determining export performance. What matters is the speed of enhancing competitive power, irrespective of the level of output per worker. G_y is just a simple rough measure of genuine competitive power. Yet, it can provide us with convincing information.[127]

Third, the macro-level observation presented above should be followed by a disaggregated analysis. A preliminary look at the productivity side is given by the sectoral productivity growth of industry (G_{yI}) and agriculture (G_{yA}). We are familiar with the behavior of these two terms in light of the discussion in Chapter 4. In forming the positive association between G_X and G_y, both

127 One may be interested in using this association in order to endorse the argument that greater export expansion brings forth higher rates of productivity growth of the economy. However, we cannot share such a view. In Chapter 1, the measures of residual growth are presented for subgroups A and B. It is definitely greater in subgroup A than in subgroup B. Roughly, subgroups A and B correspond to subgroups E and F, respectively. Thus, the G_y difference roughly approximates that of residual growth. It is our view that the difference emanates not from any single factor, such as exports, but from the basic relationship between PC and SC in the individual economies concerned.

industry and agriculture play a consistent role, without any noticeable biases. For the sake of convenience in grasping the performance of G_{yI} and G_{yA} given in the table, the simple averages of groups II, III, and IV are presented for subgroups:

	1960–1970		1970–1980		Average	
Subgroups	G_{yA}	G_{yI}	G_{yA}	G_{yI}	G_{yA}	G_{yI}
E	3.8	5.2	3.4	3.2	3.6	4.2
F	2.2	3.4	1.2	1.3	1.7	2.6

The difference between subgroups E and F is common for both industry and agriculture. That is, subgroup F tends to have the disadvantage of slower rates of industrial productivity growth but is also inferior in enhancing its rates of agricultural productivity growth. The relationship between these sectoral features to the performance of exports might be expected to be clarified by decomposing merchandise exports. The classic notion of comparative advantage is often illustrated by comparing industry and agriculture, as mentioned in Chapter 5. A theoretical model of development phasing treats a path of substituting exports of agricultural products with exports of manufactured goods as a normal shift of phases. Although we know the general trend is for an increase in manufactured goods and a decrease in agricultural products as the share of total exports, the actual performance is in fact quite complex. The actual performance of comparative rates of productivity growth in the table does not necessarily indicate a linear path of increasing industrial productivity with increasing Y/L levels in developing economies. As is pointed out in the preceding chapters, particularly in Chapter 4, realization of a higher rate of productivity growth in agriculture in postwar developing economies draws our attention.

On the export side, without setting any simple hypothesis, we will examine the existence of any regularity in the actual growth performance of merchandise exports of two major sectors: manufactures (X_M), which represent industrial products, and agricultural products (X_A), leaving out other primary commodities (such as fuels, minerals, and metals).[128] Table 6.10 summarizes the

---

128 The data for these commodities are not readily available. A simple device is made for estimating rates of export growth of these goods by using G_X with sectoral shares of merchandise exports as follows:

Table 6.10
Disaggregation of Export Growth: Manufactures and Agricultural Products, 1960–1980 [a] ***(percentage)***

Groups & Subgroups [b]		G_{XM} Simple	G_{XM} Weighted	G_{XA} Simple	G_{XA} Weighted	X_M/X 1960	X_M/X 1980	X_A/X 1960	X_A/X 1980
I		5.0	0.4	1.3	1.2	3.3	13.0	94.7	87.6
II		6.5	0.9	1.9	1.3	11.0	17.6	84.7	50.9
	E	7.7	1.3	3.4	2.3	9.2	24.2	87.0	48.3
	F	5.2	0.8	0.3	0.2	12.5	11.0	79.6	53.5
III		13.8	2.5	4.2	2.5	7.8	25.2	80.4	41.9
	E	19.0	4.0	4.0	2.1	12.3	30.0	74.7	28.0
	F	8.6	1.0	4.3	3.0	3.3	20.3	85.8	55.4
IV		10.8	3.3	2.8	1.3	15.7	44.4	69.5	40.1
	E	13.1	4.9	3.3	1.8	22.1	53.2	68.8	38.0
	F	8.5	1.9	1.0	0.7	9.3	35.5	70.0	42.2
V		7.4	4.8	4.9	1.0	60.8	70.1	27.0	16.3
	E	8.0	6.3	5.7	1.2	66.1	76.2	25.7	16.3
	F	6.9	4.1	4.2	0.8	55.5	64.0	28.3	16.3

Notes: a. The share data are in current prices. Therefore, in G_{XM} and G_{XA} the effects of commodity price changes are involved. For the weighted growth rates, the simple average of shares in 1960 and 1980 are used, taking total merchandise as 100, including non-X_M = X_A items.

b. Reliable data are not available by subgroup for group I. India from group I and Singapore from group IV are excluded. These cases are judged atypical for group observation for present purposes.

Source: World Bank, *World Development Report 1982*, Table 9.

results, together with the changes in the sectoral shares and weighted growth rates. In comparing the shares of X_M/X and X_A/A

For an interval (n years) between points 1, beginning year, and 2, ending year,

$G_{XM} = (X_M/X(2)) \;/\; (X_M/X(1)) \times (1 + G_X)^n$ and

$G_{XA} = (X_A/X(2)) \;/\; (X_A/X(1)) \times (1 + G_X)^n$.

in the two years across groups, as expected, the share of manufactures increases while that of agricultural products decreases sharply without exception towards groups of higher Y/L levels. In terms of growth rates of exports, growth of manufactures (G_{XM}) is far greater than growth of agricultural products (G_{XA}).[129] Moreover, G_{XM} tends to increase in the lower Y/L intervals, i.e., groups I–III, and turns to decrease in the higher Y/L intervals, i.e., groups III–V, thus forming an inverse U-shape pattern. G_{XA} moves along with G_{XM} in the lower Y/L intervals, but does not in the higher Y/L intervals. The weighted rates follow a similar pattern in the lower Y/L intervals, but differ in the higher Y/L intervals. The weighted G_{XM} exhibits a linear tendency of increase for the entire range, without any decrease toward groups of higher Y/L levels, whereas the weighted G_{XA} exhibits an inverse U-shape pattern. The weighted rate is an indicator of incremental "contribution." When changes are drastic, "simple" and "weighted" data present different patterns. This is the case in our examples, and the difference is substantive. For example, the share of agricultural products (X_A/X) of group IV in 1960 is still greater than that of manufactured goods (X_M/X). Yet, the weighted growth rate for manufacturing is much greater than that of agriculture. We adopt the "weighted" approach for our present purposes.

What patterns appear for these terms by subgroups? Both subgroups E and F follow virtually the same pattern in terms of the weighted growth rate of exports: there is a trend acceleration of G_{XM} and an inverse U-shape pattern for G_{XA} across groups. Within each group, both the simple and weighted G_{XM} is distinctly greater for subgroup E than subgroup F across all groups except for group I.[130] The same pattern is seen for G_{XA}, with the exception of group III. In subgroup F of group III, both the share and the rate of growth for agricultural product are very large. Since this group forms the point of turn in presenting the inverse U-shape, this performance is to be noted.

Taken together, these trends suggest the following. Through the long-term path of export growth, the so-called export

---

129 Changes in current prices are not eliminated and qualifications are made later.

130 Limited data suggest an inverse relationship for group I. Such a bias is understandable as the economy is dominated by agriculture during its preprimary phase of development.

substitution process is sustained in a broad sense in that the export of agricultural products is replaced by the export of manufactures. While one may argue that this is a mere repetition of the obvious fact, our group comparison provides distinct evidence in terms of the weighted growth rates of sectoral exports, elucidating a specific pattern: the path is produced as a combined result of a trend acceleration of G_{XM} and an inverse U-shape trend of G_{XA}. This implies that in the lower Y/L intervals both G_{XM} and G_{XA} accelerated, but in the higher Y/L intervals, the differential between the two growth rates is witnessed by a sustained acceleration of G_{XM} vs. a deceleration of G_{XA}. We know the typological difference among nations in terms of natural endowments plays a significant role, particularly for agricultural exports. Despite such possible differences among nations, we identify a general pattern of sectoral export growth in terms of both group and subgroup setting.

We interpret these findings in our framework of development phasing fully in the section that follows. For the moment, however, only a preliminary observation is made. What we conventionally call the phase of "primary export substitution" pertains to the specific interval of the long-term growth path. Specification of this phase can be made in various ways. We adopt the weighted rate of sectoral export growth, assuming this is most adequate for our purpose. The initial inequality $G_{XM}<G_{XA}$ is supposed to reverse to $G_{XM}>G_{XA}$ at a certain point along the growth path. The turning point can be defined by the equality $G_{XM} = G_{XA}$, which demarcates a shift from the phase of "traditional agricultural exports" to the phase of "primary export substitution."[131] In the cross-sectional data, the turning point may be located in group III, but by subgroup the location differs. While $G_{XM}>G_{XA}$ occurs distinctly for subgroup E, for subgroup F, $G_{XM}<G_{XA}$ is still the case. Therefore, we consider that subgroup E represents an "early" case, whereas subgroup F a "behind" case in realizing a shift from the phase of traditional agricultural exports to the phase of primary export substitution.[132] This coincides with the demarcation of the major

131 For the time being, just the export dimension is dealt with, apart from the phase of primary import substitution which was discussed in the preceding section.

132 Incidentally, an extremely low value of G_{XA} for subgroup F in group II is exceptional as $G_{XM}>G_{XA}$ is atypical.

phase proposed earlier in Chapter 4. We are convinced that this is significant in interpreting export growth performance in line with the domestic pattern of changes in productivity structure. The point of shift in exports coincides with the point of turn in the inverse U-shape path identified for aggregate performance. By sector, it now becomes clear that, after passing through that point, G_{XM} continues to sustain its path of acceleration whereas G_{XA} tends to decelerate, and the gap between the two widens.

For the higher Y/L level of developing economies, the relationship between export and productivity can be explained easily insofar as the export of manufactures is concerned. If we follow the dichotomy between machinery manufactures (Ma) and nonmachinery manufactures (NMa) adopted earlier, primary export substitution is continuously sustained further on with regard to NMa, corresponding to its productivity increase in domestic production. This export pattern may have substantive implications for the postprimary export substitution, although qualification is needed for a broader coverage of NMa than the conventional explanation centered on nondurable consumer goods. As a matter of fact, the function of substituting exports of agricultural products for exports of manufactures still goes on substantively during this interval. On the other hand, the ratio of Ma to total exports increases slightly from group III to IV. Compared with primary export substitution, the activity of secondary export substitution can only be slightly recognized between groups III and IV. A noticeable jump is identified only for the final segment from group IV to group V. As pointed out earlier, there appears to be an impressive process of discontinuity. These points will be expanded on in the section that follows.

Before moving on to the next topic, comments are needed on the following three points which are relevant to the different patterns noted between the former and the latter intervals and the related phasing. First, the associated performance of the growth rate of productivity and exports is witnessed for groups I–III, but not for groups IV–V as shown below (reproduced from Tables 6.9 and 6.10):

	I	II	III	IV	V
G_{yI}	0.5	2.9	3.3	3.6	3.8
G_{yA}	0.4	1.4	3.4	3.5	4.0
G_{XM}	0.4	0.9	2.5	3.3	4.8
G_{XA} (weighted)	1.2	1.3	2.5	1.3	1.0

How do we interpret these performances? Only an overview can be given as our data are crude. There are two major elements: one is the rate of increase in product demand which works competitively between domestic and foreign markets, and the other is the specialization of output production by subgroup within manufacturing and agriculture, respectively. These two elements are well-known factors and what we are specifically concerned with is its significance in the phasing framework.

The second element is easier to deal with. Export substitution by domestic manufacturing in developing economies occurs along the lines of comparative advantage: that is, resources are increasingly allocated more to the subsectors which can shift their supply curves at a faster pace. In terms of our simple approach, the role of productivity increase, G_{yI} (a proxy of G_{yM}) must be greater for these exporting subsectors. These cases are specified by the notation #. We have $G_{yM}\#>G_{yM}$ throughout the entire development path. The general trend of increase toward groups of higher Y/L levels found earlier for manufacturing suggests that the difference, $G_{yM}\#-G_{yM}$, tends to be greater with increasing production specialization. In the case of agriculture, we can conceive the analogous effects of specialization. This is illustrated by the well-known problem of competition between food and cash crops. However, the difference, $G_{yA}\#-G_{yA}$, tends to become smaller towards groups of higher Y/L levels, due to the smaller scope of possible relative specialization as compared to manufacturing.[133]

Demand competition, the first element mentioned, is difficult to handle beyond saying that income elasticity continues to be greater for manufactured goods as compared to agricultural products. However, we cannot emphasize too much the significance of this conventional proposition. For example, in the case of manufacturing, the magnitude of export and productivity increases differs a great deal between subgroups E and F (as was pointed out earlier). The effects of demand cannot be the major factors of interpretation here.

Second, we cannot neglect the sectoral performance of relative output prices. According to our hypothesis, differences in the speed

133 The well-known problem of the trade-off between food crops and cash crops occurs because of the difficulty in making the shift. This is because during the earlier phases of development, food crops for domestic consumption and cash crops especially for exports are in a severely competitive situation. Toward the latter phases of development, the possibility of making a shift to the production of exportable crops becomes greater. But we contend that this occurs to a lesser extent in industry.

of shifting supply curves must manifest themselves in the behavior of output prices. Is it possible to endorse the validity of our hypothesis from this aspect? Let us recall the earlier finding that the relative prices of manufacturing output in the developing economies are rather reluctant to follow a decreasing tendency in a great number of countries. Over the two decades, by group a simple average of relative prices in terms of annual rates of change is summarized together with the case of agriculture as follows (%):

	I	II	III	IV	V
Manufacturing	2.0	1.1	-0.7	-0.6	-1.0
Agriculture	0.4	0.2	0.1	-0.6	-1.4

As noted earlier, the data can be interpreted only very generally. Yet, between the lower and higher Y/L intervals, we see a certain difference. In the lower Y/L intervals, i.e., groups I–II, the rate of increase in relative output prices is greater for manufacturing as compared to agriculture. This corresponds to the inequality, $G_{XA} > G_{XM}$, observed earlier for the lower Y/L intervals. As for the higher intervals, however, a consistent correspondence appears to be recognized: the output price of manufactures decreases faster than that of agricultural products. However, the decline in prices both occur at a moderate pace. As explained earlier, a shift from the phase of traditional agricultural exports to the phase of primary export substitution occurs around group III, so that such a price behavior can also be understood without inconsistency.

For group IV, the relative price of manufactures shows a decrease of almost the same pace, whereas the price of agricultural goods tends to decline. The factors forming this trend of decrease in the price of manufactures from group III to group IV may not be simple to determine, but at least the difficulty of making a shift to producer durable equipment and related products must be relevant. On the one hand, this pertains to the problem of poor performance towards secondary export substitution. On the other hand, the rate of productivity growth in agriculture is sustained in its process of acceleration throughout the groups, starting from group I. In Chapter 4 the importance of this phenomenon was pointed out as part of the characteristics of postwar developing economies. With regard to this particular group, the advantage of productivity growth of agriculture, perhaps supported by specialization, thus appears at least not to decline compared to manufacturing.

Conceptually, the dichotomy of tradeables vs. nontradeables discussed in Chapter 5 should not be ignored here. But the data are limited and only a simple illustration is possible. In group V, prices of both manufactures and agricultural products tend to decrease concurrently to a great extent. This implies that the output price of the services sector tends to increase. If services output includes the majority of nondurables, the comparative advantage of sectors producing tradeables is, in our view, indicated by such a price behavior. If the data of export prices are available for our group and subgroup observations, this problem, as well as the specialization effects mentioned earlier, would be directly clarified.[134]

Third, concerning the wage-productivity relationship, one may make critical comments on the lack of discussion of this relationship here. One could argue that surely this is an indispensable factor in comparing developing and developed economies. As for the earlier phases characterized by flexible supplies of unskilled labor, the lag of wage increases behind productivity growth is often the case in industrial growth, particularly in the upswings of the economy. As is estimated in Chapter 1, labor's income share (β) tends to decrease in the lower Y/L levels of developing economies, particularly in manufacturing. From these observations, we speculate that during the two decades under consideration, the lag of wage increases behind productivity growth actually took place. Given the same rate of productivity increase, this must have been an advantage in raising the competitive power of developing economies in the international market. To that extent, its supply price can be lower than otherwise. We assume, however, that such effects must be included in principle in the data or prices we used, though being

134 In this regard, the findings in Chapter 5 provide a general answer to the questions regarding export prices for manufactures and agricultural products. Perhaps, contrary to one's expectation, the relative prices of agricultural products at least did not decline as compared to that of manufactures, insofar as the 1960–1980 period is concerned. The relative price of food did increase somewhat. However, the difference does not seem to be significant, so that our data of G_{XM} and G_{XA}, used in nominal terms, may not be seriously affected in real term comparisons. Substantively, this means that the phenomenon of a much smaller G_{XA}, despite greater G_{yA} found earlier for groups IV and V, can be virtually endorsed in real terms. This raises the substantive problem of the international competition in exports of agricultural products: the pressure of relative price declines in higher groups on lower groups.

mixed with certain effects of other factors, such as policies and measures regulating the market mechanism.

When viewed from these three aspects, we would like to conclude that the associated performance of productivity and exports operates as the basic force in determining the growth-trade relationship.

III. Exports and Imports: Interrelated Sequential Path Through Shifting Phases

The summary to this chapter includes discussion on two topics: (1) how to link the changes in the trade pattern with the development phases that were demarcated earlier by changes in the domestic productivity structure, and (2) how to interpret the interactive relationship in sequence between exports and imports through industrialization. The first topic provides the basic framework and the second topic will be discussed within that context.

Incorporating Trade Patterns with Phasing in the Domestic Productivity Structure

To deal with the linking of trade patterns with development phases, the export performance of the major sectors is compared with the difference indicators identified in Chapter 4. In the preceding section, a preliminary discussion on this topic was already made so that here we can focus on the aspect of demarcation of phases and the shift of phases in order to identify merchandise export performance in the phase framework.

The difference in the growth rates of exports between manufactures and agricultural goods, $G_{XM}-G_{XA}$, measures the long-term process of export expansion of substituting X_A with X_M in a general context. By group average, we identify the equality, $G_{XM} = G_{XA}$, at group III. This is the cross-point of two tendencies: a decreasing G_{XA} and an increasing G_{XM}. Subgroup observations provide more detailed information of this phasing, with subgroup E being ahead, while subgroup F lags behind. For example, in group III, subgroup E shows distinctly $G_{XM}>G_{XA}$, while the reverse is true for subgroup F; that is, subgroup E is located after the cross-point, but subgroup F is located before it. Such differences are seen also for groups IV and V, while the reverse appears to be the case for group II. Subgroup observations are not possible for group I, but we speculate that in the preprimary phase, this effect may be caused by the poor performance of agricultural exports in subgroup F. Relevant terms are listed in Table 6.11 to compare with the phase demarcation made earlier.

The major findings are deduced from $G_{XM}-D_I$ and $G_{XA}-D_A$. These indicate simply the changes in the patterns of the relationship between exports and the domestic productivity by sector. In a comparison across groups, $G_{XM}-D_I$ is near zero for groups I and II, while for groups III, IV, and V, the difference is positive and shows a distinct tendency to increase. The term $G_{XA}-D_A$ has sizeable positive values for groups I, II, and III, followed by small values for groups IV and V. The growth relationship between exports and productivity thus presents a sharp contrast in the two segments. We can say that manufactured exports expand at a faster accelerating pace than that of growth in domestic productivity, whereas agricultural goods exports increase at a slower decelerating pace than that of growth in domestic productivity. Earlier, we asked why, for the higher Y/L groups (i.e., groups IV and V), did the growth rate of agricultural goods exports become distinctly lower, despite the fact that its domestic productivity remained almost equal to the higher rates of growth in manufacturing? The answer was given earlier and is reconfirmed now. We think no further explanation is needed here.

With respect to subgroups, we know that for both D_I and D_A, subgroup E has a greater value than subgroup F for all groups without exception. This firmly characterizes the two subgroups in terms of their position in the phase demarcation. What about the dimension of exports? The same inequality manifests itself in manufacturing: that is, G_{XM} is greater for subgroup E than subgroup F. In the case of agriculture we also see this inequality. Taken together, we expect that $G_{XM}-D_I$ and $G_{XA}-D_A$ are greater for subgroup E than subgroup F, and these are shown in the table with the exception of group III for the latter difference. It is noted that in the case of manufacturing, the difference between subgroups E and F is sizeable for groups III and IV. The reason for the exception of group III is not necessarily clear, but the fact that $G_{XM}-D_I$ is near zero for subgroup F in this group may be relevant. For the most part, the subgroup observations endorse what we found from the group observations.

Next, we draw attention to the very important role played by the services sector in order to illustrate the activity of demand competition between domestic and foreign markets in terms of resource allocation. In Chapter 4, the change in the magnitude of the difference, D_S–D_I, was discussed as an indicator of subphase shifts in the secondary major phase. Here, we would like to use D_S–D_I together with D_S–D_A as a crude indicator representing the degree of competitive activities between the production of tradeable and nontradeable goods. These differences are shown to supplement Table 6.11 with a rough phase demarcation (the preprimary phase is included in the primary phase):

		Primary (I, II)	Secondary (III, IV)	Developed (V)
D_S–D_I	Average	0.1	0.3	-0.2
	E	0.1	0.2	-0.2
	F	0.1	0.5	-0.2
D_S–D_A	Average	1.2	1.2	0.4
	E	1.0	0.8	0.6
	F	1.3	1.6	0.3

Particular attention is drawn to the relatively large magnitude of the secondary phase, which is especially distinct in the case of D_S–D_A. This is primarily caused by subgroup F, particularly for group IV. In light of the supposed normal path from group III to group V, the activity of the services sector is stronger in inducing resources in group IV. To this extent, export activity is discouraged. In Table 6.11, we actually identify this phenomenon for this group, particularly for subgroup F where a greater magnitude of D_S is associated with a smaller magnitude of G_{XM} and G_{XA}.

The data are rough and crude. Industry includes public utilities and construction, and production activities of nontradeable goods are not excluded regularly. The weighting system is not the same as that used for domestic productivity growth and export expansion. The performance of output prices is not counted, although from our findings in Chapter 5, we can say for this particular interval, the rate of relative price increase is greater for the services sector which will tend to confirm our suggestion. Nonetheless, we are convinced that the role played by domestic activities as illustrated by services is one of the important elements for understanding the export growth activities in developing nations. Thus, we are convinced that it is generally possible to incorporate the trade patterns with the phase demarcation by changes in the structure of domestic productivity.

Table 6.11
Performance of Exports in the Development Phases: 1960–1980 [a]
(percentage)

Group & Subgroup	G_{XM}	G_{XA}	$G_{XM}-G_{XA}$	D_I	D_A	D_S	$G_{XM}-D_I$	$G_{XA}-D_A$
I Average	0.4	1.2	-0.8	0.5	-0.3	-0.4	0.1	1.5
II Average	0.9	1.3	-0.4	1.1	-0.3	2.1	-0.2	1.6
E	1.3	2.3	-1.0	1.4	0.1	2.2	-0.3	2.2
F	0.8	0.2	0.6	0.8	-0.6	1.9	0.0	0.8
III Average	2.5	2.5	0.0	1.2	0.7	1.4	1.3	1.8
E	4.0	2.1	1.9	1.3	1.1	1.7	2.7	1.0
F	1.0	3.0	-2.0	1.2	0.3	1.2	-0.2	2.7
IV Average	3.3	1.3	2.0	1.4	0.8	1.8	1.9	0.5
E	4.9	1.8	3.1	1.9	1.0	1.9	3.0	0.8
F	1.9	0.7	1.2	1.0	0.5	1.8	0.9	0.2
V Average	5.2	1.0	3.8	1.2	0.6	1.0	4.0	0.4
E	6.3	1.2	5.1	1.5	0.7	1.3	4.8	0.5
F	4.1	0.8	3.3	0.9	0.4	0.7	3.2	0.4

Note: a. Slight inconsistencies are due to rounding.

Sources: See sources of Tables 4.10 and 6.10.

With regard to imports, a few words will suffice. The very low level of Y_M/Y in groups I and II is relevant to the characteristics of the preprimary phase, even the import ratio of NMa increased from group I to group II. During the primary phase, which is characterized by the indicator pattern of $D_S>D_I$, the import substitution of NMa proceeds from group II to group III and from group III to group IV. However, the ratio of NMa imports to domestic production still remains large, as was shown earlier. Without specifying the indicator of individual countries, we cannot deal with the import performance and $D_S>D_I$ in the primary and the secondary phases. Yet, we suggest that the secondary import substitution must be associated with the pattern of $D_I>D_S$, which indicates the shift of structure of productivity in the economy towards fuller industrialization.[135]

Exports and Imports: Simultaneous Path in Sequence

In this section, the purpose of our discussion is to integrate the earlier discussions on imports and exports in the phasing framework by emphasizing the significance of the interactive relationship between imports and exports activities. We believe it is important to recognize this interaction given the widely prevailing view that exports and imports are alternative choices. In order to address this major issue, however, we need to first clarify the actual relationship between export and import activities. We believe it is most important to precisely identify what we call the "simultaneous process" of export expansion and import substitution. Let us begin by presenting Table 6.12. The interrelationship of imports and exports is simply treated in terms of the resource imbalance (M–X), disaggregating the imbalance into manufactures (machinery and nonmachinery) and nonmanufactures. Machinery is again a proxy for investment goods. The resource imbalance thus disaggregated is not necessarily the conventional approach, but is convenient for our purpose as it represents the resource imbalance that pertains to the foreign payment constraint. The imbalance due to nonfactor services is added, though it is roughly estimated as a residual. The primary purpose is to use M–X as a simple indicator of changes in the merchandise trade structure in the process of phase shifts through the path of industrialization. The table pertains to cross-sectional data, but useful knowledge can be drawn from it.

---

135 The Western historical experience cannot be systematically treated, but Japan's case presents some evidence of this. Japan's ratio of imports to domestic production of producer durable equipment was 40.1 percent in 1905–1909, 27.4 percent in 1925–1929, and 7.7 percent in 1935–1939 (Yuichi Shionoya, "Patterns of Industrial Development," in Klein and Ohkawa, eds., *Economic Growth*). The original annual data are taken from the Japanese version in Ohkawa and Minami, eds., *Kindai Nihon no Keizai Hatten,* Appendix Tables 13 and 14 (in current prices):600–11. As compared to the Western experience, the level of the ratio appears lower, but its trend of decline can be assumed to be similar.

Incidentally, M_A/Y_M was 2.5 in 1874–1883, 2.8 in 1892–1901 before following a trend of increase in subsequent periods (*LTES* , *Mining and Manufacturing*, Vol. 10 (by M. Shinohara), Table 1.9:13). The ratio is in gross terms for intermediate goods and yet suggests that the initial ratio might not differ much from that of contemporary developing economies.

First, in terms of average group comparisons, the magnitude of the imbalance is greater for manufactures than that of merchandise trade throughout all developing nations, with the slight exception of group IV. The other side of the coin is the contribution of nonmanufactures. The imbalance of merchandise trade tends to decrease at higher Y/L groups and approaches zero in group V. This is caused by two trends: one trend is a trend of decrease in the imbalance of manufactures and the other trend is a countermovement of nonmanufactures, that is, its trend of increase contributes to the betterment of the resource balance. Since the trend of decrease is witnessed for the total resource imbalance, this can be identified as the basic pattern of development.

One may say that these trends are all obvious and tell us nothing new. However, these trends are indispensable as preparatory information for the observation by subgroups which is of greater significance here. The magnitude of the imbalance of merchandise trade is greater in subgroup F and smaller in subgroup E for developing economies. An exception is group IV, but the imbalance of nonfactor services is atypically greater for subgroup E than subgroup F in this group. Contrarily, the magnitude of the imbalance of manufactures tends to be greater in subgroup E and smaller in subgroup F, again with the exception of group IV for the same reason. As for the imbalance of machinery, the magnitude is definitely greater in subgroup E than in subgroup F without exception. It is obvious that this is influential in giving rise to the difference in the imbalance between subgroups E and F that was found earlier. We believe this is an important phenomenon (cf. the discussion on imports in Section I). The resource balance in merchandise trade is more favorable in subgroup E than in subgroup F, except for group IV. In order to sustain this performance, subgroup E tends to rely more on the contribution of nonmanufactures and nonfactor services.

Second, the data in Table 6.12 may serve to clarify another important aspect of the phase demarcation of trade patterns. The role played by net exports of nonmanufactures increases sharply from group I to group II, but towards subsequent groups the ratio of nonmanufactures reverses and decreases rapidly. This suggests that the zero point in its contribution may be located somewhere at the midpoint between groups III and IV, as actually illustrated by subgroup F of group III. The primary phase of export expansion is centered in the range from group I to group II. As expected, a distinct trend of import substitution is witnessed, and is centered on a range from group II to group III. From group II to group III, the ratio of machinery which was less than the ratio of nonmachinery increases

Table 6.12
Resource Imbalance and Sectoral Disaggregation: The Ratio of Manufactures and Nonmanufactures to GDP, 1980 [a] ***(percentage)***

Group & Subgroup		Merchandise Trade [b]	Manufactures Total	Machinery	Nonmachinery	Nonmanufactures	Total Resource Imbalance [b, c]	Nonfuel Services [d]
I	Av.	9.3	10.7	5.9	4.8	-1.4	9.7	-0.4
	E	7.5	12.0	6.4	5.6	-4.5	10.6	-3.1
	F	10.8	9.3	5.4	3.9	1.5	8.8	2.0
II	Av.	8.5	21.6	9.8	11.8	-13.1	9.5	-1.0
	E	6.6	21.5	10.9	10.6	-14.9	5.5	1.1
	F	10.6	21.7	8.8	12.9	-11.1	13.4	-2.8
III	Av.	8.1	10.1	6.4	3.7	-2.0	8.9	-0.8
	E	7.6	11.7	6.7	5.0	-4.1	7.8	-0.2
	F	8.6	8.5	6.1	2.4	0.1	10.0	-1.4
IV	Av.	8.8	7.8	4.7	3.1	1.0	6.2	2.6
	E	13.0	8.5	5.7	2.8	4.5	8.3	4.7
	F	4.6	7.1	3.6	3.5	-2.5	4.0	0.6
V	Av.	0.9	0.2	0.3	-0.1	0.7	0.8	0.1
	E	-0.8	0.1	0.3	-0.2	0.9	1.3	-2.1
	F	-0.5	0.1	0.2	-0.1	-0.4	0.2	-0.7

Notes:
a. The difference between the data in this table and in Table 1.3 is due to a slight difference in the countries selected. Other discrepancies are due to rounding.
b. Merchandise trade is in current dollars, while total resource balance is in national currency.
c. Resource imbalance is M–X.
d. Nonfactor services are indirectly calculated by merchandise trade less the total resource balance, including the discrepancy as noted in note b.

Sources: World Bank, *World Development Report,* 1982 and 1983 issues, Tables 5, 8, 9, and 10.

to exceed nonmachinery. For the range from group I to group II, the magnitude of its imbalance does increase. We would like to call this the interval of "primary import expansion." This takes place simultaneously with the primary phase of export expansion of nonmanufactures. Conventionally, import substitution is conceived of as substituting manufactures (i.e., nondurable goods) because of the initial situation of imported manufactures brought forth during the colonial era. Accordingly, attention to the significance of such a pattern has not been duly paid. Since no difference is essentially recognized between the subgroups in this interval, the evidence regarding the primary phase of import expansion suggests the validity of specifying the earliest phase in these terms.

Let us recall the preprimary phase which was defined earlier in the discussion of the structure of productivity. The characteristic found in the cross-sectional interval can be supposed to correspond to that phase in the dimension of the trade structure, although examination using time-series data of individual countries would be needed to confirm this. With the initial low level of agricultural productivity, imports of manufactures, particularly imports of nonmachinery, are strongly depressed and tend to decline further. If this argument is accepted, a difference between groups I and II suggests that before the conventional phase of primary import substitution takes place a noticeable performance in the import expansion of manufactures is identified simultaneously with the phase of export expansion of traditional products, although the actual relationship between the two activities in terms of the time dimension can only be clarified by examining the time-series data of individual countries.

The latter interval from group III to group IV covers the major part of the secondary phase. It appears to be characterized first by the simultaneous expansion of traditional exports (i.e., nonmanufactures) which is a sustained process, then by primary import substitution, later by primary export substitution (i.e., nonmachinery), and lastly by secondary import substitution (i.e., machinery).

These simultaneous movements in export and import activities are not the statistical results of our group average treatment, but can generally be seen in the development path of individual nations and has also been pointed out by some economists. The reason for emphasizing its significance here is that it pertains to the interactive relationship between export and import activities. The financing relationship is well-known: that is, export expansion provides foreign payment funds for enhancing domestic production towards import substitution. What we are concerned with is the competitive mechanism which operates between the export industry and the industry which enhances import substitution in the domestic market. For example, let us assume a sustained export expansion of textiles in an economy that has passed through the phase of primary export substitution. The sustained export expansion of textiles is possible because this industry keeps its strong competitive power in the international market as well as in the domestic market. The basic condition for simultaneously proceeding with machinery import substitution by promoting domestic production is to be competitive with the textiles industry in terms of productivity, wages, and prices. This is the real problem of the phase which is conventionally called the secondary import substitution phase. In this phase, the rationale for protection policies is to enhance the industry's competitive power in the domestic market.[136]

Our emphasis on the significance of the simultaneous process does not intend to ignore the sequential process of phase shifts. Instead, we are convinced that this presents the essential dimension of the economic development path. It is our interpretation that the sequential process can be successfully realized by moving through several steps of the simultaneous process. However, the actual development path does not necessarily proceed along the orthodox sequence line. Of course, any mechanical application of phase shift to all developing economies is far from realistic. Allowances must be made for certain modified types of path that may be reasonable as a result of differences in initial conditions, typology, and size of country. However, we believe that unacceptable deviations from this sequence line would bring forth negative effects on development from a long-term perspective.

The sequential process discussed earlier in conjunction with the phase shifts may need some further explanation from this point of view. It is beyond the scope of our study to investigate the complexity of reality in detail, and we want to focus our discussion on secondary import substitution activities as the most important illustration. This can be seen by comparing machinery and nonmachinery in Table 6.12: nonmachinery decreases from group II to group IV by 8.7 percentage points but machinery also decreases by 5.1 percentage points. This performance by the group average comparison is essentially the same as that derived by the subgroup comparison, although subgroup E shows a greater difference for machinery imbalance. This may seem contradictory to the conventional notion of a sequential path of movement from the primary to the secondary levels. The data for nonmachinery indicate the net result of primary import substitution and primary export substitution, so that taken together with the data on machinery, it implies that the activity of secondary import

136 This is a simplified explanation for illustrating the mechanism. For a detailed analysis of the actual process, see Japan's experience which is described in Ohkawa and Kohama, *Lectures on Developing Economies*, Lecture 2.

substitution occurs during the phase of primary export substitution. When we discussed the import dependency of investment goods, M_I/I, earlier in section I of this chapter, the ratio M_I/I was shown to have a tendency to slowly decrease for the majority of developing economies. This certainly supports the validity of the sequence of substitution, although the direct indicator, the ratio of imports to domestic production, is not available.[137] Thus, with regard to the substitution of investment goods, we would like to describe this pattern as the "earlier start" phenomena.

The findings for import performance in Section I are useful in answering this question of the substitution sequence.

One may say that the data represent the average performance for country groups of which the simultaneous process may be the result. We do not think so, though we do allow for such a possibility. Yet, it does suggest that even for the development path of individual countries, a clear-cut phase demarcation between the primary and the secondary phase is not always possible. The reason for referring to this as an "earlier start" is that it cannot be interpreted as being different from the orthodox sequential path, but it should be accepted as an extended modification of our phasing framework. It goes without saying that the earlier start case requires more intensive protectionist policies than the conventional case because of the borrowing of foreign technology, in the sense discussed in Chapter 1, in the context of an economy which is not yet mature, and a domestic market which is weak in terms of competitive ability. In fact, this "earlier start" has been attempted in a number of nations. The situation differs between nations and sweeping generalizations are risky. Yet, we can say that an acceptable economic reason for such protectionist policies (apart from the industries related to defense) is the long interval required to absorb the technological knowledge borrowed from abroad.

---

137 This ratio would show the earlier start more directly. For example, in Japan's case:

1905–1909	30.0	1915–1919	19.6	1925–1929	25.7	1935–1939	8.6
1910–1914	22.9	1920–1924	23.7	1930–1934	15.3		

(Ohkawa and Kohama, *Lectures on Developing Economies,* Table 2.8:72–73). Conventionally, secondary import substitution is demarcated in the period 1920–1924 to 1935–1939, but a decrease in the ratio did start earlier, although we have to consider the specific influence of World War I.

Now, let us sum up the discussion of the sequence in the interactive relationship between exports and imports. This notion includes a wide area and indeed all of the preceding discussion in this chapter pertains to this aspect. A comprehensive treatment is not possible here; rather we focus on the performance of investment goods.

In shifting phases in sequence in the long-term path of upgrading production capacity, developing economies need more and more imports of investment goods which are the carriers of advanced technological knowledge. The process of realizing this technological requirement is actually constrained by the ability to make foreign payments. A greater rate of export expansion contributes to weaken the pressure of this constraint. Export promotion is thus needed but depends upon a number of factors, both internal and external. Apart from the effects of external factors, we have emphasized the significance of augmenting the rate of productivity growth of the exporting industry. The production activity of tradeable goods in the domestic economy is linked to the input-output relationship and thus, the requirement of augmenting productivity growth must be of a wider scope in the domestic economy. This claims, in turn, greater amounts of investment goods imports.

Import substitution of investment goods can take place simultaneously by absorbing advanced technological knowledge and this would contribute to weakening the pressure of import requirements. However, this is only possible after the economy has achieved the phase of primary import substitution. Primary import substitution is made possible, in turn, by using the foreign payments resources provided through the primary export expansion. The process is highly simplified and further explanation on other relevant factors would be required to more accurately reflect reality. Yet, we are convinced that recognition of such interactive relationships is indispensable in light of the actual observations of the sequential path of exports and imports as demarcated by the phases.[138] From this long-term perspective, the policy choice of

---

138 In this regard, the conclusions of Chapters 6 and 7 in Chenery et al., *Industrialization and Growth,* are cited here as these are valuable empirical studies focused on nine semi-industrial nations.

> A period of import substitutions and deepening of input-output relationships is required before an economy can compete successfully in world market and so shift to either an export-led or open development strategy (Yuji

import substitution vs. export promotion is not an issue of choosing one or the other. The problem is currently important and debatable but its dimension differs. Nobody denies the possible contribution of export promotion in easing the pressure of foreign payments constraints, so long as this is brought forth by a shift in the supply curve. For example, if the exchange rates are artificially fixed and are distorted so that export activities are discouraged, elimination of such distortions are no doubt desirable. The same is true for import substitution as the measures to encourage it artificially also lead to distortions, and there is no accompanying rate of productivity growth in the industry concerned. These examples illustrate the issue of adjustments in the short-term dimension.

In our view, the strategy problem at issue is how to realize comparative advantage in the long-term interactive sequence, not only in terms of export expansion, but also in terms of import substitution. In order to achieve this, it is vitally important to identify the real problems at issue which change through the various development phases. The basic criterion is the possibility of enhancing the rate of productivity growth in the industries concerned from a long-term perspective. What was proposed in Chapter 1 with regard to the problem of maintaining balance between productive capacity and social capability in the macro-setting is basically relevant here in terms of sectoral components.

Kubo, Jaime de Melo, and Sherman Robinson, "Trade Strategies and Growth Episodes," in Chenery et al., *Industrialization and Growth*, 187).
There is also some sequencing in that the economies successfully pursuing open development strategies did so after a period of significant import substitution (Yuji Kubo, Jaime de Melo, Sherman Robinson, and Moshe Syrquin, "Interdependence and Industrial Structure," in Chenery et al., *Industrialization and Growth*, 224).

Appendix I

Initial Conditions of Japan:
Economic Level and Structure[139]

This appendix examines the economic level and structure of Japan during its inauguration to modern economic growth. Discussions of the social aspects may be found in other studies.

I. Comparisons of Japan and Europe

Income per capita or product per worker is the conventional measure of economic level. Qualifications and reservations are of course needed in applying modern, market-oriented concepts to the analysis of the initial conditions when the market economy is only partially developed. However, we find no reason for rejecting the use of these concepts,[140] and, in fact, find that this is the only operational way to link the initial conditions with the subsequent pattern of development. Our primary objective is to compare Japan and developing economies, but a comparison of Japan with Europe is indispensable because Japan can be viewed as a bridge between the developed and developing economies and can improve our understanding of the general implication of the measurement with these concepts.

---

139 Reproduced from the International Development Center of Japan, *Japan's Historical Development Experience and the Contemporary Developing Countries,* CA Project Report, with minor editorial changes.

We would like to acknowledge useful comments received from the project members at the Hakone Meeting as well as the Planning Conference and the computational help rendered by Messrs. Shokichi Motai, Yutaka Shimizu, and the late Nobukiyo Takamatsu.

140 For example, see the arguments put forth by Gunnar Myrdal, "Difference in Initial Conditions," Chapter 14, in *An Approach to the Asian Drama* (New York: Vintage Books, 1970).

Table A1.1
Comparisons of Product in the Nonagricultural and Agricultural Sectors; Europe and Japan in the Initial Period[a] (1965 US$)

	Per Capita GNP at the Initial Dates	Initial Dates	Labor Force[b] /Total Population (%)	Years	Per Worker GNP[c]	Percentage of Nonagri. Prod. Per Worker to Countrywide Prod. Per Worker[d]	
United Kingdom	227	1765-1886	45.0	1851	504	0.93	
France	242	1831-1840	39.1	1856	618	1.42	
Germany	302	1850-1859	38.3	1882	788	(1.15)	
Denmark	370	1865-1869	46.6	1882	794	1.27	
Norway	287	1865-1869	45.0	1875	637	1.66	
Sweden	215	1861-1869	39.8	1870	540	1.64	(1.77)
Italy	261	1861-1869	58.8	1881	632	1.81	(1.34)
Japan	154	1876-1887	59.9	1881-82	274	1.95	(2.14)

	Years	Per Worker Product of Nonagri.[d,e]		Percentage of Agricultural Prod. Per Worker to Countrywide Prod. Per Worker[d]		Per Worker Product of Agriculture[d]
United kingdom	1801-1811	469		1.16		
France	1896	878		0.55		
Germany	1850-1859	906		0.87		
Denmark	1870-1879	1,008		0.80		
Norway	1865	1,057		0.61		
Sweden	1861-1870	886	(956)	0.75	(0.71)	(383)
Italy	1861-1870	746	(847)	0.88	(0.77)	(487)
Japan	1887	535	(586)	0.60	(0.59)	(162)

Notes:
a. The coverage of sectors is not necessarily consistent between product and labor for some countries. The following is particularly noted for Kuznets' product data from Table 45:290-91: (1) exclusion of income from dwellings for the United Kingdom, France, Germany, and Norway; (2) exclusion of finance and income from dwellings for Sweden and Italy; and (3) no exclusions for Denmark. Incidentally, our data for Japan exclude income from dwellings for NDP but income is included in the GDP data.
b. Labor force are from unadjusted data.
c. Per worker GNP = (per capita GNP at the initial dates)/(labor force/total population).
d. Figures in parenthesis are in constant prices.
e. Per worker product of nonagriculture = (per worker GNP)*(percentage of nonagricultural product per worker to countrywide product per worker).

Sources: Kuznets, *Economic Growth of Nations,* Table 2:24, 53-54; Table 45:290-91; Ohkawa et al., *LTES*, M. Umemura in *LTES*, Vol. 2, Vol. 1, Table 24:226 and Table 25:227.

On the one hand, Japan's initial level of per capita GNP has long been considered to be "exceptionally low" compared to that of the Western nations. Much of the facile discussions, on the other hand, cast doubt on this notion simply because it cannot be conceived that Japan's development began from such a low economic level. In this appendix, the results of our research in clarifying this issue is summarized in terms of: (1) per capita GNP measured in US$; (2) a breakdown of product per worker in agriculture and nonagriculture; (3) per capita GNP measured in terms of purchasing power parity (ppp); and (4) distribution of labor in the industrial sector.

Our measure of per capita GNP of Japan for 1876–1887 is US$154 as compared to US$74 for 1874–1879 by Kuznets (both are expressed in 1965 US$).[141] The rather large difference between the two estimates is partly due to our revision of the previous underestimation of GNP in the constant price series for the early Meiji years, for which we are responsible. Per capita GNP of European countries in the initial dates are US$230–300. Thus, even the revised levels of per capita GNP for Japan can still be said to be "exceptionally low."

The breakdown of per worker product into agriculture and nonagriculture shown in Table A1.1 is a crude approximation. The number of countries used in obtaining the data is limited. Yet, the figures suggest a substantive significance of such measures. The average per worker product of nonagriculture and of agriculture for six European countries is US$841 and US$488 (both in 1965 prices, 100:58.1, respectively). The corresponding figures for Japan are US$535 and US$164 (100:30.7). Note that the data are in current prices; while a constant-price series would be more adequate, the number of countries is more limited. In terms of constant prices, the average per worker product of three European countries is US$903 for nonagriculture and US$519 for agriculture (100:57.5) as compared to US$586 and US$162 (100:27.6) for the two sectors, respectively, in Japan. Japan's level of per worker product relative to Europe is thus 63.6 for nonagriculture and 33.6 for agriculture in terms of current prices and 64.9 for nonagriculture and 31.2 for agriculture in terms of constant prices.

---

141 Kuznets, *Economic Growth of Nations,* Table 2:24.

Though the coverage is narrow, it may be enough to indicate that the "exceptionally low" level of per capita output in Japan is due to very low levels of per worker product in agriculture. Japan's relative level of output in nonagriculture is almost two-thirds the average European level and seems to be reasonably acceptable in view of the comparison of labor's sectoral distribution shares (this will be discussed later). The warning that the use of per capita product as an indicator of economic level requires some qualification, particularly as there may be possible distortions arising for agricultural product due to the varied restraint of land endowments; a phenomenon that is not new but is, in fact, rather well known.[142] Yet, the magnitude of these "distortions", as indicated by our measures, seems to be very impressive. The relative level of per worker product in the nonagricultural sectors is a much more acceptable indicator of economic activity than are macro-measures.[143]

The use of exchange rates, especially for conversions of values to be expressed in US$, needs qualification. An alternative measure of using ppp has a long history and its unavoidable arbitrariness in assuming "equivalence" of commodities and services between countries of varied preference scales is well known. Yet, recent developments involving this measurement deserve attention, particularly with respect to making meaningful qualifications of what was mentioned above.[144] Using ppp measures, the measures of per capita GDP of other nations relative to the United States are greater than the measures using exchange rate measures. The gap between the two tends to be greater for developing economies than for developed economies. Such findings in the international cross-sectional approach cannot be exact substitutes for the historical trends of nations, but may be suggestive for our purposes. Direct

---

142 The importance of this was first discussed by Simon Kuznets, "Present Underdeveloped Countries and Past Growth Patterns, 1958," in *Economic Growth and Structure* (New York: W. W. Nortons, 1965).

143 Incidentally, Hayami's comments on the effects of possible statistical discrepancies between output and labor in making such a sectoral breakdown is accepted as a qualification (underestimation of agricultural productivity for backward countries).

144 See for example Irvin B. Kravis et al., *A System of International Comparisons of Gross Product and Purchasing Power* (Baltimore and London: Johns Hopkins University Press, 1975). Further discussion of this is found in Chapter 5.

measures, however, are impossible for the initial periods. Taking the gaps between ppp and exchange rate measures (i.e., the ratio of the former to the latter) in 1970 for Japan (1.46), an adjustment ratio of 16 percent is assumed for raising Japan's economic level. This is an approximation for the ppp–exchange rate ratio for the same base year used in the domestic constant-price series. To this extent, Japan's lower economic level may be discounted.

The shares of the total labor force employed in agriculture are 34.4 percent in the United Kingdom (1801/1811), 51.7 percent in France (1856), 64.1 percent in Germany (1852/1855/1858), 51.7 percent in Denmark (1874–1875), 63.7 percent in Norway (1865), 64.0 percent in Sweden (1860), and 57.5 percent in Italy (1861/1871); these are to be compared with 72.0 percent in Japan (1881–1882).[145] Compared to the shares of output, these shares are more convincing indicators of the degree of economic development in the initial periods of modern economic growth (MEG). Japan's economic backwardness is clear. Even in Sweden and Norway, which were latecomers in Europe, labor's share in agriculture is 64 percent, which is still much smaller than that of Japan. It was not until 1907–1908, near the ending year of the initial phase of development in our demarcation, that Japan's agricultural labor's share reached 64 percent. The relationship between labor's share and per worker product cannot be found to be uniform and stable in a historical comparison of Japan and Europe. However, an illustration may be useful. Japan's product per worker (on a GDP basis) in the nonagricultural sector in 1907–1908 is estimated to be 1–1.5 times that of 1876–1887, while per worker product of nonagriculture in Table A1.1 shows Sweden's level is 1.6 times that of Japan in the initial periods. The difference in the level can be largely explained by Japan's economic backwardness.

Finally, the greater difference found in terms of per worker product in agriculture between Japan and Europe is, of course, essentially caused by Japan's unfavorable man-land ratio. For example, agricultural land area per male worker is 7.18 ha for France as compared to 0.89 ha for Japan in 1880.[146] Taking into account differences in land quality would not alter the comparison.

---

145 The figures for the European countries are from Kuznets, *Economic Growth of Nations,* Table 38:250–51. A slash (/) denotes that data are for single years. Japan's figures are from data in Ohkawa et al., *LTES,* Vol. 2, *Labor Force.*

146 According to Hayami and Ruttan, *Agricultural Development,* agricultural output per male in 1965 is 357 for an average of 10 European countries, where Japan is assumed to be 100 (calculated from Table 4.2:73),

Table A1.2
Comparisons of Product Per Worker in Nonagriculture and Agriculture: Asia and Japan in 1953–1957 and 1968–1972 (1965 US$)[a]

	Per Capita GDP	Per Worker GDP	By Sector			
			Agri.[b]	Nonagri.	Industry[c]	Services
Philippines						
1953-1957	136	379	254	649	708	590
1968-1972	172	596	364	915	857	973
Thailand						
1953-1957	95	186	80	1,006	1,229	783
1968-1972	158	350	135	1,178	1,349	1,006
India						
1953-1957	63	202	153	324	398	249
1968-1972	158	232	135	485	512	457
Pakistan						
1953-1957	91	258	222	304	232	375
1968-1972	113	394	259	598	477	718
Korea, Rep. of						
1953-1957	158	503	304	986	655	1,316
1968-1972	303	970	540	1438	1,644	1,234
Taiwan						
1953-1957	175	617	317	992	943	1040
1968-1972	394	1,261	518	1,680	1,792	1,567

	Share of Labor (%)		
	Agri.[b]	Industry[c]	Services
Philippines			
1953-1957	66.9	14.2	18.9
1968-1972	57.8	21.2	21.0
Thailand			
1953-1957	87.2	3.9	8.9
1968-1972	78.2	8.7	13.1
India			
1953-1957	68.6	13.1	18.3
1968-1972	72.2	13.8	14.0
Pakistan			
1953-1957	56.0	22.0	22.0
1968-1972	58.1	23.8	18.1
Korea, Rep. of			
1953-1957	72.5	11.6	15.9
1968-1972	50.7	21.1	28.2

while the corresponding figure is 224 for per capita GNP, calculated from Kuznets, *Economic Growth of Nations,* Table 2:24. The former is in terms of wheat unit while the latter is in terms of 1965 US$ and is not exactly comparable. Yet, a comparison of the two is informative for what is stated in the main text.

Taiwan			
1953-1957	55.6	20.7	23.7
1968-1972	36.7	33.5	29.8

Notes: a. Because of either lack of data or irregularities, the periods were taken as follows: (1) three-year average of 1953, 1965, and 1957; (2) two-year average of 1956 and 1957; and (3) three-year average of 1955, 1956, and 1957.
b. Includes forestry and fishery.
c. Includes manufacturing, mining, construction, transportation, and communication.

Sources: United Nations, *Yearbook of National Accounts Statistics*, various issues; International Monetary Fund, *International Finance Statistics*, various issues; International Labour Organization, *Yearbook of Labour Statistics*, various issues; Economic and Social Commission of Asia and the Pacific, *Statistical Yearbook of Asia and the Far East*, various issues; Directorate-General of Budget, Accounting and Statistics, *Statistical Yearbook and Explanatory Notes* (Taipei: DGBAS, 1975).

Even in the premodern period, the selection of crops and a farming system had adapted to the given conditions with land-saving occurring together with technological progress in Japan. As a result of these activities, product per land did increase, but the increase in product per worker was limited. The difference in resource limitation between Europe and Japan is thus important; this is not only the major explanation of the "exceptionally low" level of Japan when statistically estimated in the initial periods, but also continued to operate as a restraint on the effects of technological progress in agriculture, to which we will come back later.

II. Comparisons Between Japan and the Asian Developing Economies

There has been much superficial talk about the "equivalence" (i.e., no difference) of economic levels between mid-nineteenth century Japan and the postwar initial conditions of the Asian developing economies. Superficially, there seems to be many similarities, including such factors as the natural resource endowments and the subsistence level, but no comprehensive study has yet been done thus far because of the data limitations. However, we believe that a general observation is now possible for selected Asian countries. Having discussed the Japan-Europe comparison in the preceding section, the problem here can be categorized into two elements: one element is Japan's economic backwardness as seen by labor's share of sectoral distribution and the other element is Japan's much lower level of per worker product of agriculture. If the notion of equivalence between Japan and the other Asian countries is accepted, then these characteristics can also be applied to the other Asian countries, using Japan as a bridge between Europe and Asia.

Relevant statistics are shown in Table A1.2 for the average of 1953–1957 and 1968–1972. The former period approximates the initial period and the latter period, which is immediately before the year of the oil crisis, is added for reference. To be directly comparable to the figures in Table A1.1, the data are all expressed in 1965 US$.

Note first that per capita GDP in 1953–1957 in the Asian countries, except Taiwan, is lower than that of Japan (154 in 1876–1887, GNP). Even in 1968–1972, per capita GDP of Thailand, India, and Pakistan is still lower than that of Japan. The notion of equivalence appears to be accepted. According to Kravis and others, the ratio of ppp to the exchange rate measure between India and the United States is 2.99 for 1970; this gap is much larger than the gap between advanced countries. If we use this together with the Japan–U.S. ratio of 1.46, India is 188 (63 for 1953–1957 in the table) vs. Japan 225. We believe that the ppp measure is more realistic as a measure of the economic level, in view of the fact that the expenditure composition and the price system in the Asian countries differ considerably from those of the United States. Because we do not have direct ppp measures between Japan and India or the other Asian countries, it is not possible at present to qualify them in a reliable way.[147] Yet we speculate that the average per capita

147 Another comparison between India and Japan with respect to personal consumption is as follows (India's values are relative to Japan which is 100):

		Japan		
		1885	1892	1909
India	1967	118.3	90.6	73.1
	1970	124.0	95.6	76.8

product of the Asian economies in the initial periods may be closer to that of Japan than was the case mentioned earlier.

Sectoral distribution of labor in 1953–1957 in the Asian economies is very close to that of Japan, except for Taiwan: on average, the share of agriculture is 70.9 percent. However, the variation in the labor share in agriculture requires further comment. Thailand's share is exceptionally high, while that of India and Pakistan are low: in particular, in India, labor in agriculture increased from 69 percent to 72 percent in 1968–1972. This trend over time does not follow the usual pattern and may be the result of factors such as a varying female participation rate in the labor force and the different ways of counting the labor force in relation to industrial classification. This would require further examination if we were concerned with the study of individual countries. However, it is beyond the scope of our study.[148]

GDP per worker of the other Asian countries appears to be closer to that of Japan on average than was the case of GDP per capita. Of importance here is the differences between nonagriculture and agriculture. In 1953–1957, the ratio of agriculture to nonagriculture (where nonagriculture = 100) was 39 in the Philippines, 9 in Thailand, 47 in India, 73 in Pakistan, 31 in Korea, and 32 in Taiwan. Except for Thailand, where the ratio is

---

For the 1967 and 1970 ppp measures, data are from Kravis et al., *A System of International Comparisons of Gross Product and Purchasing Power*. For Japan, quantity indexes of 1967 and 1970 are linked with prewar selected years using prewar personal expenditure series (*LTES*, Vol. 6) and the BSO, Prime Minister's Office, *Annual Report on the Retail Price Survey* (Tokyo: BSO, 1970). This procedure is not possible for individual items and is applied to five major broad items: food, clothing, housing, fuel and light, and miscellaneous. The above figures are derived from geometric means. The tentative measures seem to support what is stated in the main text.

148 It is reported that the labor shares in India were 74.8 percent in agriculture, 12.2 percent in industry, and 13.0 percent in services in 1911. In 1951, these shares had changed to 75.7 percent, 11.9 percent, and 12.4 percent, respectively (J. Krishnamurty, Some Aspects of Structural Change in the Indian Economy Since 1881, unpublished mimeo, September 1977). Note that agriculture includes ancillary activities and services excludes income from house property. The initial share of agriculture seems larger than the figure in Table A1.2. (Incidentally, Table A1.2 contains, as a reference, the share of labor for industry and the services sectors.) The share of services is larger than that of industry for most of the countries in 1953–1957. This pattern is similar to Japan, including India.

extremely small, two general groups can be distinguished: one group is represented by South Asia, where the ratio is relatively larger, and the other group is represented by East Asia (the Philippines appears to be between the two groups). The East Asian pattern is similar to that of Japan, whereas the South Asian pattern is not. The ratio tends to decline to 28 in India and to 43 in Pakistan in 1968–1972; these ratios are closer to that of the former group. Together with the previous discussion of the distribution of labor by sector, we would like to refrain from making any definite statements.[149]

With respect to per worker product of agriculture, however, we can be more definite. The immediately preceding discussion is based on national income data. It is difficult to expect statistical consistency between the expenditure approach and the production approach in deriving constant-price series data for the industrial sector in developing economies and in the earlier years of Japan. Therefore, another measure—the wheat unit (WU)—is examined below. In Table A1.3, the relevant statistics are summarized. The major indicator is output per worker but output per unit of land is also included in the table. Because of the varied rate of female participation in the labor force, even among the Asian nations, the male worker labor force participation rate is used.

The table reveals a simple but important fact. On average, whether it be for the region or for individual countries, product per worker in the Asian developing economies is not very different from the level of Japan in the initial periods. The level of South Asia is slightly lower than Japan's and that of Southeast Asia is close to the initial level of Japan. The corresponding figures in Table A1.2 may not be contradictory with this finding, if Thailand is excluded as an exceptional case. Unfortunately, with respect to the nonagricultural sector, no such definite observation can be made. However, we are inclined to say that (1) compared to South Asia, Japan's initial economic level is higher in the nonagricultural sector, particularly in the services sector, while for agriculture the notion of equivalence may be generally valid, and (2) compared to

---

149 By the same token, we refrain from discussing the varied levels of nonagricultural product per worker in comparison with Japan. Both South Korea and Taiwan show a pattern similar to that of Japan in that at the initial periods, per worker product of services is higher than that of industry and at a certain point in the subsequent development, the relationship is reversed. In other countries, except Pakistan, the relationship is reversed in the initial periods and in 1968–1972, it is mixed. This needs further examination to help overcome the reservation.

Table A1.3
Agricultural Output Per Worker and Per Unit Land (WU): Comparisons Between Japan and the Asian Developing Economies [a]

(A)	Japan, 1880	SE Asia, 1963	South Asia, 1963
Output per worker	2.50	2.33	1.82
Output per land (ha)	2.90	1.43	1.70

(b)	Sri Lanka	India	Pakistan	Philippines
Added value per male worker, 1960	3.7	2.1	2.4	3.7
Added value per hectare, 1960	2.71	1.05	-	1.83

Note: a. The measure of WU (wheat unit) differs in its construction between Yamada and Hayami and others. The latter study is specifically designed for both historical and international cross-sectional comparisons (see Hayami and Ruttan, *Agricultural Development*), while the measures in the former study are based on FAO's Far East weights. However, the results of the measures seem to differ only very little. In addition, Yamada's figures are gross of current input, but may not differ much from added values because the proportion of current input is small. The population engaged in agriculture often includes those engaged in forestry and fishery as a principal occupation in Yamada's estimates, while these are excluded for Japan's figures. This may imply some underestimation of other countries. Lastly, land does not exclude permanent pasture.

Sources: Yujiro Hayami et al., *A Century of Agricultural Growth in Japan: Its Relevance to Asian Development,* Table 1.2:8 (Tokyo: University of Tokyo Press, 1975); Saburo Yamada, "Ajia Nogyo no Tonyu Sanshutsu Kozo to Hatten no Pataan" [Input-output structure and its pattern of development], Parts I, II, and III. *Ajia Keizai* [Asian economics], 16 (June 1975) and 16 (August 1975); Yujiro Hayami in association with others, "An International Comparison of Agricultural Production and Productivities," *Technical Bulletin 277,* Agricultural Experiment Station, University of Minnesota (1971).

Southeast Asia, Japan's initial economic level is higher for nonagriculture.

Output per unit of land in Japan is much higher than that of other Asian countries. The difference is larger in the initial periods (mid-1950s for the developing economies). Though to a lesser extent, this is also seen in Southeast and South Asia. The factor that is directly responsible for this is, of course, the difference in the worker-land ratio, the indicator of which is given by the ratio of the two indicators. This is relevant to what was found in the preceding section between Japan and Europe. In this regard, the notion of equivalence between Japan and the other Asian countries in terms of the initial conditions cannot be accepted. In fact, we see a sharp contrast in terms of an intensive vs. an extensive type of farming.[150] With its intensive farming supported by improved infrastructure, particularly the irrigation facilities and progress in technology and/or institutions of the indigenous type, Japan's per hectare output surpasses those of the developing Asian countries. Nevertheless, it is important to note that Japan's per worker product barely reaches the Asian level. In other Asian nations, output growth had mostly been attained by land expansion in the past. Only quite recently have they been faced with the problem of severe limitation of land. In this sense, their agricultural development phasing can be said to be different from Japan, as seen from the initial periods of modern economic growth. The underdeveloped irrigation facilities in contemporary Asia has often been discussed in comparisons of their development experiences with Japan. Particular attention should be paid to human adaptation to given natural conditions, which differ between Japan, East Asia, and the other Asian countries in order to attain more or less the same level of product per worker, which is the basis of the Asian subsistence level.

III. Concluding Remarks

The preceding discussion still leaves a number of inconclusive points because of data limitations. Nevertheless, it is possible to say a few words in conclusion and to discuss the implications.

First, the revision of the statistical estimates, the adequate qualification by considering real purchasing power, the examination of labor shares by sector, and a two-sector (i.e., nonagriculture and agriculture) comparison of per worker product have all contributed

---

150 For a detailed discussion on this point, see Kazushi Ohkawa, "Asian Agricultural Development," Part IV, in *Differential Structure and Agriculture* (Tokyo: Kinokuniya, 1972).

to clarifying the "economic backwardness" of Japan, which was explicitly or implicitly assumed in the comparison with Europe. To this extent, the magnitude of the gap has been reduced. Even by Western standards, Japan's initial economic level and structure can be understood to be "not exceptionally low" insofar as the economic level is represented by the activities of the nonagricultural sector. The restraints stem from natural endowments in relation to population and can explain a great part of Japan's lower economic level. However, this does not necessarily imply a lower level of social capability.

Second, on the reinterpretation of the notion of equivalence about the initial conditions between Japan and the developing Asian countries, the similar approaches carried out for this comparison, as for the previous case, revealed two major points.

1. The initial economic level and structure of the developing Asian countries can generally be recognized by the notion of equivalence, but Japan's level of nonagricultural activities is higher, though this varies between nations and regions. In this sense, Japan's case can serve as a bridge between Europe and Asia. One of the important points deduced from these observations is that the so-called barriers which stem from the low level of initial conditions may not be as serious as has often been asserted.

2. On the other hand, there is another aspect to which the notion of equivalence cannot be applied; this is the different performance of obtaining a similar level of agricultural product per worker. Other Asian countries in the South and Southeast regions had no accumulation of premodern technological and/or institutional progress comparable to that which had been accumulated in Japan. As noted earlier, this is rather a result of their adaptation to the given conditions so that one cannot merely describe their capability and potential as being backward. The difference we would like to emphasize is the lack of historical experience in adapting to the restraints of land resources in the initial periods.[151]

---

151 In this respect, we have the impression that the different performances in the economic and agricultural activities recognized between states in contemporary India is an interesting approach from which to explore the problem; that is, to what extent can such differences be attributed to those of the initial conditions. Suggestions are contained in Krishnamurty, Some Aspects of Structural Change in the Indian Economy Since 1881.

Third, we should comment on the effects of the initial conditions in shaping the performance of growth and structural change during the subsequent years of development. This is too big a problem to be discussed here, but again the pattern of agricultural development provides a representative illustration. The data shown in Table A1.2 do not extend far back enough to fully discuss this aspect. However, Japan's historical experience tells us that product per worker in agriculture increased at a slower pace than that of the industrial sector, resulting in a growing gap of productivity between the two sectors through the entire process of modern economic growth, though with long-swing fluctuations. The problem of convergence or divergence in the trend of sectoral differences in output per worker is a classic one. Our research revealed that the gap in productivity between agriculture and industry has widened at a much faster rate in Japan than in the Western countries when a process of divergence occurred there.[152] This finding implies that the initial conditions, particularly the man-land ratio, have long had a great, sustained effect on the pattern of structural changes in subsequent development. This may suggest a more general study analyzing the problems of contemporary developing countries, where the man-land ratio is much more unfavorable than it was in the Western case.

A general setting of such a problem would lead us to a discussion on the final aspect of the initial conditions. By definition, these initial conditions include the historical heritage at the initial dates and in this sense the conditions are "given," and are essentially indigenous and traditional. However, these conditions should not be thought of in too rigid a fashion. Development is an interrelated change in the indigenous components together with the process of introducing foreign and modern components into the economy. As illustrated in the preceding paragraph, the traditional components may operate as a resistance to modernization, but they can also be advantageous if the interrelated development of the two components is appropriately pursued in terms of technology and institution as well as the conventional economic mechanism. Japan's historical experience illustrates such a possibility. Thus, the initial conditions should be interpreted more flexibly in terms of developing the social capability of nations, particularly in terms of adapting to these conditions.

---

152 Refer to Chapters 3 and 4.

Appendix **II**

Sectoral Residual Growth and Reallocation Effects

The residual macro-growth measurement cannot be fully substantiated without sectoral breakdowns. The available data, however, are limited even if the scope is confined to the three major sectors. Nevertheless, what we can do is to make preliminary observations using the estimates derived somewhat "hypothetically" on certain presumptions. The presumptions are drawn from our knowledge obtained in the preceding chapters, and we hope that the results are generally acceptable. Future efforts in arranging the relevant data will test the validity of this approach.

The normal macro-level pattern was that the effect of the nonconventional factors (G_R) tends to increase compared to that of the conventional factors ($\alpha(G_K - G_L)$) in the path of productivity growth. In terms of sectoral breakdowns, will this show a uniform or nonuniform performance? The uniform pattern implies that all three sectors exhibit some similarity in the macro-level comparison. We believe this is far from reality, and the problem here is to clarify the varied patterns by sector. Since investment data are not available by sector, we will estimate G_{RA}, G_{RI}, and G_{RS} (i.e., the residual growth of sectors, A, I, and S) using the output approach. The actual procedure is as follows: (1) the macro-level measurement for 1960–1980 is taken, excluding the 1980–1985 downswing interval; (2) the performance of the industrial sector is assumed to play a central role in shaping the macro-level pattern of increasing G_R; (3) agriculture requires specific treatment because the contribution of land plays an influential role; and (4) for the services sector, data are least available and we are obliged to derive G_{RS} as a residual (where G_R is the weighted sum of G_{RI} and G_{RA}).[153] First, the summary is presented and the details are described later.

I. Summary Presentation

In Table A2.1, a summary of the estimates is presented. In Panel (A), G_{RI} shows a path of trend acceleration interrupted by group IV; this is the same pattern that was observed for the aggregate performance. The term G_{RA} increases dramatically from a small magnitude in the lower Y/L groups to a large magnitude in the higher Y/L groups. In fact, G_{RA} is even greater than G_{RI} in groups III and IV, but G_{RA} decreases in group V. The term G_{RS} is of a much smaller magnitude than either G_{RA} or G_{RI} throughout the entire range of countries.

The aggregate G_R, after taking into account the reallocation effects to be estimated later, is decomposed in Panel (B) into three sectoral components and shows the sectoral contribution to the aggregate. The data suggest an interpretation that is relevant to the development phasing, although the data pertain to cross-sectional observations. It goes without saying that a phase shift is led by enhancing the residual growth of industry, but in this path the residual growth of agriculture also plays a much more influential role than is usually expected. The inverse U-shape pattern of the residual growth of agriculture draws particular attention: this pattern is primarily due to the advance of technology of the biochemical (BC) type. Direct estimates of the residual growth for the services sector are beyond our reach due to the paucity of available data, particularly for G_{KS}. From the conditional estimates of G_K tried earlier, it is not possible to derive even a rough estimate of G_{RS} as the great part of capital for infrastructure must be included in this sector. Thus, we are obliged to estimate G_{RS} as a residual, assuming that the estimate of G_R is acceptable. To the extent that the reallocation effects of capital would not be minor, our estimate is biased. The estimates of G_{RS} in Table A2.1, Panel (A) are to be understood with these reservations in mind, but we believe its magnitude, which seems relatively small, is acceptable. One may legitimately question the magnitude which

153 The reallocation effect is included in the macro-level residual (cf. Chapter 1). To this extent, the residue can be deduced.

Table A2.1
Sector Residual Growth by Group, 1960–1980 (percentage)

(A)

	G_{RA}	G_{RI}	G_{RS}	G_R(Aggregate)
I	0.08	-0.20	-1.01	0.23
II	1.01	1.16	-0.22	1.03
III	2.90	1.42	0.81	1.89
IV	2.48	1.17	0.75	1.36
V	1.44	1.72	0.74	1.57

(B) Weighted [a]

	$w_A G_{RA}$	$w_I G_{RI}$	$w_S G_{RS}$	Reallocation Effects
I	0.04	-0.03	-0.41	0.63
II	0.33	0.28	-0.10	0.52
III	0.82	0.37	0.38	0.32
IV	0.37	0.27	0.37	0.35
V	0.07	0.67	0.42	0.41

Note: a. In panel (B) the weights, w_A, w_I, and w_S are from Table 4.10 on average for 1960–1970 and 1970–1980. Since $G_{RA} = \Delta R_A/Y_A$, the weight should be Y_A/Y, etc., to link the sectoral values with the aggregate.

Sources: See sources of Table 2.8 and sources in other appendix tables.

is considerably smaller than is usually expected and may question whether this is brought forth by an underestimation of one sort or another? Actually, the data of G_{yS} used in Chapter 4 suggest the possibility of a higher G_{RS}. However, on the other hand, the plausibility of a smaller magnitude for G_{RS} cannot be denied because the rate of technological advance is least in the services sector.[154] One may assert that, in fact, a portion of G_{yS} may be better

154 Although the growth pattern of services is different from those of the postwar developing economies, the historical records of Japan draw attention in this regard. Throughout the long prewar period, G_{RS} is estimated to be near zero with positive values in the upswings but negative values in the downswings. See Kazushi Ohkawa and Nobukiyo Takamatsu,

interpreted as pertaining to the reallocation effects. We are inclined to share this view and believe that this is shown by the greatest part of the reallocation effects, which is relevant to services.

In Chapters 1 and 2, the factors associated with the level of Y/L and the factors not associated with it were distinguished from each other in dealing with the residual growth. What is described above pertains to the operation of the factors associated with the Y/L level. It is, however, also important to know the effects of the operation of the nonassociated factors by subgroup observations. But, because of data limitations, it is not possible for us to apply the same procedure as that used for the group treatment. Here, a semi-hypothetical approach is used for analysis of subgroups E and F in estimating sectoral rates of residual growth by applying the ratio, G_{RA}/G_{yA}, etc., measured on average for the groups to subgroups E and F, assuming the difference may not be significant. The relevant figures are listed in Table A2.2 and are arranged to correspond to the format of Table A2.1.

Despite the crude measurement procedures, we are convinced that several significant points are revealed. First, the subgroup difference in the sectoral residual growth is not uniform but appears to change through Y/L groups. The dominant sector in the macro-level difference of the residual growth is agriculture in group II, industry in group III, industry cum agriculture in group IV, and services in group V. This seems to support our view that the component for upgrading the level of social capability differs in the path of shifting phases of development. The sector which plays the dominant role can be called the "major sector" in this sense. The figures in Panels (A) and (B) in the table, which are drawn from the cross-sectional data, suggest a normal path of agriculture —> industry—>services with a few distortions. For example, for group IV, the "major sector" is not simply industry but is a combination of industry and agriculture. We believe this to be relevant to the dip pointed out earlier.

An important implication of this finding is that we cannot fully accept the widely prevailing assertion that industrialization proceeds through rises in the level of per capita income for two reasons: (1) the role played by nonassociated factors must be influential, and (2) much more attention should be drawn to the performance of agriculture.

"Productivity of Commerce and Services," Chapter 12, in Ohkawa and Minami, eds., *Kindai Nihon no Keizai Hatten*.

Table A2.2
Sectoral Residual Growth by Subgroups, 1960–1980 [a] ***(percentage)***

(A)

Group [b]	Subgroup	G_{RA}	G_{RI}	G_{RS}
II	E	1.7	1.3	-0.5
	F	0.2	0.9	-0.4
III	E	3.4	1.9	1.2
	F	2.2	1.0	0.5
IV	E	3.4	1.6	0.9
	F	1.8	0.8	0.7
V	E	1.5	1.9	1.0
	F	1.3	1.6	0.4

(B)

Group [b]	Subgroup	$w_A G_{RA}$	$w_I G_{RI}$	$w_S G_{RS}$	Sum [c]
II	E	0.50	0.32	-0.23	0.59
	F	0.06	0.21	-0.18	0.09
III	E	0.98	0.51	0.53	2.02
	F	0.60	0.26	0.24	1.10
IV	E	0.61	0.54	0.43	1.58
	F	0.27	0.27	0.14	0.68
V	E	0.13	0.65	0.57	1.35
	F	0.12	0.55	0.23	0.90

Notes: a. The ratio (G_{RA}/G_{yA}, etc.) by group is as follows:

Sector	Group II	III	IV	V
Agriculture	0.72	0.83	0.75	0.35
Industry	0.40	0.44	0.33	0.46
Services	0.24	0.77	0.34	0.46

b. For group I, the data are not sufficiently available for classification into subgroups.

c. The averages of sums do not necessarily coincide with those in Table A2.1, because the selected countries differ slightly between the two tables due to data availability.

Sources: Tables 4.5, 4.6, and 4.7; and Table A2.1.

II. Detailed Description of the Procedures Used to Derive the Estimates

The estimating procedure is described for the two sectors, respectively, with interpretations of the assumptions and results.

Industry

The presumption is that industrialization plays the leading role in absorbing advanced technological knowledge and the records of successful development show an impressive trend of increase in the ratio. That is, $G_{RI}>G_R$: the rate of residual growth of industry is sustained at a higher rate than that of the macro-economy. In looking at preliminary historical records, unfortunately, only the historical data for the United Kingdom and Japan can be used. As was discussed in Chapter 2, the United Kingdom illustrates a case of slow growth in currently developed nations, and Japan's faster growth pattern has both similarities and dissimilarities with the growth pattern of contemporary developing nations. Table A2.3 presents the historical patterns for both the United Kingdom and Japan. In both cases, the rate of residual growth of industry (of manufacturing for the United Kingdom) is sustained at a higher rate than the rate of aggregate residual growth. The ratio G_{RI}/G_R (G_{RM}/G_R for the United Kingdom) is much greater for Japan, the latecomer. Yet, we must note that the prewar records of the United Kingdom presents a trend of increase in the ratio except for the period 1937–1951.

These historical patterns differ from the patterns of the developing economies which are derived from Table A2.1. The ratio G_{RI}/G_R in the cross-sectional observations is -0.87 for group I, 1.17 for group II, 0.75 for group III, 0.86 for group IV, and 1.10 for group V. For the majority of developing economies, the ratio is less than unity. Why is it that the presumption that industrialization plays the lead role in development does not fit for the case of the developing economies? Is the procedure of estimation responsible for this difference? These approximations are crude and yet may produce no substantive distortions. The value of the ratio, G_{RI}/G_R, of less than unity, we believe, represents the characteristics of postwar developing economies insofar as the group average is concerned, although its value greater than unity can be expected for a number of nations, including those which have grown vigorously. It is our view that the difference in the pattern is caused by the patterns of the other sectors, namely, agriculture and services.

Table A2.3
Historical Residual Growth of the United Kingdom and Japan: A Comparison of Industry with Aggregate Measures (percentage)

	1856-1873	1873-1913	1924-1937	1937-1951	1951-1973
United kingdom					
G_{RM}	0.9	0.6	1.9	0.9	2.4
G_R	0.9	0.5	0.7	1.4	2.3
G_{RM}/G_R	1.00	1.20	2.71	0.64	1.04
	1887-1904	1914-1917	1917-1938	1954-1965	1965-1976
Japan					
G_{RI}	1.1	2.1	2.9	6.7	5.5
G_R	0.8	1.8	1.6	5.0	4.9
G_{RI}/G_R	1.36	1.16	1.79	1.34	1.17

Sources: R. C. O. Mathews et al., *British Economic Growth,* Table 17.4:532; Kazushi Ohkawa and Nobukiyo Takamatsu, "Capital Formation, Productivity and Employment: Japan's Historical Experience and its Possible Relevance to LDCs," *IDCJ Working Paper Series,* No. 26 (March 1983); and Table 2.4.

Actually, in estimating G_{RI} in Table A2.1 with the formula, $G_{RI} = \beta G_{yI}+\alpha(G_{YI}-G_{KI})$, we depend heavily on the data of G_{yI} (cf. Table 4.6) that was observed earlier. The simplest link between G_{RI} and G_{yI} would be $G_{RI} = \beta_I G_{yI}$, assuming that $G_{KI} = G_{YI}$. In reality, however, the capital-output ratio changes and hence the ratio G_{KI}/G_{YI} is estimated as follows: 1.10 for group I, 1.00 for group II, 0.95 for group III, 1.05 for group IV, and 1.10 for group V. The data shows that in the lower Y/L intervals, the capital-output ratio tends to decrease and in the higher Y/L intervals, it increases. No data are available for industry and the aggregate pattern is used as a general approximation (cf. the technical notes to Chapter 1). With regard to the data of $\beta(\alpha)$ for the industrial sector, β_M (for manufacturing), presented in Table 1.3, is used as a proxy for β_I. The relevant data are shown in Table A2.4.

Agriculture

In applying the conventional procedure to the agricultural sector, we do not try to derive our own estimates, but rather, relevant data derived by experts in this field of study are borrowed with the

Table A2.4
Estimates of Residual Growth in Industry of the Developing Economies by Group [a, b] ***(percentage)***

Group	G_{yI}	β_I	$\beta_I G_{yI}$	$\alpha_I(G_{YI}-G_{KI})$	α_I	$G_{YI}-G_{KI}$	G_{RI}
I	0.5	50.6	0.25	-0.46	47.2	-0.10	0.21
II	2.9	49.9	1.45	-0.29	50.1	-0.18	1.16
III	3.2	45.6	1.46	-0.14	54.4	-0.02	1.42
IV	3.6	47.5	1.76	-0.59	52.5	-1.12	1.17
V	3.7	61.6	2.28	-0.56	38.4	-1.46	1.72

Notes: a. $G_{yI} = G_{YI} - G_{LI}$.
b. $G_{RI} = \beta_I G_{YI} + \alpha_I(G_{YI} - G_{KI})$.

presumption that the data can be consistent with G_{YA}, G_{LA}, and hence GyA which were measured in Chapter 4. The relevant data are presented in Table A2.5.

Our attention is drawn to the large difference in G_{RA} between groups I and II, and groups III through V. The data exhibit a most distinctive inverse U-shape pattern. Changes in the factors which contribute to forming this pattern show a general regularity: that is, land exhibits the same shape in a milder form, capital follows a reverse U-shape, and labor tends to increase throughout the entire range of countries. Although the estimates are rough, we are convinced that these patterns are largely acceptable and can be understood as being generally consistent with the types of technological advance in agriculture described in Chapter 4.

One may ask why the magnitude of G_{RA} in groups III and IV is so large? Is this due to the estimate bias? The answer is "no." Essentially, the postwar agricultural revolution—in particular, the advances of the BC type that were discussed in Chapter 4—explain this large magnitude. The case of Japan may be illustrative of this as the data are much more reliable. Japan's postwar G_{RA} is so large as to be almost 3.0 percent.[155] Due to the advances in scale-

155 For the corresponding periods in Table A2.2, G_{RA} is estimated for historical Japan as follows:

G_{RA}	0.95	1.36	0.67	2.97	1.87

Source: Ohkawa and Takamatsu, "Capital Formation, Productivity and

Table A2.5
Estimates of Residual Growth in Agriculture of the Developing Economies by Group [a] *(percentage)*

(A)

	$\alpha_A(G_{YA}-G_{KA})$	$\beta_A(G_{YA}-G_{LA})$	$\gamma_A(G_{YA}-G_B)$	G_{RA}
I	-0.60	0.23	0.45	0.08
II	-0.44	0.87	0.54	0.97
III	-0.10	1.81	1.38	3.09
IV	-0.42	1.88	1.01	2.47
V	-1.09	2.05	0.48	1.44

(B)

	G_{YA}	G_{KA}	G_{LA}	G_B	α_A	β_A	γ_A
I	1.7	6.8	1.3	0.2	11.8	58.4	29.8
II	3.2	6.8	1.7	1.4	12.1	57.7	30.2
III	4.5	5.3	1.1	0.5	12.4	53.2	34.4
IV	2.8	5.8	-0.7	-0.3	13.9	53.6	32.5
V	1.1	6.3	-3.0	-0.5	21.0	49.9	31.0

Note: a. The group average for G_B and G_{KA} are drawn from the countries in the Hayami and Ruttan source. The number of countries are 1 for group I, 3 for group II, 7 for group III, 6 for group IV, and 12 for group V. Because of the limited number of countries in the first two groups, particularly for group I, strong reservations are required. G_{KA} is replaced by the value for group II. G_{KA} is a weighted average of the average annual rate of growth of livestock and tractor horsepower (the weight is the output elasticity for livestock and machinery in this measurement of the intercountry production function of group II, Table 6.1:145–46). G_{KA} is arbitrarily increased by 1 percentage point for groups I, II, and III, and 1.5 percentage points for groups IV and V to include infrastructure facilities such as irrigation and drainage. The income shares (α, β, and γ) are assumed to be on average roughly represented by the average of the historical data (since 1880) of the United States and Japan calculated from the original data in Hayami and Ruttan, *Agricultural Development,* Table 7.8:204. The figures in Panel (B) are somewhat modified by the Asian data. For details, see the main text.

Sources: Hayami and Ruttan, *Agricultural Development;* and Table 4.1.

dependent technology in group V, G_{RA} is much lower despite its higher G_{yA}.

Some explanation is needed with regard to the estimates of the output elasticities (which are approximated by income shares) of the three factors. (As mentioned earlier, the basic data are from Hayami and Ruttan.) First, the shares including those of intermediate inputs (power and fertilizer) are converted to shares of value added. In doing so, for the United States one-half of the share of "power" is assumed to be intermediate inputs. Second, the average of the shares of the two countries for the selected years is applied as follows for the groups in the cross-sectional data: 1885 for group I, 1885–1910 average for group II, 1910 for group III, 1935 for group IV, and the average for 1955 and 1970 average for group V. However, as mentioned earlier, in light of our knowledge about the Asian developing economies, the shares thus derived appear to be somewhat higher for labor and lower for land. Though the available data are limited even for the postwar period, the share of labor in the Philippines, Korea, and Taiwan are added in deriving the group averages (the Philippines for group II and the latter two cases for group III). The share of land is modified accordingly.[156]

Reallocation Effects

Because of the procedure mentioned earlier, residual growth of the services sector can be estimated only after measuring the reallocation effects. However, beyond this specific requirement, it is generally indispensable to quantify the pattern of reallocation effects in our conceptual framework of the residual approach, as pointed out in Chapter 1. A rough approximation of its magnitude was presented earlier in Table A2.1, Panel (B).

Before looking at the reallocation effect in depth, a preliminary approach using a simplified formula may be attempted to deal with the general subject of the contribution of structural changes to aggregate productivity growth. Our rationale for believing this to be important is detailed in Chapter 4. In Chapter 3, in the analysis of trend convergence or divergence, the two-sector approach—agriculture (A) vs. nonagriculture (N)—was used. Along

---

Employment."

156 Kazushi Ohkawa, "Implication for Agricultural Development," Part IV, in Yujiro Hayami, Vernon W. Ruttan, and Herman M. Southworth, eds., *Agricultural Growth in Japan, Taiwan, Korea and the Philippines* (Tokyo: University of Hawaii Press, 1979).

this line, a simplified formula is applied here regarding the two sectors A and N.

$$G_y = (w_A G_{yA} + w_N G_{yN}) + \{(w_A - v_A) G_{LA} + (w_N - v_N) G_{LN}\}$$

where y = labor productivity, and w and v are weights (for the output share and the labor share, respectively). The macro-level of productivity growth, Gy, is decomposed into two terms: a productivity term and a structural term. The productivity term is the weighted sum of the productivity of the two sectors and the structural term (σ) is the weighted sum of employment terms in parenthesis { }. The magnitude of σ, on average, for 1960–1970 and 1970–1980 is as follows:

	I	II	III	IV	V
σ	0.75	1.03	0.83	0.82	0.10

Note that σ increases from group I to group II, but then decreases moderately through subsequent groups followed by a sharp drop from group IV to group V. Unlike the performance of aggregate productivity growth, G_y, σ does not exhibit an inverse U-shape. The ratio, σ/G_y, exhibits a trend of decrease:

	I	II	III	IV	V
σ/G_y	0.63	0.40	0.28	0.28	0.03

This is a simple measurement in terms of the average productivity of labor that provides preliminary knowledge about the shape of the effects of structural changes. In particular, two points are noteworthy. First, for the developing economies, the ratio σ/G_y tends to decline distinctly in the lower Y/L range, and second, we see a drastic drop to group V, where σ/G_y has a very minor effect. These appear to be acceptable insofar as the average productivity is concerned.

Now, the effects of structural change measured in terms of marginal productivity or factor prices are conventionally referred to as the "reallocation effects" because the sectoral differences in the marginal productivity of factors is assumed to narrow as a result of factor reallocation. To this extent, productivity at the macro-level is assumed to increase even if there is no technological and/or organizational advances. Thus, this pertains to the measurement of residual growth. The actual estimates cannot be precise enough, but useful approximation is possible.[157] The result of applying the conventional formula to labor is described below. The reallocation effects of capital cannot be estimated due to data limitations, but we assume these effects are less significant than that of labor. The relevant data are presented in Table A2.6. Only the sum of the reallocation effects has substantive meaning, but the sectoral figures are added for reference as these may support our interpretation of the results.

The results exhibit a U-shape pattern: that is, the sum of the reallocation effects decline from the lowest group to group III, then increases somewhat in group V. One may wonder why the reallocation effects do not follow the pattern that is expected in light of the inverse-U shape pattern of growth found earlier. This is a legitimate question because the reallocation effects are expected to also follow an inverse U-shape pattern because the sectoral differential of the marginal product of labor is supposed to widen in the lower Y/L intervals and narrow in the higher Y/L intervals. Is the unexpected shape pattern due to our estimates bias? We do not think so, although we admit some possible distortions are involved due to the rough procedures, including, in particular, the use of a constant value for β_S.

We believe the pattern can be explained by two mutually related factors: one factor is the large effect of changes in the product shares and the other factor is the significant role played by the services sector. In addition to the differentials in labor productivity, changes in the product shares affect the measurement of the reallocation effects to a considerable extent. The formula, if converted to the incremental level to correspond with the treatment

---

157 With regard to the Japanese historical records, see Kazushi Ohkawa, "Effects of Structural Change on Productivity Growth: A Long-term Measurement," in R. F. Kosobud and Ryoshin Minami, eds., *Economic Studies of the Japanese Economy* (Chicago: Chicago University Press, 1977). For the theoretical relationship between the residual and the structure term, see Kazushi Ohkawa, "Sangyo Kozo Henka to Makuro Seisansei" [Industrial structural change and macro-productivity growth], Chapter 4 in *Nihon Keizai no Kozo.*

The formula used is

$$G_R = \Sigma X_I + G_{RI} + [\Sigma X_I \alpha_I (G_{KI} - G_K) + \Sigma X_I \beta_I (G_{LI} - G_L)]$$

where X = output share and I designates sectors. The terms in parenthesis are the reallocation effects in the aggregate growth of the residual, G_R. The term $\Sigma X_I G_I$ represents the residual growth net of reallocation effects.

Table A2.6
The Reallocation Effects of Labor, 1960–1980 [a] (percentage)

	Agriculture			Industry		
Group	$w_A\beta_A$	$G_{LA}-G_L$	$w_A\beta_A(G_{LA}-G_L)$	$w_I\beta_I$	$G_{LI}-G_L$	$w_I\beta_I(G_{LI}-G_L)$
I	0.28	-0.3	-0.08	0.08	2.7	0.22
II	0.18	-0.6	-0.11	0.11	1.5	0.17
III	0.15	-1.8	-0.27	0.14	1.9	0.27
IV	0.08	-2.7	-0.22	0.16	1.0	0.16
V	0.03	-4.2	-0.13	0.24	-0.5	-0.12

	Services		
	$w_S\beta_S$	$G_{LS}-G_L$	$w_S\beta_S(G_{LS}-G_L)$
I	0.17	2.9	0.49
II	0.22	2.1	0.46
III	0.20	1.6	0.32
IV	0.24	1.7	0.41
V	0.42	1.6	0.67

	Sum of the Reallocation Effects (Re) [b]	G_R Net of the Reallocation Effects	Re/G_R
I	0.63	-0.35	273.9
II	0.52	0.51	50.5
III	0.32	1.57	16.9
IV	0.35	1.01	22.1
V	0.42	1.16	26.1

Notes: a. w_A, w_I, and w_S are the output shares of the agricultural, industrial, and services sectors, respectively. These shares correspond to X_I in the preceding footnote.

b. $Re = w_A\beta_A(G_{LA}-G_L) + w_I\beta_I(G_{LI}-G_L) + w_S\beta_S(G_{LS}-G_L)$.

Sources: See sources of Tables 4.10 and 4.1.

in Chapter 1, implies $1/Y[\Delta L_I(w_I-w)]$ assuming equality between the marginal product of labor and wages, w_I (w is the macro-level average wage). Table A2.7 summarizes the relevant terms by drawing from the preceding estimates.

Table A2.7
Services and Industry Performances Relevant to the Reallocation Effects of Labor: 1960–1980 [a] (percentage)

(A): Weighted growth rates of labor and output

	$v_S G_{LS}$	$v_I G_{LI}$	$w_S G_{YS}$	$w_I G_{YI}$
I	0.50	0.15	0.15	0.75
II	0.75	0.40	2.85	1.50
III	1.40	0.90	3.00	2.05
IV	1.45	0.75	2.90	2.20
V	1.50	0.50	2.50	1.70

(B): Proxies of the differentials (ratio) in the marginal product of labor

	A	I	S			
	$\theta_A \cdot \beta_A/\beta$	$\theta_I \cdot \beta_I/\beta$	$\theta_S \cdot \beta_S/\beta$	A/I	A/S	S/I
I	0.67	4.48	3.38	0.15	0.20	0.75
II	0.55	2.88	2.95	0.19	0.19	1.02
III	0.55	1.36	1.74	0.40	0.32	1.28
IV	0.47	1.31	1.33	0.36	0.35	1.02
V	0.44	0.92	1.19	0.48	0.37	1.29

Note: a. For agriculture, $\theta_A = Y_A/L_A \,/\, Y/L \text{x} \beta_A/\beta = L_A w_A/Y_A \,/\, Lw/Y = w_A/w$. Likewise, we have w_I/w for industry and w_S/w for services.

Sources: Tables 4.10, 3.7, and 1.7.

Panel (A) simply illustrates that in the major route of labor's sectoral migration from agriculture to nonagriculture, the services sector plays a much more influential role as compared to industry in absorbing labor. Since the weighted growth rate of output is also greater for services than industry, we can suppose that the role played by services must be much more influential than industry in reallocating labor. The panel shows that this is the pattern witnessed for all groups except group I for which data reliability is questionable.

Our data are not reliable enough to measure the differences in the marginal productivity of labor by sectors, but in Panel (B) of the table a rough proxy is presented to illustrate the order of magnitude. Since the sectoral differences in the product per worker are drastic (i.e., it is enormous in the lower Y/L range and even larger in the higher Y/L range), as was discussed in Chapter 3, the effect of differences in the sectoral β is relatively less influential. To a large degree, an approach which uses average productivity makes sense, as illustrated by our preliminary approach. What we are particularly concerned with here, however, is the higher level of marginal productivity of labor, in terms of the sector average, that is sustained for the services sector (despite underemployment) as compared to that of industry (note that the assumed magnitude of β_S is conservative).

If the pattern of reallocation effects that is expected stems from knowledge based on the historical experience of presently developed nations, the pattern of the reallocation effects derived here may be "unexpected" because of the characteristics of the structural changes in the contemporary developing nations (this was pointed out repeatedly in preceding discussions).

The performance of the services sector is symbolic in this respect. Let us recall the pattern of phase shifts which is characterized by a sustained $D_S > D_I$ as shown in Panel (A). Special attention is required for group V. The major reason for the larger magnitude of services is given by its higher level of marginal productivity of labor as compared to that of industry. The estimate is rough and possible distortions should be allowed for, but note that this takes place with the inequality, $G_{LI} < G_L$, in this group; that is, services becomes the sole sector for absorbing increased labor.

What about the significance of the pattern of the reallocation effects? The measurement is simple and rough and does not take into account the differences in quality of labor, including unskilled vs. skilled labor, the reallocation effects within each major sector, and the relationship between labor reallocation and the pattern of sectoral allocation of capital. Yet, we are convinced that the measurement is useful in light of our original hypothesis of technological advance which suggests that the pattern of the G_R net of the reallocation effects should be the genuine pattern pertaining to technological and/or organizational advance. In this regard, we are particularly concerned that the G_R net of reallocation effects exhibits essentially the same pattern as G_R. For reference, the ratio Re/G_R is shown in Table A2.6. Although the abnormally high figure for group I is doubtful, the ratio appears to follow a U-shape pattern with group III marking the turning point. In the lower Y/L range, the ratio decreases, whereas in the higher Y/L range, the ratio increases. This indicates the changes in the role played by the two terms: reallocation effects are dominant in the lower Y/L range and less influential in the higher Y/L range.

This is a significant finding because we are now able to say that the reallocation effects have a different pattern and yet all the discussions on the residual growth pattern that have been developed thus far can be kept broadly intact by taking into account the reallocation effects. The interpretation can be fully given by direct observation of changes over time, but it may be speculated that the pattern of reallocation is a property that accompanies the shift in development phases: put more simply, the reallocation effects are dominant in the primary phase and less influential in the secondary phase.[158]

Explanation of the conceptual reason and the substantive evidence for what we call the "concomitant" property is warranted here. In this regard, let us recall two concepts contained in the preceding discussions: (1) the characterization of technological property by sector concerning the relationship between productivity growth and increase in employment, and (2) the different growth performances between nations in subgroups E and F. These two concepts are interrelated. Table A2.8 summarizes the results of the subgroup approach; the table is arranged to correspond to Table A2.6.

The magnitude of the sum tends to be greater for subgroup E than for subgroup F, with the exception of group IV. In addition, the subgroup differences are not uniform. However, with regard to the two goods-producing sectors, agriculture and industry, the inequality where subgroup E>subgroup F appears regularly throughout all groups. This is the pattern which we expect from the preceding proposition which suggested a "concurrent" pattern for industry and a "reverse" pattern for agriculture. Nations with higher rates of productivity growth show a larger positive magnitude for the I term

158 In terms of the pattern of long-term changes over time, available data are limited. Japan's case suggests a trend of increase, instead of decrease, in the reallocation effects of labor (the ratio of the reallocation effects to G_R in percentage terms is shown in parenthesis):

1905–1917	0.51	(27.7)	1932–1938	0.60	(17.6)
1918–1931	0.50	(74.6)	1955–1964	1.45	(21.6)

With an abnormally high figure for the downswing interval, it is difficult to discern the trend pattern of the ratio, but at least we can say that the accompanying reallocation changes are likely to have taken place.

Table A2.8
Reallocation Effects of Labor by Subgroup, 1960–1980 [a] (percentage)

Group [b]	Subgroup	$w_A\beta_A(G_{LA}-G_L)$	$w_I\beta_I(G_{LI}-G_L)$	$w_S\beta_S(G_{LS}-G_L)$	Sum
II	E	-0.10	0.25	0.56	0.71
	F	-0.07	0.11	0.31	0.35
III	E	-0.35	0.28	0.39	0.32
	F	-0.19	0.14	0.33	0.28
IV	E	-0.35	0.23	0.25	0.13
	F	-0.15	0.11	0.31	0.27
V	E	-0.20	0.06	0.76	0.62
	F	-0.18	-0.28	0.59	0.13

Notes: a. No estimates are possible for the sectoral β by subgroup and the group average value is used for both subgroups E and F (cf Table 1.7). The simple averages of these terms do not necessarily coincide with the figures in Table A2.5 because of slight differences in country selection.
b. For group I, the data are too limited to make subgrouping meaningful.

Sources: Tables 4.10, 4.5, 4.6, and 4.7.

combined with a larger negative magnitude for the A term, while nations of lower rates of productivity growth exhibit a lower positive magnitude for the I term with a lower negative magnitude for the A term. The difference is sizeable and distinct. As for the services sector, what we have called a "mixed" pattern appears to be relevant to its performance. The pattern of difference between but the size of the difference is much smaller in group III and is even reversed for group II. It is easy to see that this is the major cause for the irregularity in the difference in the reallocation effects (i.e., the sum) of the two subgroups. It can be concluded that the parallel occurrence of the two effects is essentially brought forth by the technological property of the goods-producing sectors and that some deviation is due to the performance of services.

Syrquin's Estimates

The results of our rough approximation in measuring the sectoral residual growth as well as the reallocation effects should be interpreted with due qualification and reservations. The problems left unsolved cover a wide range and further research is desirable. The magnitude of the reallocation effects measured by other studies does not necessarily show a trend of decline, but rather exhibit a trend of increase for the range of developing economies. Thus, further scrutiny is needed. In this regard, Moshe Syrquin's work deserves particular attention.[159] A brief review would contribute to our analysis by suggesting the actual problems at issue in our study.

A simplified summary of Syrquin's estimates is shown in Table A2.9. The estimates are comparable with our estimates as the concept and methodology used to derive the estimates do not differ, although the procedure of the data treatment does differ. In addition to agriculture (A), manufacturing (M), and services (S), mining and social overhead are treated in Syrquin's work. For simplicity, agriculture, manufacturing, and services are taken up here, and manufacturing is assumed to generally represent the pattern of industry (I). Both similarities and dissimilarities are witnessed with regard to the measured results. It is not our intent to discuss the performance of aggregate G_R but let us just say that the two estimates of G_R appear to be similar except that in Syrquin's case the inverse U-shape pattern is milder without showing a dip in groups corresponding to group IV in our case. A very similar pattern is identified between G_{RM} in Syrquin's estimates and G_{RI} in our study, where both exhibit trend acceleration, though again a dip is witnessed in the latter.

For agriculture, G_{RA} exhibits a distinct inverse U-shape pattern in Syrquin's estimates and its magnitude is always much smaller than that of G_{RM}. This is both similar and dissimilar to our estimates. Syrquin's estimates of G_{RA} is similar to ours in that both exhibit an inverse U-shape pattern, but Syrquin's G_{RA} is dissimilar to our estimates in that the magnitude is distinctly smaller than the level of G_{RI}. This requires further scrutiny, though we are convinced that $G_{RA}>G_{RI}$ can be possible because of the specific contribution of agricultural technology. As for G_{RS},

159 Moshe Syrquin, "Productivity Growth and Factor Reallocation," Part III, in *Industrialization and Growth* (Washington, D.C: World Bank, 1986) and "Resource Reallocation and Productivity Growth," in Kelley Williamson, ed., *What Drives Third World City Growth* (Princeton: Princeton University Press, 1984). I wish to express my deepest gratitude for his kindness in extending valuable comments on my earlier draft and for sending me his papers for my research.

Table A 2.9
Syrquin's Estimates of Sectoral Residuals and Reallocation Effects (percentage)

Group [a]	G_{RA} (A)	G_{RM} (M)	G_{RS} (S)	Aggregate G_R	Reallocation Effects (Re) [b]	Ratio of (Re/G_R-Re)
0	0.00	0.53	0.72	0.44	0.04	9.09
1	0.11	0.91	0.88	0.72	0.15	20.83
2	0.23	1.61	1.20	1.40	0.29	20.71
3	0.86	2.11	1.66	2.28	0.56	24.56
4	1.46	2.52	2.19	2.92	0.75	25.68
5	1.55	2.67	2.99	3.11	0.40	12.86
6	1.49	2.79	2.91	2.80	0.08	2.86

Notes: a. Classification of groups in terms of per capita income (U.S. dollars) is used for the "simulation". The per capita income of each group is as follows: 0, US$100–140; 1, US$140–280; 2, US$280–560; 3, US$560–1,120; 4, US$1,120–2,100; 5, US$2,100–3,360; and 6, US$3,360–5,040.

b. Note that the magnitude of Re is slightly different from that mentioned in the footnote.

Sources: Moshe Syrquin, "Productivity Growth and Factor Reallocation," Part III, in *Industrialization and Growth* (Washington, D.C: World Bank, 1986).

there is a distinct dissimilarity in terms of both magnitude and pattern, and further scrutiny would be required not only for statistical measurements but also for substantive interpretation.

Finally, with regard to the reallocation effects, Syrquin's research is much more intensive and includes the unique device of treating the capital-labor ratio. The value of Re shown in the table includes the effects of capital, but can be generally comparable to our estimates. Its pattern parallels that of the macro-level of residual growth and the path of narrowing the differentials in the marginal productivity of labor towards higher income levels is clearly shown. A dissimilar pattern for group V in our case seems to be an issue for further examination related to the problem of G_{RS}.

Appendix **III**

Labor Productivity, Investment Efficiency, and the Residual with Sectoral Estimates of Three Selected Countries

I. Introduction

The purpose of this appendix is to illustrate the macro-sectoral approach of residual growth and the performance of relevant terms by selecting a few countries for which the required data are available. At present, this is not possible for all countries because of the lack of data on the sectoral breakdown of gross domestic investment, which is not usually available from the conventional national accounts. We hope this preliminary attempt will lead to further research in this area.

The concept, framework, and estimating procedure are essentially the same as in Chapter 1 and the basic formula is (using the same notation):

$$\Delta R/I = \Delta Y/I - w\Delta L/I - r\varepsilon$$

where $\varepsilon = I/\Delta K$.

This basic formula is applied to the macro-economy and to two major sectors, agriculture (A) and manufacturing (M). Additionally, $\Delta R/Y$ will be dealt with, and for the non-AM sector, a residual estimate will be attempted.

Three countries in Asia were selected: India (group I), Thailand (group II), and the Republic of Korea (group III). All of these nations belong to subgroup A in Chapter 1 and subgroup E in Chapter 4. Although the number of developing countries for which sectoral investment data are available is limited, an alternative selection of countries is possible. Nevertheless, these three cases were chosen in view of other factors such as data continuity and comparability, and the relative convenience of data availability of the relevant terms. After all, our aim is to present an illustrative example for the possibility and usefulness of applying a sectoral approach to individual developing economies, instead of attempting a substantive analysis of these economies themselves for which the available literature is rich in each case insofar as the conventional approach is concerned.

Changes of growth performance over time of the individual countries will be observed in some detail in order to supplement the contents of the latter part of Chapter 2, where this subject was discussed lightly. Finally, with regard to the possible shift of development phases, a brief discussion is given in light of the framework developed in Chapter 4, in addition to incorporating the sectoral performance with the macro-performance.

II. Growth Performance

Growth Rates of Output, Labor Force, and Labor Productivity

Table A3.1 shows the growth performance of the three countries. While the data are not meant to reveal any new findings (as these are widely known from the conventional data), they are presented here as a preliminary step that links the performance of the key terms in our framework (and which will be dealt with later). First, let us note that the macro-level term G_Y for the selected Asian countries follows the acceleration path of the inverse U-shape that was identified by the group comparisons in Chapter 1. The data also indicate that a greater magnitude of G_Y in the manufacturing sector is associated with the pattern of the aggregate G_Y, but a smaller magnitude of G_Y in the agricultural sector is not necessarily associated with the pattern of aggregate G_Y. This is also discussed in Chapter 4.

With respect to changes over time, in Korea the aggregate G_Y declines gradually over the entire period. A similar declining trend of G_Y is observed in the manufacturing sector as well. With regard to the agricultural sector, G_Y in the 1970s was only one-half that of the other periods. Why did the G_Y of Korean agriculture reveal a U-shape pattern during the observation period? We do not have an immediate answer and yet it should be noted that G_Y of the agricultural sector was kept considerably low and that agriculture did not contribute very much to the high aggregate growth, as will be discussed later in Section III. The aggregate growth rate of output in Thailand also shows a trend of decrease similar to that of Korea. The sectoral growth performances, however, are dissimilar; because of the relatively vigorous growth of agricultural output, G_Y of the

Table A3.1
Growth of Output, Labor Force, and Productivity [a] *(percentage)*

Country/ Period	G_Y[b] Aggr.	G_Y[b] Agri.	G_Y[b] Manu.	G_L[c] Aggr.	G_L[c] Agri.	G_L[c] Manu.	G_y[d] Aggr.	G_y[d] Agri.	G_y[d] Manu.
Korea, Rep. of									
1960s	9.6	4.0	19.8	2.7	-0.5	10.4	6.9	4.5	9.4
1970s	8.8	1.9	16.1	3.6	-0.4	8.9	5.2	2.3	7.2
1980–1985	7.3	4.1	8.8	1.8	-4.3	3.5	5.5	8.4	5.3
1963–1985	8.8	3.2	16.1	2.9	-1.2	8.4	5.9	4.4	7.7
Thailand									
1960s	7.9	5.4	10.6	1.7	1.4	3.1	6.2	4.0	7.5
1970s	6.9	4.2	10.0	3.4	2.2	11.7	3.5	2.0	-1.7
1980–1985	5.0	5.0	5.0	1.6	3.3	2.6	3.4	1.7	2.4
1963–1985	6.9	4.8	9.2	2.4	2.1	6.4	4.5	2.7	2.8
India									
1960s	4.1	2.1	5.5	1.5	1.4	1.5	2.6	0.7	4.0
1970s	3.3	1.7	4.1	1.7	0.3	2.5	1.6	1.4	1.6
1980–1985	5.4	2.7	7.5	2.0	—	—	3.4	—	—
1963–1985	4.0	2.1	5.3	1.7	—	—	2.3	—	—

Notes: — = Not available.

a. The average annual growth rate for each period is calculated simply by bridging the beginning and the ending year, and the estimate for the entire period is a weighted average of these (weight=duration, hereafter the same for all tables).

b. Y refers to GDP in 1975 prices for Korea, 1972 prices for Thailand, and 1970 prices for India.

c. For Korea, data refer to employment; for Thailand, data refer to economically active population in 1960 and employment after 1970; for India, data refer to economically active population.

d. $G_y = G_Y - G_L$.

Sources: International Labour Organization, *Yearbook of Labour Statistics,* various issues; United Nations, *Yearbook of National Accounts Statistics,* various issues.

1980s is almost identical between the aggregate, and the agricultural and manufacturing sectors in Thailand. In the case of India, the aggregate growth rate of output increased sharply from the 1970s to 1980–1985 (the interval of downswing, cf. Chapter 2), which is different from the case of Korea and Thailand. With respect to sectoral growth, it is specifically noted that G_Y of both sectors attained higher growth rates in the 1980s which must have enhanced the aggregate G_Y.

Looking at the labor force data, the aggregate G_L on average is of a smaller magnitude than that of the aggregate G_Y. Changes in the aggregate G_L over time generally follow similar patterns among the three countries: particularly notable are the higher values of G_L in the 1970s. With respect to the sectoral performance, G_L of the agricultural sector is negative in Korea; this is in sharp contrast to the positive values for agricultural G_L of Thailand and India. This illustrates well the changes in employment patterns through the phases discussed in Chapter 4. In this regard, two points draw our attention. The first is the sustained increase in agriculture's G_L in Thailand and the second is the extraordinary magnitude of manufacturing's G_L in Korea in the 1960s and in Thailand in the 1970s. Despite the oil crises in the 1970s, the G_L of manufacturing in the entire period on average in Korea and Thailand indicates maintenance of very high rates of absorbing labor into the manufacturing sector.

Finally, let us examine the growth performance of labor productivity. The significance of the inverse U-shape pattern of increase in labor productivity was noted in Chapter 1. In the earlier years, the accelerating path is illustrated vividly here in the three Asian countries. This is obviously the combined result of G_Y and G_L which was explained above. On average for the entire period, the aggregate G_y is 2.3 percent in India, 4.5 percent in Thailand, and close to 6 percent in Korea; these are extremely high rates in the international perspective. However, with regard to changes in G_y over time, after having the highest records in the 1960s, we see the maintenance of fairly high rates of G_y in Korea and Thailand. Together with the high G_y of India in the 1950s, this pattern draws our attention because these trends took place despite a downswing in the economy. With regard to sectoral performance, agriculture and manufacturing show a contrasting pattern in Korea; that is the highest value for the agricultural sector and the lowest value for the manufacturing sector both occur in the 1980s. In Thailand, the manufacturing sector experienced negative growth in G_y in the 1970s largely as a result of the large G_L. This is an unexpected finding and must be investigated further.

The Investment Ratio and ICOR

The estimates of the investment ratio (I/Y) and the incremental capital-output ratio (ICOR) are presented in Table A3.2. A relatively higher proportion of gross domestic investment is

Table A3.2
The Investment Ratio and ICOR

	I/Y(%)			ICOR(I/ΔY)[a]		
	Aggr.	Agri.	Manu.	Aggr.	Agri.	Manu.
Korea, Rep. of						
1960s	19.2	4.4	31.4	2.0	1.1	1.6
1970s	29.0	8.3	25.1	3.3	4.4	1.6
1980–1985	35.6	12.7	22.1	4.9	3.1	2.5
1963–1985	26.4	7.6	26.9	3.1	2.8	1.8
Thailand						
1960s	21.2	—	—	2.7	—	—
1970s	24.2	8.3	24.7	3.5	2.0	2.5
1980–1985	22.1	—	—	4.4	—	—
1960–1985	22.6	—	—	3.4	—	—
India						
1960s	14.8	6.8	24.9	3.6	3.2	4.5
1970s	15.7	7.8	23.5	4.8	4.6	5.7
1980–1985	18.2	9.2	26.9	3.4	3.4	3.6
1960–1985	15.8	7.7	24.7	4.0	3.8	4.8

Notes: — = Not available.
a. ICOR = I/Y / G_Y. Inventory changes are included in Table 1.1, and the ICOR estimated in this table is somewhat smaller. The magnitude of ICOR for the entire period is the weighted average of the three periods.

Sources: United Nations, *Yearbook of National Accounts Statistics*, various issues.

one of the important characteristics of contemporary developing economies as compared to presently developed economies. The implications of this significant fact have repeatedly been mentioned in Chapters 1 and 2. The data of our selected Asian countries confirm this as the typical pattern. For the entire period even in India, I/Y is close to 16 percent, in Thailand it is 22.6 percent, and in Korea it is over 26 percent (note that inventory changes are excluded). What about changes in I/Y over time? The ratio I/Y is generally characterized by a tendency of acceleration for the three countries, with the exception of the 1980s in Thailand.

Korea increased its capital investment ratio (I/Y) of the macro-economy in a remarkable way since the early 1960s and reached the level of 35 percent in the 1980s. In Thailand, a rising trend of I/Y is also seen but after 1980 it is interrupted, and contrary to our expectations, I/Y declines thereafter. The ratio I/Y in India shows a gradual increase through the entire period, and yet its level is much lower than that of the other two countries.

Features of I/Y by sector are different among the three countries. In Korea, I/Y in the agricultural sector increases from the 1960s to the 1980s, while in the manufacturing sector, the ratio decreased. In India, I/Y in the agricultural sector follows a rising trend and I/Y in the manufacturing sector remains almost unchanged at around 25 percent in the entire period. As for Thailand, the data are available only for the 1970s. However, the difference between I/Y of the agricultural and manufacturing sectors is noticeably large, and one may say this is a common phenomenon to many other countries. Despite these features, taken together, one notes that, regarding both agriculture and manufacturing, the ratio I/Y does not appear to differ much over time. At least we can say that the variance is much narrower than would be expected from the wider variance of growth performance identified earlier. This poses a challenge for our analysis.

The measures of the incremental capital-output ratio or ICOR ($I/\Delta Y$) is a rough attempt to respond to the challenge. In our basic formula we treat its reciprocal, $\Delta Y/I$, which we refer to as investment efficiency for simplicity. On average, for the entire period ICOR is 4.0 in India, 3.4 in Thailand, and 3.1 in Korea. The data of our Asian countries seem to confirm exactly the pattern of ICOR decrease from group I to group III identified in Chapter 1, implying that the variance between individual nations may be rather wide.

With regard to changes in ICOR over time, both Korea and Thailand show signs of an increasing ICOR, reaching levels close to 5 in the 1980s. This implies that the efficiency of capital investment for these countries is becoming worse during the downswing. In India, an increasing trend of ICOR emerged from the 1960s to the 1970s and then its tendency changed to one of decrease. It should be noted that the ICOR of India was rather high during the 1970s when the country suffered from low growth rates of both output and labor productivity. Therefore, its turn to decrease during 1980–1985 is worthy of particular attention.

Let us now look at the sectoral differences in ICOR. It is an important fact that ICOR in the manufacturing sector in Korea has been kept considerably low in contrast to the big increase of ICOR in the agricultural sector from the 1960s to the 1980s. Korea's high economic growth and rapid industrialization could be accomplished under the favorable condition of efficient capital investment in the manufacturing sector. In Thailand, the ICOR in both sectors is low during the 1970s. When the ICOR of Thailand is compared with that of India, the Thai figures for the agricultural and manufacturing sectors are distinctly smaller. The ICOR in both sectors in India became smaller in 1980–1985 which is consistent with the macro-performance mentioned above.

Employment and the Residual

In estimating the employment effect in this study, we need data on labor's income share (β). Our concern with the employment effect is to look into the magnitude of $w\Delta L/I$, and this can be measured by $\beta \cdot G_L$ / I/Y in each period. The term $w\Delta L/I$ is the incremental term of wL/K and indicates not only the effect of employment change, but also the characteristics of technology adopted in the process of production. Therefore, β is a critical term and we have tried to estimate it in various ways. The values that were finally obtained for labor's income share by sector are shown in Table A3.6.

For now, however, let us turn our attention to Table A3.3, where $w\Delta L/I$ of the macro-economy in Korea shows a trend of decline throughout the observation period. Particularly noteworthy is the large drop in the magnitude of $w\Delta L/I$ from the 1970s to the 1980s. This dramatic decline implies a shift from labor-intensive to capital-intensive production, in association with a rise of the wage rate for workers. On the other hand, India shows an increasing trend of $w\Delta L/I$ suggesting an expansion of labor absorption capacity in the country's macro-economy. The magnitude of $w\Delta L/I$ in the Thai economy increased from the 1960s to the 1970s, but decreased in the 1980s. Whether Thailand has come to the turning point to a trend of decline is still to be seen, but at least we may say that the country is coming closer to that point.

As expected, the sectoral performance of $w\Delta L/I$ differs significantly between countries. In Korea, $w\Delta L/I$ of the agricultural sector is negative in each period and is exceedingly large in 1980–1985. In contrast, the manufacturing sector realized a high $w\Delta L/I$. Rapid industrialization was accomplished by absorbing a lot of the labor force through the 1960s and the 1970s. Another interesting fact is that in Thailand the $w\Delta L/I$ of both sectors are larger than that of the macro-economy in the 1970s. It should be emphasized here that the absorption capacity of the labor force (in our specific sense) in agriculture was as large as that of manufacturing. In the case of India, $w\Delta L/I$ was comparatively high for the agricultural sector in the 1960s and for the manufacturing sector in the 1970s. In addition,

Table A3.3
Employment and the Residual Increase Per Investment

	wΔL/I			ΔR/I		
	Aggr.	Agri.	Manu.	Aggr.	Agri.	Manu.
Korea, Rep. of						
1960s	7.0	-4.4	14.2	27.2	79.5	32.9
1970s	6.2	-1.9	16.0	11.9	12.6	35.9
1980–1985	2.8	-13.5	7.9	9.1	37.2	23.3
1963–1985	5.8	-6.4	13.6	17.5	44.3	32.2
Thailand						
1960s	3.6	—	—	19.3	—	—
1970s	7.2	13.5	21.3	11.8	27.5	9.7
1980–1985	4.3	—	—	10.7	—	—
1960–1985	5.2	—	—	16.6	—	—
India						
1960s	6.1	10.5	2.5	11.4	10.5	9.4
1970s	6.5	2.0	5.0	4.7	10.1	2.9
1980–1985	6.6	—	—	15.2	—	—
1960–1985	6.4	—	—	9.5	—	—

Notes: See notes of Table A3.6.

Sources: See sources of Table A3.6.

the manufacturing sector tends to use more labor force than it did before.

Several interesting points emerge from the estimates of the residual increase with regard to capital investment (which is estimated using the basic formula stated earlier). In the case of Korea, ΔR/I of the macro-economy exhibits a sharp declining trend from the 1960s to the 1980s. This appears to be similar to the pattern of the growth rates of output and labor productivity. As expected, a lower ICOR is connected with a higher residual increase. With respect to sectoral performance, agriculture attained an exceedingly large ΔR/I in the 1960s. It should be noted that the ICOR of the agricultural sector was as low as 1 in the 1960s. The ΔR/I of the manufacturing sector was the largest in the 1970s. Features of residual increase are quite different between the two sectors. The ΔR/I of the manufacturing sector follows an inverse U-shape pattern which is in striking contrast to the U-shape pattern of the agricultural sector during the period under observation. This implies a different type of contribution from the two sectors to macro-economic growth in Korea. In Thailand, the $\Delta R/I$ of the macro-economy shows a decline that resembles the case of Korea. The difference between Korea and Thailand is the magnitude of the residual increase: the level of the Thai $\Delta R/I$ is distinctly lower than that of Korea in each period. Regarding this point, it should be recognized that there appears to be a big difference in the ICOR between the two countries. With respect to the sectoral performance of residual growth, it is important to note that the agricultural sector achieved a larger $\Delta R/I$ than the manufacturing sector in the 1970s, although the G_Y of agriculture was smaller than that of manufacturing in the same period. This confirms our argument in Appendix II.

India's case is dissimilar to the other two countries. The $\Delta R/I$ of the macro-economy shows a U-shape pattern with a very small figure in the 1970s. There appeared a big shift of residual growth from the 1970s to recent years. With respect to sectoral performance, the agricultural sector did not show any distinct change from the 1960s to the 1970s, but the manufacturing sector revealed a big decrease in $\Delta R/I$ during the same period. The reason for the change in the manufacturing sector may be suggested from its increase in $w\Delta L/I$.

Finally, further observation is made on the residual growth and employment-technology. In Table 1.1, $\Delta R/I$ shows a sharp increase from group I to group III. This is distinct for subgroup A, but its pattern is mixed for subgroup B. In light of this international perspective, the $\Delta R/I$ estimated for the three nations shows the same pattern but with higher magnitudes, even compared with subgroup A of each group. Two points are noteworthy. One is the effect of a downswing where we see for 1980–1985, the aggregate $\Delta R/I$ in India is 9.5, in Thailand it is 16.6, and in Korea it is 17.5. But for 1960–1980 we have 8.1, 15.6, and 19.6, respectively. The greater value for India and Thailand is largely affected by its relatively high value during the downswing, 1980–1985, as discussed earlier. The other finding concerns the varied effects of sectoral contribution; this is also important, though the available data are limited. A suggestive illustration is given by comparing Korea and Thailand in the 1970s. In Korea, $\Delta R/I$ in the manufacturing sector is much greater than that of the agricultural sector, whereas in Thailand the pattern is exactly reversed. This no doubt pertains to typological features.

As was revealed in Chapters 1 and 2, the residual growth per investment is greatly influenced by the performance of investment efficiency, $\Delta Y/I$, the reciprocal of ICOR. The estimates for the three nations present essentially the same picture as shown in the figures of Table A3.3. Reservation is required because of the reliability of our estimates, in particular, regarding the term r, as will be discussed later. However, we are convinced that in looking for broad patterns, the possible distortions that may arise because of this are not very large.

The performance of $w\Delta L/I$ needs special comment in this regard. First, in light of the general international pattern presented in Table 1.1 by group (i.e., 6.4 for group I, 5.7 for group II, and 5.8 for group III), the figures of the three nations in Table A3.3 appear to show a similar pattern: they are 6.4 for India, 5.2 for Thailand, and 5.8 for Korea for the period 1960–1985, although the magnitude of Thailand is somewhat biased. Second, with regard to changes over time, the case of Korea presents a distinct pattern for the manufacturing sector: $w\Delta L/I$ increases slightly from the 1960s to the 1970s and turns sharply to decrease during 1980–1985. In our specific sense, we see a tendency of a labor-intensive pattern in the former period and a new movement to a capital-intensive pattern in the latter period. The difficulty of making such a turn efficiently has been repeatedly discussed in the main text regarding the phase shift from group III to group IV. Finally, an increase in $w\Delta L/I$ of the manufacturing sector in India is understandable. However, its low level is somewhat questionable, though it may be acceptable in comparison with agriculture. Further scrutiny would be needed.

III. Incorporating Macro-sectoral Estimates

We could compare the features of the growth performance of the three countries by summing all the results of our estimates, but this merely repeats what has already been discussed. Rather, we prefer to focus on our group-subgroup analysis and try to incorporate the macro-sectoral estimates into this analysis.

Two topics are considered: (1) productivity growth and phasing, and (2) the measurement of residual growth using the output approach.

Let us begin with an examination of sectoral labor productivity growth. In addition to what was pointed out earlier regarding its macro-level performance, the results of a two-sector comparison (1960–1980, excluding the downswing) are shown below:

	G_{yM}	G_{yA}	$G_{yM}-G_{yA}$
India	2.8	1.1	1.7
Thailand	2.9	3.0	-0.1
Korea	6.1	3.4	2.7

The corresponding figures for subgroup E are as follows (note that the value for group I is the group average as the subgroup average is not dealt with in Chapter 4).

Group	G_{yI}	G_{yA}	$G_{yI}-G_{yA}$
I	0.5	0.4	0.1
II	3.4	2.4	1.0
III	4.4	4.0	0.4

Assuming no sizeable difference between G_{yM} and G_{yI}, we can say that the acceleration path of both agriculture and industry is clearly revealed by the three-country comparison, but with an important qualification due to typological factors. In both India and Korea, the pace of industrialization is much higher than growth of the agricultural sector, while Thailand seems to show a near equivalent path of two-sector growth in the international perspective. This illustrates that a direct comparison of productivity growth rates of individual nations when looking at the phases of development is often misleading. To be realistic, we need a yardstick of phasing that is modified by typological features.

This problem was dealt with in Chapter 4 by the "difference indicator" for demarcating development phases in terms of sectoral productivity growth. The idea behind this approach is to associate the phase shift with changes in the sectoral shares of output and labor in macro-economic growth. This is a way of incorporating our estimates of the aggregate or macro-level on the one hand, and the two major sectors, agriculture and manufacturing, on the other. In order to do this, however, a gap is to be filled by the residue estimates for the S' sector (i.e., the non-AM sector). This is actually composed of additional subsectors to the services sector, S, in our classification: that is, it includes public utilities, mining, and construction. The major element can be called facilitating industries (services and infrastructure) in the broadest sense, but the unavoidable inclusion of other industries makes the meaning of this residual somewhat vague. The results are shown in Table A3.4.

Table A3.4
Sectoral Contribution to Aggregate Productivity Growth and Difference Indicators (D)

(A) Weighted growth rates

Output	Agriculture			Manufacturing			Non–AM (S')			
	G_{YA}	w_A	$w_A G_{YA}$	G_{YM}	w_M	$w_M G_{YM}$	$G_{YS'}$	$w_{S'}$	$w_{S'} G_{YS'}$	Sum
Korea, Rep. of										
1960s	4.0	37.2	1.5	19.8	13.9	2.8	16.2	48.9	7.9	12.2
1970s	1.9	22.4	0.4	16.1	26.4	4.3	8.4	51.2	4.3	9.0
1980–1985	4.1	14.2	0.6	8.8	35.4	3.1	7.3	50.4	3.7	7.4
1963–1985	3.2	26.7	0.9	16.1	23.2	3.7	11.3	50.1	5.7	10.3
Thailand										
1960s	5.4	36.5	2.0	10.6	13.8	1.5	9.0	49.7	4.5	7.9
1970s	4.2	28.6	1.2	10.0	18.1	1.8	7.4	53.3	3.9	7.0
1980–1985	5.0	24.1	1.2	5.0	20.7	1.0	5.6	55.2	3.1	5.3
1960–1985	4.8	30.9	1.5	9.2	16.9	1.6	7.1	52.2	3.7	6.7
India										
1960s	2.1	47.2	1.0	5.5	12.2	0.7	6.0	40.6	2.4	4.1
1970s	1.7	40.0	0.7	4.1	13.5	0.6	4.4	46.5	2.0	3.3
1980–1985	2.7	34.6	0.9	7.5	14.8	1.1	6.6	50.6	3.3	5.4
1960–1985	2.1	41.8	0.9	5.3	13.2	0.7	5.5	45.0	2.5	4.1

Labor	Agriculture			Manufacturing			Non–AM(S')			
	G_{LA}	v_A	$v_A G_{LA}$	G_{LM}	v_M	$v_M G_{LM}$	$G_{LS'}$	$v_{S'}$	$v_{S'} G_{LS'}$	Sum
Korea, Rep. of										
1960s	-0.5	56.9	-0.3	10.4	10.6	1.1	6.0	32.5	2.0	2.8
1970s	-0.4	42.3	-0.2	8.9	17.4	1.5	5.8	40.3	2.3	3.7
1980–1985	-4.3	29.5	-1.3	3.5	22.5	0.8	4.9	48.1	2.4	1.9
1963–1985	-1.2	45.6	-0.5	8.4	15.5	1.3	5.7	38.7	2.2	3.0
Thailand										
1960s	1.4	80.6	1.1	3.1	3.7	0.1	3.0	15.8	0.5	1.7
1970s	2.2	75.0	1.7	11.7	6.0	0.7	6.1	19.1	1.2	3.5
1980–1985	3.3	70.3	2.3	2.6	7.8	0.2	5.3	22.0	1.2	3.7
1960–1985	2.1	76.3	1.6	7.4	5.4	0.4	4.7	18.4	0.8	2.9
India [a]										
1960s	2.0	73.1	1.5	0.8	10.3	0.1	4.4	16.7	0.7	2.3
1970s	1.7	71.6	1.2	3.3	10.3	0.3	1.6	18.2	0.3	1.8
1960–1979	1.9	72.5	1.4	2.2	11.0	0.2	3.1	16.5	0.5	2.1

(B) Difference indicators

	D_A	D_M	$D_{S'}$	$D_M-D_{S'}$
Korea, Rep. of				
1960s	1.8	1.7	5.9	-4.2
1970s	0.6	2.8	2.0	0.8
1980–1985	1.9	2.3	1.3	1.0
1963–1985	1.4	2.4	3.5	-1.1
Thailand				
1960s	0.9	1.4	4.0	-2.6
1970s	-0.5	1.1	2.7	-1.6
1980–1985	-1.1	0.8	1.9	-1.1
1960–1985	-0.1	1.2	2.9	-1.7
India [a]				
1960s	0.0	0.6	2.0	-1.4
1970s	0.5	0.4	0.7	-0.3
1960–1979	0.3	0.5	1.4	-0.9

Note: a. Data were not available for India 1980–1985.

Sources: See sources of Table A3.1.

We are specifically concerned with two points: one is the issue of the positive or negative value of D_A, the indicator of agriculture, and the other is the variation in the difference between the two indicators, $D_M-D_{S'}$, which relates to the manufacturing and facilitating sectors.

The indicator D_A in Thailand is near zero on average, though its value in India is slightly positive. This requires careful interpretation in relation to our proposal of the preprimary phase in Chapter 4. In Table 4.10, Panel (C), D_A is negative for groups I and II. The comparable value of D_A for 1960–1980 in Thailand is 0.2 on average, and this, similar to India, shows its superior performance in terms of an international perspective. Of course, the concept of the preprimary phase is not applicable for Thailand. However, attention is drawn to the sizeable negative value of D_A in 1980–1985 for Thailand. This illustrates the operation of backward forces in the downswing period for a number of countries belonging to group II as well as group I. It goes without saying that this phenomenon stems from the fact that, despite the favorable growth of agricultural output, G_{YA}, the share of labor engaged in agriculture is sustained at extremely high levels.

The difference indicator of manufacturing, D_M, cannot be directly compared with D_I in Table 4.10, but we can generally say that its contribution to aggregate productivity growth is superior in both Korea and India to the average performance of subgroup E in each group and is equivalent with the average performance of subgroup E, group II, and Thailand. On average for 1960–1980, the difference indicators are:

D_M, Korea	2.25	D_M, Thailand	1.25	D_M, India	0.50
D_I, E, III	1.50	D_I, E, II	1.25	D_I, I	0.35

The data are comparable to the data in Table 4.10, Panel (C) and may serve to quantify the role played by industrialization. What we are specifically concerned with is the performance of $D_M-D_{S'}$, shown in the last column of Panel (B) in Table A3.4. Except for one case in Korea, $D_M-D_{S'}$ is negative and is negative for the entire period on average. According to our phasing framework, the turning point where the difference in the indicators turns from negative to positive (i.e., $D_I = D_S$) is of significance in the path of industrialization. Korea seems to have arrived at this point in the beginning of the 1970s, which is much earlier than the arrival dates of countries in subgroup E, group III. Both in Thailand and India, the negative magnitude of the difference tends to decrease over time and a turn in the difference of the indicators to positive values is in the distant future, though the negative magnitude is smaller than the average of subgroup E for each group.

The output approach, discussed in Chapter 1, is attempted here with sectoral residual growth measurements, as this is the conventional way of incorporating macro-economic and sectoral performance. The same procedure that was used in Appendix II is applied. Again, we encounter difficulty in treating the services sector, but a rough approximation is achieved by using S', which is derived as a residual. Another difficulty is the measurement of the reallocation effects for labor which is discussed in some detail in Appendix II. Because of the limited data available, only an average adjustment for the 1960–1980 period is tried to indicate the order of its magnitude. Table A3.5 presents the (unadjusted) residual growth.

The value of aggregate $\Delta R/Y$ exhibits the expected pattern: that is, it is largest in Korea, smallest in India, and is in-between in Thailand. In light of the international perspective shown in Table 1.2, the order of $\Delta R/Y$ is as expected but needs an important qualification: that is, the value of ΔR/Y of these nations are all greater than that of subgroup A for each group (1.1 for group I, 1.9 for group II, and 3.0 for group III). In making technological-organizational advances, all of these nations show a performance that is superior even to that of subgroup A. This judgment varies from the previous case of ΔR/I and the reason for this stems from the performance of I/Y.

By sector, the greatest ΔR/Y in manufacturing is noted for Korea, but this is not necessarily so for Thailand and India. Note that in Appendix II, its ratio to the macro-level value tends to be smaller than unity for the group average of developing economies and this is confirmed by Thailand and India which show the characteristics of contemporary developing economies. We are inclined to say that Korea's pattern is similar to that of the historical records of presently developed nations.

However, we must note an important reservation. As touched upon earlier, the reallocation effects are reflected in the magnitude of ΔR/Y for the non-AM sector. To the extent that this is true, the value of S' is overestimated. The outcome of applying the same method used in Appendix II appears to vary between periods, and we have taken the average for 1960–1980 (1960–1970 for Thailand). The reallocation effects of labor thus measured and the adjusted values are:

	Korea	Thailand	India
Reallocation effects	0.95	1.24	0.62
Adjusted ΔR/Y of S'	1.50	1.20	0.20
Weighted	0.75	0.57	0.07

These adjusted magnitudes of DR/Y for the S' sector appear to be comparatively smaller than those of the agricultural and manufacturing sectors (see Table A3.5).

A comparative judgment between the estimates in Appendix II and the estimates above would require appropriate reservations. Yet, the two estimates appear to be generally consistent with each other and we assert that the contribution of the facilitating sector in promoting aggregate residual growth is rather conservative. Technological and/or organizational advance is chiefly brought forth by manufacturing and agriculture.

In this regard, let us mention a few final words on the role played by agriculture in the path of shifting development phases. To increase aggregate residual growth, the contribution of

Table A3.5
Residual Growth: Output Approach [a] *(percentage)*

	$\Delta R/Y$				$w_I \Delta R/Y$ (weighted)		
	Aggr.	Agri.	Manu.	Non–AM	Agri.	Manu.	Non–AM
Korea, Rep. of							
1960s	5.2	3.5	10.3	5.0	1.30	1.43	2.49
1970s	3.5	1.0	9.0	1.8	0.22	2.38	0.90
1980–1985	3.2	4.7	5.1	1.4	0.67	1.81	0.72
1963–1985	4.1	2.7	8.7	3.0	0.74	1.89	1.50
Thailand							
1960s	4.0	—	—	—	—	—	—
1970s	2.9	2.3	2.4	3.4	0.66	0.43	1.81
1980–1985	2.4	—	—	—	—	—	—
1960–1985	3.1	—	—	—	—	—	—
India							
1960s	1.7	0.7	2.3	2.7	0.33	0.28	1.09
1970s	0.7	0.8	0.7	0.6	0.32	0.09	0.29
1980–1985	2.8	—	—	—	—	—	—
1960–1980	1.4	—	—	—	—	—	—

Notes: — = Not available.

a. $\Delta R/Y$ is calculated by $\Delta R/I \cdot I/Y$. The term w_I indicates the output share of each sector.

Sources: See sources of Table A3.3.

agriculture is sustained higher than that of manufacturing through a long path before arriving at the point where $\Delta R/Y$ between agriculture and manufacturing becomes equal. Although the effects of typological differences cannot be ignored, the data in the table suggest that the point would be located somewhere between Thailand and Korea.

IV. Notes on the Estimates

All of the data of relevant terms used in estimating the residuals per investment using the basic formula presented in the Introduction are listed in Table A3.6.

The estimates were targeted to span 25 years, from 1960 to 1985, in order to be comparable with the data in Table 1.1. However, some deviation is unavoidable. In the case of Korea, the interval is 23

Table A3.6
Relevant Terms in Estimating the Residuals

	ß (%)				D/Y	(I–D)/I		r	ε
	Aggr.	Agri.	Manu.	Non–AM	(%)	(%)	K/Y	(%)	(%)
Korea, Rep. of									
1960s	50	39	43	60	5.1	63.1	2.0	25.0	15.8
1970s	50	40	45	57	7.8	73.0	3.9	16.7	12.2
1980–1985	55	40	50	63	9.7	73.1	3.8	11.8	8.6
1963–1985	51	40	45	59	7.1	69.0	2.8	19.0	12.9
Thailand									
1960s	45	—	—	—	5.8	73.6	2.8	19.6	14.4
1970s	51	51	45	53	7.3	69.9	3.6	13.6	9.5
1980–1985	59	—	—	—	7.4	66.4	3.6	11.4	7.6
1960–1985	52	—	—	—	6.7	70.7	3.3	14.8	10.9
India									
1960s	60	51	42	76	4.8	67.0	2.6	15.4	10.3
1970s	60	51	47	72	5.0	67.4	2.8	14.3	9.6
1980–1985	60	51	49	69	5.8	67.8	3.2	12.5	8.5
1960–1985	60	51	45	73	5.1	67.7	2.8	14.5	9.7

Notes: — = Not available.
β = labor's income share, D = depreciation of capital (K), r = rate of return on capital, and ε = coefficient of capital replacement.

Sources: United Nations, *Yearbook of National Accounts Statistics*, various issues; Hayami, Ruttan, and Southworth, eds., *Agricultural Growth in Japan, Taiwan, Korea and the Philippines*, Table K.56; Government of Korea, *Report on Mining and Manufacturing Survey* and *Input-Output Table of Korea;* Government of Thailand, *Report of the Industrial Census;* Government of India, *Annual Survey of Industries.*

years from 1963 to 1985 due to data limitations for the labor force in each sector. In the case of Thailand, investment data for agriculture and manufacturing are available only for the period from 1970 to 1978. Nevertheless, the final results of the sectoral estimates are presented by demarcating three equal periods: the 1960s, the 1970s, and 1980–1985.

Beginning with $w\Delta L/I$, to apply $\beta G_L \cdot Y/L$, β of the macroeconomy can be measured by using data from the national accounts of each country. The problem is how to measure β for the various sectors. We try to estimate the β in the manufacturing sector on the basis of manufacturing census data which are fortunately available for the three countries selected. For the agricultural sector, estimation has been conducted in the case of Korea by agricultural economists (see the sources of Table A3.6), but no exact data are available for Thailand and India. Under these conditions, we attempt to estimate β in the agricultural sector for these two countries by utilizing data on daily wage rates of agricultural workers, working days, and the number of laborers.[160] Finally, β for the S' sector is obtained as a residual, that is, the aggregate level of β minus the weighted values of β_A and β_M.

With respect to capital, data on capital replacement are not directly available. We assume that ΔK can be approximated by (I–D) where D is the value of consumption of the fixed capital stock and is estimated by depreciation allowance.

In order to estimate the rate of return on capital (r), we use data on capital's income share (α) and the capital-output ratio (K/Y), as r is equal to $\alpha \cdot Y/K$. The term α is measured assuming $\alpha + \beta = 1$. The ratio K/Y is estimated using the formula $K/y = K/D \cdot D/Y$, following the procedure used in Chapter 1, Technical Notes. The average duration of capital life is assumed to be 39 years for Korea, 49 years for Thailand, and 55 years for India. These are the same values used for duration of capital life for groups III, II, and I, respectively, in Chapter 1.

Finally, a few words are required for qualification and reservation. First, with respect to the aggregate β, the estimated values generally seem to be consistent with those of groups I, II, and III in Chapter 1. According to the World Bank, *World Development Report 1978,* Indicators, Table 8, β in the manufacturing sector in the three Asian countries are:

	Korea	Thailand	India
1970	25	25	47
1985	27	24	49

160 These data are obtained by our own field work conducted in 1978. For detail, see Katsuo Otsuka, "The Transfer of Technology in Japan and Thailand: Sericulture and the Silk Industry," *Development and Changes* (July 1982), Table 5. The β in the agricultural sector for India is estimated in the same way. Data from the Ministry of Labour, Government of India, *Rural Labour Equity 1977–78,* were also used.

The sizeable differences for Korea and Thailand, in our view, may stem from the underestimates due to the undercoverage of very small firms. The term β in agriculture is generally consistent with the estimates derived in Appendix II, but the term is not very reliable and further scrutiny would be needed. Second, with respect to capital, our estimates are crude and rough in two ways: (1) land is not treated separately from capital, and (2) no specific estimate is possible for the sectoral values of capital life duration. The aggregate value of capital life duration may appear longer than is usually expected, but this is due to the greater weight attached to structures and buildings. Therefore, for manufacturing this duration is too long, and further scrutiny would be required. Finally, comparisons of K/Y with ICOR regarding the aggregate estimates on average for 1960–1980 excluding the downswing intervals, are shown below:

	Korea		Thailand		India	
K/Y	2.8	(3.5)	3.3	(3.2)	2.8	(2.7)
I/ΔY	3.1	(2.7)	3.4	(3.1)	4.0	(4.2)

The values of the two terms are fairly close. Our estimates can be largely accepted insofar as the macro-level performance is concerned, although perhaps the value of India may be somewhat underestimated.

References

Chapter 1

Kazushi Ohkawa and Hirohisa Kohama, *Lectures on Developing Economies: Japan's Experience and Its Relevance* (Tokyo: University of Tokyo Press, 1989).

Kazushi Ohkawa and Henry Rosovksy, *Japanese Economic Growth: Trend Acceleration in the Twentieth Century* (Stanford: Stanford University Press, 1973).

Angus Maddison, *Phases of Capitalist Development* (London: Oxford University Press, 1982).

Simon Kuznets, "Notes on Japan's Economic Growth," in Lawrence Klein and Kazushi Ohkawa, eds., *Economic Growth* (Durham, N.C.: Duke University Press, 1955).

Lawrence Klein and Kazushi Ohkawa, eds., *Economic Growth: The Japanese Experience Since the Meiji Era* (Homewood, Illinois: Richard D. Irwin, 1968).

Edmond Malinvaud, ed., *Economic Growth and Resources,* Vol. I, The Major Issues, Proceedings of the Fifth World Congress of the International Economic Association held in Tokyo, Japan (London: Macmillan, 1979).

Moses Abramovitz, *Thinking About Growth* (Cambridge: Cambridge University Press, 1989).

Simon Kuznets, *Modern Economic Growth: Rate, Structure, and Spread* (New Haven, Yale University Press, 1966).

Nicholas Kaldor, *Causes of the Slow Rate of Economic Growth of the United Kingdom: An Inaugural Lecture* (London: Cambridge Univ. Press, 1966).

Simon Kuznets, *Economic Growth of Nations: Total Output and Production Structure* (Cambridge, Mass.: Harvard University Press, 1972).

John W. Kendrick and Ryuzo Sato, "Factor Prices, Productivity and Growth," *American Economic Review* (Nashville, T.N.: American Economic Association, December 1967).

M. Ishaq Nadiri, "Some Approaches to the Theory and Measurement of Total Factor Productivity: A Survey," *Journal of Economic Literature* 3 (Nashville, T.N.: American Economic Association, December 1970).

Hollis Chenery, Sherman Robinson, and Moshe Syrquin, *Industrialization and Growth: A Comparative Study* (New York: Oxford University Press, 1986).

World Bank, *World Development Report* (Washington D.C.: World Bank, 1982, 1987).

World Bank, *World Tables* (Washington D.C.: World Bank, 1983).

International Labour Office, *Economically Active Population, 1950–2025,* Volumes 1–5 (Geneva: ILO, various years).

Hollis Chenery and Moshe Syrquin, "Typical Pattern of Transformation," Chapter 3, in Hollis Chenery, Sherman Robinson, and Moshe Syrquin, *Industrialization and Growth: A Comparative Study* (New York: London: Oxford University Press, 1986).

Kazushi Ohkawa, "Capital-Output Ratio and the 'Residuals': Issues of Development Planning," *IDCJ Working Paper Series,* No. 28 (Tokyo: IDCJ, March 1984).

Simon Kuznets, "Distribution of Product and Income," Chapter 4, in *Modern Economic Growth.*

World Bank, *Statistical Tables,* various issues (Washington, D.C.: World Bank, various years).

Yujiro Hayami and Vernon W. Ruttan, *Agricultural Development: An International Perspective,* Revised edition (Baltimore: Johns Hopkins University Press, 1985).

Ryoshin Minami and Akira Ono, "Yoso Shotoku to Bumpairitsu no Suikei" [Estimates of factor income shares], *Keizai Kenkyu* [The economic review] 29 (Tokyo: Iwanami Shoten, April 1978).

Saburo Yamada and Yujiro Hayami, "Agriculture," Part II, Chapter 4, in Kazushi Ohkawa and Miyohei Shinohara, eds., *Patterns of Japanese Economic Development* (New Haven: Yale University Press, 1979).

Shokichi Motai and Kazushi Ohkawa, "Small-scale Industries: A Study on Japan's 1966 Manufacturing Census," *IDCJ Working Paper Series*, No. 11 (Tokyo: IDCJ, December 1978).

Edward F. Denison, *Why Growth Rates Differ: Postwar Experience in Nine Western Countries* (Washington, D.C.: The Brookings Institution, 1967).

United Nations, *National Accounts Statistics* (New York: United Nations, 1985).

Kazushi Ohkawa et al., *Long-term Economic Statistics (LTES)*:
Vol. 1, *National Income*; (Tokyo, Toyokeizai Shimposha, 1974)
Vol. 2, *Labor Force*; (Tokyo, Toyokeizai Shimposha, 1988)
Vol. 3, *Capital Stock* ; (Tokyo, Toyokeizai Shimposha, 1976)
Vol. 6, Personal Consumption Expenditures; (Tokyo, Toyokeizai Shimposha, 1977).
Vol. 8, *Prices*. (Tokyo, Toyokeizai Shimposha, 1971).

Chapter 2

Ryoshin Minami, *The Economic Development of Japan: A Quantitative Study* (London: Macmillan, 1986).

B. R. Mitchell, *European Historical Statistics, 1750–1970* (London and Basingstoke: Macmillan, 1978).

John W. Kendrick, *Productivity Trends in the United States* (Princeton, N.J.: Princeton University Press, 1961).

J. J. Carre, P. Dubois, and Edmond Malinvaud, *French Economic Growth* (Stanford: Stanford University Press, 1976).

R. C. O. Mathews, C. H. Feinstein, and J. C. Odling-Smee, *British Economic Growth, 1856–1973*, (Stanford: Stanford University Press, 1983).

Alexander Gerschenkron, *Economic Backwardness in Historical Perspective* (Cambridge: Harvard University Press, 1962).

Kazushi Ohkawa, "Investment Spurts and Resource Allocation," Chapter 5, *Nihon Keizai no Kozo* [Structure of the Japanese economy] (Tokyo: Keiso Shobo, 1974).

Toshiyuki Mizoguchi and Mataji Umemura, eds., *Kyu Nihon Shokuminchi Keizai Tokei 1895–1938* [Basic economic statistics of former Japanese colonies] (Tokyo: Toyokeizai Shimposha, 1988).

Kazushi Ohkawa, *Shokuryo Keizai no Riron to Keisoku* [Theory and measurement of food economy] (Tokyo: Nihon Hyoron-sha, 1945).

Kazushi Ohkawa and Miyohei Shinohara, eds., *Patterns of Japanese Economic Development: A Quantitative Appraisal* (New Haven: Yale University Press, 1979).

Japan Government, Economic Research Institute, Economic Planning Agency, *Gross Capital Stock in Private Enterprises* (Tokyo: EPA, 1965).

Japan Government, Economic Planning Agency, *National Economic Accounts Quarterly*, No. 45, 60, 70, and 82 (Tokyo: EPA, various years)

Ryoshin Minami and Akira Ono, "Factor Incomes and Shares," Part 4, Chapter 11, in Kazushi Ohkawa and Miyohei Shinohara, eds., *Patterns of Japanese Economic Development.*

Kazushi Ohkawa, "Initial Conditions: Economic Level and Structure," in *Japan's Historical Development Experience and the Contemporary Developing Countries: Issues for Comparative Analysis* (Tokyo: IDCJ, 1978).

John C. H. Fei, Kazushi Ohkawa, and Gustav Ranis, "Economic Development in Historical Perspective: Japan, Korea and Taiwan," Part II, in Kazushi Ohkawa and Gustav Ranis, eds., *Japan and the Developing Countries* (Oxford: Basil Blackwell, 1985).

Moses Abramovitz, "Rapid Growth Potential and Its Realization: The Experience of Capitalist Economies in the Postwar Period," in Edmond Malinvaud, ed., *Economic Growth and Resources.*

Chapter 3

Kazushi Ohkawa, "Keizai Hatten to Kozo" [Economic development and structure], in Kazushi Ohkawa and Ryoshin Minami, eds., *Kindai Nihon no Keizai Hatten* [Economic development of modern Japan] (Tokyo: Toyokeizai Shimposha, 1975).

Kazushi Ohkawa and Nobukiyo Takamatsu, "Shogyo to Sabisu no Seisansei" [Productivity of commerce and services], Chapter 12, in Kazushi Ohkawa and Ryoshin Minami, eds., *Kindai Nihon no Keizai Hatten.*

Akira Ono and Tsunehiko Watanabe, "Changes in Income Inequality in the Japanese Economy," in Hugh Patrick, ed., *Japanese Industrialization and Its Social Consequences* (Berkeley and Los Angeles: University of California Press, 1976).

Chapter 4

David Ricardo, "An Essay on the Influence of a Low Price of Corn on the Profits of Stock", 1815, and "On Protection to Agriculture" in Pierro Staffa, ed., *The Works and Correspondence of David Ricardo* (Cambridge: Cambridge University Press, 1966).

W. W. Rostow, "Notes on the Take-off," Chapter 2, in W. W. Rostow, ed., *The Economics of Take-off into Sustained Growth* (London: Macmillan, 1963).

Chapter 5

Kazushi Ohkawa, "Long-term Changes on the Price of Agricultural Products," Chapter 2, in *Nogyo no Dotai Bunseki* [Dynamic analysis of agriculture] (Tokyo: Taimeido, 1954).

Enzo R. Grilli and Alan Chang Yang, "Primary Commodity Prices, Manufactured Goods Prices and the Terms of Trade of Developing Countries: What the Long Run Shows", in the World Bank, *Economic Review* 2 (Washington, D.C.: World Bank, January 1988).

Rudiger Dornbusch and F. Leslie C. H. Helmers, eds., *The Open Economy* EDI Series in Economic Development, (New York: Oxford Univ. Press, 1988).

United Nations, *World Product and Income,* United Nations International Comparison Project, Phase III (New York: United Nations, 1982).

United Nations, Commission of the European Communities, *World Comparisons of Purchasing Power and Real Product for 1980,* Part II (New York: United Nations, 1987).

Chapter 6

Bela Ballasa, "Essay in Development Strategy," *Occasional Papers,* No. 5, International Center for Economic Growth (San Francisco: ICEG, 1988).

Yuichi Shionoya, "Patterns of Industrial Development," in Lawrence Klein and Kazushi Ohkawa, eds., *Economic Growth.*

Yuji Kubo, Jaime de Melo, and Sherman Robinson, "Trade Strategies and Growth Episodes," in Chenery et al., *Industrialization and Growth.*

Yuji Kubo, Jaime de Melo, Sherman Robinson, and Moshe Syrquin, "Interdependence and industrial Structure," in Chenery et al., *Industrialization and Growth.*

Appendix I

International Development Center of Japan, *Japan's Historical Development Experience and the Contemporary Developing Countries: Issues for Comparative Analysis* (Tokyo: IDCJ, 1978).

Gunnar Myrdal, "Difference in Initial Conditions," Chapter 14, in *An Approach to the Asian Drama* (New York: Vintage Books, 1970).

Simon Kuznets, "Present Underdeveloped Countries and Past Growth Patterns, 1958," in *Economic Growth and Structure* (New York: W. W. Nortons, 1965).

Irvin B. Kravis et al., *A System of International Comparisons of Gross Product and Purchasing Power* (Baltimore and London: Johns Hopkins University Press, 1975).

Bureau of Statistics, Prime Minister's Office, *Annual Report on the Retail Price Survey* (Tokyo: Bureau of Statistics, 1970).

J. Krishnamurty, "Some Aspects of Structural Change in the Indian Economy Since 1881," September 1977. Unpublished mimeo.

United Nations, *Yearbook of National Accounts Statistics*, various issues (New York: UN, various years).

International Monetary Fund, *International Financial Statistics,* various issues (Washington, D.C.: IMF, various years).

International Labour Organisation, *Yearbook of Labour Statistics*, various issues (Geneva: ILO, various years).

Economic and Social Commission of Asia and the Pacific, *Statistical Yearbook for Asia and the Pacific*, various issues (Bangkok: ESCAP, various years).

Republic of China, Directorate-General of Budget, Accounting and Statistics, *Statistical Yearbook and Explanatory Notes* (Taipei: DGBAS, 1975).

Kazushi Ohkawa, *Differential Structure and Agriculture* (Tokyo: Kinokuniya, 1972).

Yujiro Hayami et al., *A Century of Agricultural Growth in Japan: Its Relevance to Asian Development* (Tokyo: University of Tokyo Press, 1975).

Saburo Yamada, "Ajia Nogyo no Tonyu Sanshutsu Kozo to Hatten no Pataan" [Input-output structure and its pattern of development], Parts I, II, and III, in *Ajia Keizai* [Asian economics] 16(June 1975) and 16(August 1975) (Tokyo: Institute of Developing Economies)

Yujiro Hayami in association with others, "An International Comparison of Agricultural Production and Productivities," Technical Bulletin 277, University of Minnesota, Agricultural Experiment Station (1971).

Appendix II

Kazushi Ohkawa and Nobukiyo Takamatsu, "Capital Formation, Productivity and Employment: Japan's Historical Experience and Its Possible Relevance to LDCs," *IDCJ Working Paper Series*, No. 26 (Tokyo: IDCJ, March 1983).

Kazushi Ohkawa, "Implication for Agricultural Development," Part IV, in Yujiro Hayami, Vernon W. Ruttan, and Herman M. Southworth, eds., *Agricultural Growth in Japan, Taiwan, Korea and the Philippines* (Tokyo: The University of Hawaii Press, 1979).

Kazushi Ohkawa, "Effects of Structural Change on Productivity Growth: A Long-term Measurement," in R. F. Kosobud and Ryoshin Minami, eds., *Economic Studies of the Japanese Economy* (Urbana: Univ. of Illinois Press, 1977).

Kazushi Ohkawa, "Sangyo Kozo Hatten to Makuro Seisansei" [Industrial structural change and macro-productivity growth], Chapter 4, in *Nihon Keizai no Kozo.*

Moshe Syrquin, "Productivity Growth and Factor Reallocation," Part III, in *Industrialization and Growth* (Washington, D.C.: World Bank, 1986).

Moshe Syrquin, "Resource Reallocation and Productivity Growth," in Kelly Williamson, ed., *What Drives Third World City Growth* (Princeton: Princeton University Press: 1984).

Appendix III

Katsuo Otsuka, "The Transfer of Technology in Japan and Thailand: Sericulture and the Silk Industry," *Development and Change* (London and Beverly Hills: SAGE, July 1982).

Ministry of Labour, *Rural Labour Equity 1977–78.*

Korea Government, *Report on Mining and Manufacturing Survey* (Seoul: Economic Planning Board, various years)

Korea Government, *Input-Output Table* (Seoul: Bank of Korea. 1980)

Thailand Government, *Report of the Industrial Census,* (Bangkok: National Statistical Office, various years).

India Government, *Annual Survey of Industries,* (New Delhi: Ministry of Planning, various years).

Index